URBAN WATER INFRASTRUCTURE

URBAN WATER INFRASTRUCTURE:
Planning, Management, and Operations

NEIL S. GRIGG
Colorado State University,
Fort Collins

A Wiley-Interscience Publication
JOHN WILEY & SONS
New York Chichester Brisbane Toronto Singapore

Library of Congress Cataloging in Publication Data:

Grigg, Neil S.
Urban water infrastructure.

"A Wiley-Interscience publication."
Includes bibliographical references and index.
1. Municipal water supply—Management.
2. Municipal water supply—United States—Management.
I. Title.
HD4456.G75 1986 363.6′1′068 85-26580
ISBN 0-471-82914-5

10 9 8 7 6 5 4 3 2 1

DEDICATION

This book is dedicated to Murray B. McPherson who was a pioneer and advocate for the field of urban water research. The field is devoted to improving the welfare of humanity, and "Mac" was responsible for many advances.

Mac's primary professional contribution was some 15 years of dedicated service as director of the Urban Water Resources Research Program of the American Society of Civil Engineers. During his tenure there he relentlessly pursued America's urban water managers and engineers in an attempt to get them to put their research discoveries and ideas down on paper for the benefit of the profession. The result of this effort is an outstanding collection of technical reports and papers that describes the state-of-the-art of many of the most important urban water problems. Because these papers are very valuable, a complete list of them appears in Appendix 2. Most of them can be ordered from the National Technical Information Service (NTIS).

Mac died too early on August 20, 1981. He had been honored several times by his colleagues, and one rather complete tribute is contained in the *Proceedings of the Conference on Stormwater Detention Facilities*, published by the American Society of Civil Engineers, New York, 1982.

PREFACE

The ideas that have gone into this book have come generally from two sources: those that have to do with the "infrastructure problem" and those that come from the research that has been done on improving operations in urban water systems. In the case of infrastructure I became aware of the problem years ago in conjunction with courses I was teaching about urban engineering problems, and I realized recently that what we were hearing about as a national "crisis" was the same set of public works management problems that we have been working on for a long time. What they need for solution is the same approaches we have been following, but with more money and better management. That is to say, with adequate finance the solutions needed are combinations of engineering and management; the same combinations that gave rise to the profession of public works management.

This book was written to describe potential solutions to the infrastructure problem to engineers, planners, and managers. I hope that those who take the time to read it will find some new ideas or at least different ways to present some old ones. At the least, the combination of ideas may serve as a good review of the literature on modern management principles applied to urban water systems. Many of the concepts have appeared in government documents and professional journals of different fields, but the way they are assembled here is new. Since some of the material presented here is from difficult-to-find research reports, the reader will find a great deal of material that appears to be new even though it is available in reports of previous research. There is also a good bit of original research reported in the book, some that has not yet appeared in any journal.

The professionals that should find the book of interest are water managers in the local, state, and federal governments; boards of governors of water organizations; consulting engineers; water and wastewater product developers in industries that serve utilities; and associated professionals such as planners, developers, and attorneys who have occasion to work on urban water problems.

I would like to acknowledge the research programs of the federal and state governments that have given me the opportunity to study urban water problems over the past 15 years. My first research grant as a professor was one entitled "legal constraints on urban water systems," one of the engineering initiation grants of the National Science Foundation. Without that grant and subsequent others I would not have been able to spend the time to research the subject of managing urban water systems leading to the courses I teach and to the writing of this book.

Several persons have helped me over the years in my work in urban water systems. I mention first Murray B. McPherson, who as the director of ASCE's research program in urban water helped many of us. Some of the members of his advisory council and of the Urban Water Resources Research Council were also very influential in advancing the cause of urban water. I learned a lot from these men.

During the late 1960s Scott Tucker worked with "Mac" and became an expert in urban water problems. I subsequently replaced Scott at Colorado State University and was able to take over research projects concerned with the automation of urban water systems. This led to many valuable experiences in urban water for me, and I have had the benefit of continued professional contact with Scott over the years. He has performed in an exemplary manner since 1972 as the executive director of the Urban Drainage and Flood Control District located in Denver. I would also like to acknowledge the help I received from my past partner in the consulting engineering business, Dave Sellards. We worked on a number of urban water problems together after we founded Sellards & Grigg, Inc., in Denver in 1968 and many of my early experiences were due to his help.

Finally, I would like to acknowledge the positive influence of many of America's public works and water utility managers. I admit that I am always impressed by the contributions of dedicated public servants, and the field of urban water has a large share of them. They more than others are responsible for the continued effective operation of our urban water facilities.

NEIL S. GRIGG

Fort Collins, Colorado
February, 1986

CONTENTS

URBAN WATER INFRASTRUCTURE

CHAPTER ONE

MANAGING URBAN WATER SYSTEMS

Urban water management is a general term that means taking care of the water supply, wastewater, and flood protection needs in urban areas. Without a doubt these are among the most important of all urban public services. Access to a safe, reliable water supply is an important goal for citizens and political leaders in every country. And with accelerated urbanization, wastewater and stormwater management have taken on far greater importance than in former years.

The importance of improving the management of systems was highlighted during the early 1980s when the "infrastructure crisis" received so much attention. This "crisis" was not something new but a refocusing on the badly needed investment and maintenance needs of public infrastructure systems, which have traditionally been neglected. The reasons for the neglect are complex, but what seems apparent is that solutions to the infrastructure crisis must involve better management of existing and new public systems.

This text discusses the factors that lead to effective total management of water systems in urban areas, including water supply, wastewater, and stormwater. It is intended for those managers, engineers, and planners who work with the planning, management, and productivity parameters of urban water systems as opposed to only technical design matters. This includes professionals who seek to get into management or who have a need to know how urban water systems work and students who need a place to start learning about the overall aspects of water systems before they study the systems in detail. The text follows the assumption that the future will bring different management challenges than the past and that engineers will need to know more than just design practices to develop and maintain effective water systems. Managers will feel a need to know the whole picture before getting into the details. This text attempts to paint the whole picture of urban water management, following the assumption that as the whole story comes into view, then the pieces can be better planned and designed.

The text discusses technical design questions as they relate to management and planning, with many of them concentrated in a chapter on engineering essentials. The text also includes many observations about the business of water, following the assumption that improving productivity in water systems is generally the same kind of problem as in other businesses. Also, improving productivity presents some interesting business opportunities for manufacturers and service industries.

The organizing concept of the book is to present the principles of management that lead to raising productivity and improving service through management of urban water systems. "Management" is taken here to include the tasks that are under the direct control of those who have responsibility for ensuring the success of the water undertakings, and includes planning, engineering, and operational management, with special emphasis on the use of computers. The text covers enough of the principles of planning to relate them to the success of the water undertakings. It presents the main points of design engineering from a management viewpoint and it covers the control of the engineering function to show how to direct its outputs toward higher productivity. The book covers some techniques of administrative management and focuses on the responsibilities of managers to direct operations so that higher productivity is achieved. And what is higher productivity in a water program? It means the same thing as in a business: better performance and service at lower cost. As a spinoff from good urban water management, we also get the benefit of a better environment in a direct sense when we meet the requirements of a wastewater discharge permit and in an indirect sense when we enhance an area through a better water supply system or stormwater project.

In addition to some general principles of planning, engineering, and management, the text focuses on improving operations through using management information and decision support systems better. The chapters on operations and maintenance, automation, and decision support systems are all aimed at presenting the state of knowledge of using computer-based information-handling techniques for improving management.

A final feature of the text is a careful consideration of the institutional and political factors that constitute constraints for the water manager. These are important since the gains in productivity and the improvements in management must be won in an environment that is largely adversarial and charged with risk. The title of the 1970 book by Robert Moses, the famous public administrator in New York City explains the hazards: *Public Works: A Dangerous Trade* (1).

The general thesis of the book is that water is an essential service, part of the urban infrastructure, and essential for the city to survive and the people to prosper. If the infrastructure is adequate, then the city has the opportunity to support a high quality of life; therefore, it is of utmost importance that the urban water system be given the care and attention it needs (including good leadership and adequate funds!).

Notice that the implication is not that the furnishing of a good water infrastructure will by itself bring a high quality of life. This is an important fact to recognize for effective total urban management: it is primarily the social, economic, and intangible factors in the city that lead to the quality of life; the infrastructure constitutes a necessary, but not sufficient condition. But the infrastructure is a starting point to make the other factors possible. This is important since the water manager must be able to articulate the reasons why programs need support. A lack of salesmanship results in inadequate budgets and leads to a loss of good personnel, which leads to a breakdown of the systems. It is important that the manager have a good understanding of these underlying facts.

The assumption behind the book is that if urban water managers know the principles of management as they apply to urban water systems, then management will improve. Accordingly, the following specific objectives are set forth:

1. To understand the general principles of management as they apply to urban water systems, including planning, organizing, directing, and controlling.
2. To understand the financial dimensions of all three urban water systems, including the cost factors and sources of finance.
3. To know about technological opportunities for improving urban water systems.
4. To know the productivity measures that can be useful in managing urban water systems.
5. To see the possibilities for integrating the three services.
6. To see the role of computers and decision support systems in raising productivity in the water services.
7. To fully appreciate the importance and principles of operations and maintenance in fulfilling the objectives of urban water systems.

When the above objectives are achieved, the manager should be able to do a better job of running an urban water system, be able to serve effectively in a regulatory organization, be able to carry out feasibility studies, see business opportunities in the water industry, and know where the productive places for further study are.

LIFE CYCLE MANAGEMENT PERSPECTIVE

In preparing this book I constantly asked myself, what are the most important points to include if our objectives are to gain an overall management perspective, one that will lead to excellence in the management of urban water systems? The answer came back that there were two concepts that were absolutely vital. One was an emphasis on making life cycle management work in the urban water organization, and the other was the practice of effective management principles all through the life cycle of the systems. The purpose of this section is to introduce these concepts, and the rest of the text is dedicated to explaining the procedures that can make them happen.

"Life cycle management" simply means effective management all through the useful lifetime of a system. The concept can be illustrated from the cycle of human life. If we take care of our physical and mental health properly, we can be useful, functioning citizens well into old age. The same is true of a physical water system, and when it finally reaches the end of its useful life, it should be replaced, just as a human dies and new life begins.

Although we need to pay attention to life cycle management concepts, we should not get too enamoured that it is a totally new or self-sufficient concept. It was explained in a paper at the 1983 Public Works Congress and Equipment Show by N. M. Hoyt (2), who pointed out the procedure in use by Public Works Canada, and cautioned that having some kind of automated total system to evaluate systems through their entire lives would be unwieldy. The management subsystems they used were a project delivery system, a real estate investment analysis system, and an automated works management system. Hoyt related a comment about the systems by a business executive who stated that he liked the system, but that with all the unknowns and the need to make judgment calls, he wasn't sure it added anything; that as an investor he wouldn't rely on it too much; and that as an operator he wouldn't have the time and patience for it, due to the constant need to perform in the short term.

Life cycle management has a precise meaning to industrial engineers, of course, as any project or piece of equipment has a definite useful life, and the economic analysis must be made with appropriate interest rates and operation and maintenance costs over the lifetime, as well as considering the salvage value and replacement cost in an appropriate way. With this viewpoint, we have two ways to consider life cycle management: one a broad philosophical approach and the other a specific economic approach.

I have a simple model for life cycle management in urban water systems. It is based on the concepts described above, but it shows how effective management is needed all through the life cycle of a system (Figure 1.1). The useful service life of the system is preceded by the

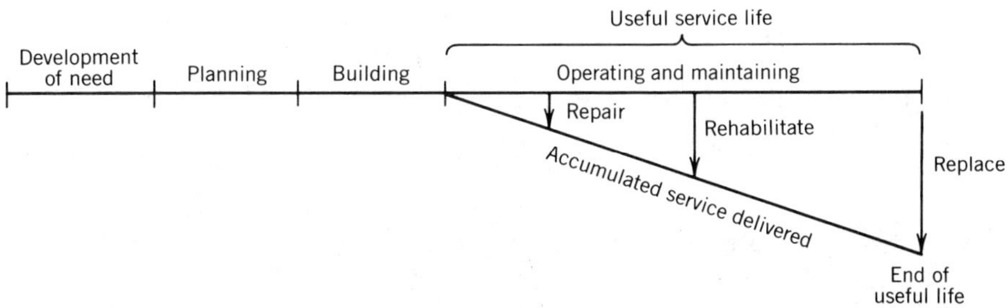

FIGURE 1.1 The life cycle of an urban water system.

three phases of developing the need, planning, and building the system. Each of these has its unique requirements for success and it is explained later in the text. Figure 1.1 shows three events during the lifetime: repair, rehabilitation, and replacement. These are shown to indicate in general that these measures will be necessary and are critical to the effective management of infrastructure systems. They come in degrees; repair is the minor event, and rehabilitation and replacement are major events, essentially falling into the category of major investments rather than routine work. Procedures for determining the timing of these three "R's" are given in Chapter 9, which covers operations and maintenance.

The first phase shown in Figure 1.1, the development of the need, has always been a critical one in public works, especially for water facilities, since it is always easy to postpone needed facilities for visible public projects that enhance the image of the political leadership. A new stadium, for example, or a civic center, might get funded before a needed wastewater treatment plant. Chapter 6 (planning and management essentials) points out that developing the need will end with a commitment to proceed with a project. Only then can we be sure that our planning will be taken seriously. The techniques for planning are also discussed in Chapter 6. Building is an activity that requires close attention to cost and quality, and procedures are discussed in Chapter 6 under quality control and in Chapter 7 as part of the essential aspects of engineering for urban water systems.

In concluding this section, I have asked myself what are the rules for success in life cycle management? They are the same as for the management of any capital-intensive enterprise. Although this is essentially the subject of the entire text, the rules seem to fall into just a few large categories.

1. Make sure of the need and commitment for the system before undertaking to plan and build it.
2. Plan it well.
3. Build it with quality construction, equipment, and operating arrangements.

4. Operate it with the best available strategies.
5. Get the most from the system through life cycle management, especially with effective maintenance strategies.

The latter two points are especially important. They include financing the system to generate enough revenues for effective life cycle management, making sure that both preventive and corrective maintenance is done well, and having effective capital planning and budgeting in place.

Although these points seem simple, those experienced in management of urban water systems will recognize that there are many roadblocks and obstacles in the path to success. Hopefully, the remainder of this book will offer some useful suggestions for overcoming them.

CRITICAL ROLE OF INFORMATION IN MANAGEMENT

If there is anything unique about the goals of this book, it is that they attempt to place a perspective on the roles of automation and information management technology in raising the productivity of urban water systems. In searching for the possibilities for information management to play such a decisive role, the ways that information management connects to decision making and to the effectiveness of management have been considered. This connecting point seems to be through the data base and the design and management of information systems. The overall concept for this is described in Chapter 8, which covers decision support systems, considering that the decision support system is, above all, a way to manage and enhance information to improve decision making and management.

The key role of information is previewed in the observations made above about the critical determinants of success in urban water systems. The first one, obtaining a commitment, is a combination of political consensus building and management planning. Since this process is so important, it is illustrated in Figure 1.2, which illustrates the combination of the political and management factors in deciding what to do. The planning aspect of management illustrated in the figure depends, of course, on the adequacy of the information data base.

The key role of information in the remainder of the processes is illustrated in Figure 1.3, which shows the central role of the data base and

FIGURE 1.2 The process of developing commitment to build or make major changes in a system.

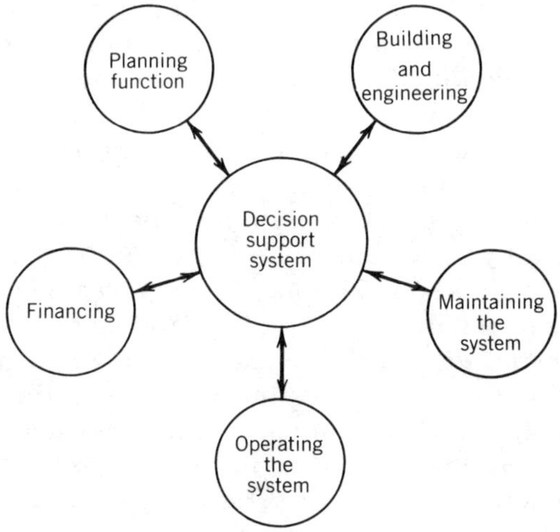

FIGURE 1.3 Key role of decision support system in management activities.

the decision support system in handling the functions of planning, financing, building, operating, and maintaining the system.

The shifts toward more and better use of information in management are already evident. The remainder of the text will present the factors and developments of new approaches to management of urban water systems using information technology. This will be the main subject of Chapter 8, on decision support systems, and Chapter 10 on automating operations. Since the field of information management is moving so fast, it has proven difficult to present every development, but the most important ones will be presented.

INFRASTRUCTURE CRISIS AND URBAN WATER SYSTEMS

The infrastructure crisis received considerable attention in the early 1980s, but it is inevitable that this attention will recede. The water components of the crisis have not been as dramatic as the total needs estimates, which have been dominated by the highway repair and replacement problems, which reach 20-year needs levels of nearly $2 trillion. Water needs for the same period are more "modest," with needs ranging only in the hundreds of billions of dollars. Specifically, water supply system needs seemed to be in the range of about $125 billion for expansion, replacement, and the handling of deferred maintenance; wastewater needs seem to be in the range of $100 billion, according to Environmental Protection Agency (EPA) needs estimates, not including combined sewer overflows or stormwater. The needs for combined sewer systems were fixed by EPA at $35 billion in 1982 and for stormwater pollution control at $93 billion. However, these figures relate to pollution control only and do not include needs that might be generated by urbanization or to solve flooding problems. Also, the need for stormwater systems to fix local nuisance problems could be vast, perhaps exceeding $200 billion. Since there is no accepted definition of "need," no one knows what the requirements really are.

Exactly what the infrastructure crisis will mean to urban water system managers is not fully clear, but some trends are certainly evident. On the water supply side, the $125 billion figure quoted above had its origins in two government reports of the early 1980s, both which seem to essentially agree on the extent to which new treatment plants will be needed (3). The $125 billion was an estimate for the 756 largest cities and included treatment plants, distribution systems, and storage capacity. The original estimate was by the water supply study group of President Carter's Intergovernmental Water Policy Task Force, which was working in the late 1970s, so the estimate needs some correction for inflation. Also, the needs of the many small water systems may not be adequately accounted for in the studies. A market researcher, William Lorenz, reports that over 100 new water treatment plants will be needed each year in the 1980s and that by the year 1990, the market for water treatment chemicals will have increased by two-thirds over the 1981 figure of $300 billion (4).

In the case of wastewater, the retreat of the federal government from the financing responsibility will be the biggest determinant for the 1980s. In an interview with a water newsletter the newly elected president of the Water and Wastewater Equipment Manufacturers Association was predicting in 1985 that this retreat—the new emphasis on privatization—would mean new opportunities for equipment companies due to the increased emphasis on performance and quality. The Wall Street Journal reported in early 1984 that there were still some 4700 wastewater plants to be built. This is probably the number on the funding list awaiting federal funds, so it is still unknown how many will be built.

In the case of combined sewer systems and stormwater, there is really not any reliable way to know what will happen, since the need for stormwater systems is not really known. I made an estimate in a recent paper that the total needs to construct stormwater and combined sewer systems was on the order of $150–200 billion (5). This estimate was meant to represent the amount that can reasonably be expected to be spent over the next 20 years, and would represent an investment of around $5–10 billion per year, which seems about right, especially when systems to meet urbanization demands are taken into account.

What we do know about the infrastructure crisis is that it will only be handled with good management. The funds to construct hundreds of billions of dollars of new facilities will not be made available, except incrementally. This is a period of fiscal austerity in government, and it seems destined to continue; this means we need more attention than ever on good management.

HISTORY OF URBAN WATER SYSTEMS

The linkages between the three components of the urban water system are apparent when reviewing the history and stages of development of the subsystems. For example, the story of the development of the water supply for the City of Rome is a starting point for this history (6). When Rome was at its peak, its water supplies were quite adequate, with the sources being the aquaducts that are famous today. There were, of course, no treatment plants or regulatory agencies such as we are familiar with today. When the grandeur of Rome ended with the sacking by the barbarians, the water supply system was destroyed and this ended the high quality of life in the city for hundreds of years.

While it was historically necessary to provide drainage for cities, the use of sewers to remove household and other wastes is a comparatively new phenomenon. There is quite a bit of literature about sanitation and related public health issues in the 19th century, but little from earlier times. Fair, Geyer, and Okun mention a bathroom of King Minos of Greece that was purported to have a latrine that was flushed by rainwater (7). This standard of engineering for sanitation apparently did not return until the 19th century, at least as far as Europe and the United States are concerned.

There was little organization for common action for services such as water supply and wastewater, which were considered private responsibilities for the most part. In earlier times when impressive water systems are mentioned, it is usually in connection with a palace or some other special place. The idea of expending public funds for systems that serve the entire population is comparatively new.

The construction of water supply systems in the United States dates from 1754 when the system for the Moravian settlement of Bethlehem, Pennsylvania, was built. It consisted of spring water forced by a pump through bored logs. In this same period Philadelphia was developing a water supply system, with combined public and private pumping facilities. The pumps were for the most part horse driven since this was before the invention of the steam engine. These were the early days of water supply in America and every city has a unique story (8).

In all of history the incidence of waterborne disease has been a problem, but it was in the late 19th century that we started to see the solutions we know today. During the 19th century advances were made in halting waterborne disease. There were a number of factors, including the protection of water supplies, improvements in wastewater disposal, the beginning of regulation, and an awareness of the importance of sanitation to public health.

In the 19th century there were sewers developed to remove wastewater from the central cities. Storm sewers were developed to drain cities and pressure mounted to use the same storm sewers for the disposal of sanitary wastes. This led to the combined sewer that we see in large cities in the United States and Europe today. By the middle of the 20th century those combined sewers were recognized to be a problem and are still under study today.

In the United States, it was during the 19th century that attention began to be directed to the connections between supplies, water courses, and the management of wastewater. Many communities were served with water from nearby sources such as springs and wells that were easily contaminated. New York City had to finally go a great distance for water with the construction of their Croton Aqueduct which opened in 1842 (8).

There was a growth in the development of both water supply and wastewater management programs during the early part of the 20th century. The U.S. Public Health Service was created, parts of it evolving into today's Environmental Protection Agency, a regulatory agency that promises to be with us for the forseeable future. After World War II there was heightened interest in wastewater treatment, and successive legislation was passed that led to the general system we have today for regulating and treating wastewater. The system continues to evolve to address new and more serious problems such as trace organic pollutants. The Clean Water Act of 1972 and the Safe Drinking Water Act of 1974 have proved to be the most important determinants of policy at all three levels of government. For the most part, policies relating to stormwater have evolved more as a result of concern about water pollution than local drainage and flood control, this also affirming the correctness of addressing the three water services together.

In the 1980's there is a lot of attention given to water management at all three levels of government, and with most of the legislative questions settled, attention seems directed at doing a better job of managing within the system we have. That is the subject of this book.

FUTURE WATER MANAGEMENT

The future holds some very important developments in water management. They are described throughout this text. One of the interesting paradoxes will be the divergence in water management needs caused by two phenomena: urbanization and the development of the automatic society. On the one hand the explosive urbanization in the developing countries will place demands on urban water systems such as have never been witnessed. We are already seeing cities with populations of tens of millions with very inadequate water services, and almost no attention to wastewater management in some areas. Just to gain some control over this situation will require some really imaginative approaches to management and some real commitment, as well as considerable political will. On the other hand there looms the prospect of "telematics," the development of the automatic society where professionals will work at home on their computer terminals and shop by video-based systems, with transportation and infrastructure requirements being dramatically altered.

These changes in the structure of life will have dramatic effects on water management needs. Already the small aspects of the changes are felt. In my own city the transformation of the city from primarily single-family customers to multifamily is proceeding apace. Many of the affluent families are opting to live in dispersed patterns, presaging the future. Some resort areas are touting themselves as workplaces of the future.

The shifts that are occurring are placing high demands on management to respond with solutions that will maintain good water service at reasonable cost. Problems that are anticipated confirm the trends. Managers see the management of the small water system as one of the primary challenges of the future. This is not a question of technology but of finance, training, and incentives. Toxic chemicals are defying the efforts of water managers to develop ways to detect and control them. The prospects of sabatoge loom over water system managers. Finance is a continuing problem as customers are not accustomed to paying the full cost of water service; the luxury of government subsidies takes a long time to wear off. In some parts of the world there simply exists no mechanism to charge for water. In fact, in some societies there is no tradition of credit, and water bills would have to be collected daily if they were used. Management operations will continue to be a challenge. How to operate the increasingly complex systems and plants with inadequately trained and financed work forces is a reality that is faced daily, and solutions are needed. This will especially be a problem in the complex large-scale urban areas where systems are interdependent.

Chapter 11 concludes the text with a discussion of the main issues that will emerge for urban water managers in the remainder of the century. The bottom line is that management and technical operations must continue to improve, and the achievement of this goal will depend on engineers and managers understanding the importance of management, especially how information systems and automation can be put to better use, and committing to improving their personal capabilities and those of their staff members.

REFERENCES

1. Robert Moses, *Public Works: A Dangerous Trade*, McGraw-Hill, New York, 1970.
2. N. M. Hoyt, Life Cycle Management of Public Works, *APWA Reporter*, April 1984.
3. Congressional Budget Office, Public Works Infrastructure: Policy Considerations for the 1980's, Washington, 1983.
4. William Lorenz, New Business Opportunities, *Pollution Engineering*, September 1983.
5. Neil S. Grigg, *Urban Storm Drainage Investment Problems: Policy Questions, International Conference on Urban Storm Drainage*, Goteborg, Sweden, 1984.
6. Sextus Julius Frontius, *The Two Books on the Water Supply for the City of Rome, AD* 97 (trans. by Clemens Herschel), republished by the New England Water Works Association, Boston, 1973.
7. Gordon M. Fair, J. C. Geyer, and D.A. Okun, *Water and Wastewater Engineering*, Wiley, New York, 1966.
8. American Public Works Association, Historical Society, *History of Public Works in the United States*, American Public Works Association, Chicago, 1976.

CHAPTER TWO

THE INTEGRATED URBAN WATER SYSTEM

The term *urban water system* implies that there is a single system, rather than several different ones that provides water service in urban areas. It is true, there is a single urban water system, and the result of this reality is a need to view it as an integrated whole; this can lead to the solution of problems and to large cost savings. This text discusses many ways to improve urban water management, but the unrealized opportunities for improving urban water management are greater in the areas of integration and in cooperation than in any of the other areas of management and technology. The reason for putting this chapter early in the book is to highlight these opportunities and to describe how they can be exploited.

In setting the stage for this chapter I would like to point out that water organizations and management practices are the way they are because of history. Often the current way of doing business is recognized as inferior to other ways, but we continue since it is often difficult and maybe impossible to make substantive changes, especially in organizational and political arrangements. The material in this chapter provides examples of how we might be able to improve. If we could wipe the slate clean and design new and "optimal" ways to do business, things could go much better. If we cannot, what do we do?

Urban water managers and engineers sometimes think only of the parts of the system where they have responsibility for management or design; they do not consider the advantages of approaching management from an "integrated viewpoint." Sometimes they do not think of what can be accomplished even through cooperation with other nearby water organizations. This is an expected situation, since we often do not have the luxury of "blue sky" thinking. However, if we are to make progress in the future, we must introduce innovations, and the concept of integrated management of urban water systems is an attractive way to look for possible improvements.

This chapter covers first the hydrologic connectivity of the natural water systems and the resulting need to protect our water from abuse and contamination that results from unexpected effects of growth and development. The hydrologic effects of urbanization have made the interdependence and connectivity of urban systems painfully obvious. Such problems as groundwater contamination and flooding that result from urban development have raised the public's awareness of water problems in recent years.

Recognizing the interdependence of natural systems, the discussion then shifts to focus on the possibilities for improvement of service through integration and cooperative management of urban water systems. Integration and cooperation offer three main classes of benefits. First, there is the economy of scale that comes from joint development of projects and services. Next, there is the possibility of offering service where there was none before, by taking advantage of centralization. This can be particularly useful in rural areas and in developing countries. The third category of benefit is the protection of the quality and abuse of the water resource by integrated management and protection of source development and wastewater disposal. The benefits available from cooperative management are the same as for integration, but cooperation offers a way to achieve some of the benefits without giving up the autonomy of decision making that becomes necessary when systems are integrated.

INTEGRATION AND COOPERATION IN URBAN WATER MANAGEMENT

Integration can be by service (water, wastewater, flood control) and by area. We refer to these as functional integration and areawide integration. In other words, if the water and wastewater utilities are combined under the same management, that is functional integration. If a regional service authority is organized to offer water service, that is areawide integration. Cooperation refers to working together without combining the management organizations. Integration can be in part or in whole, and it can involve different levels of the same organization.

There are many variations to the main themes of integration and cooperation. They lead to many interesting management and political scenarios. These range from buyouts to political struggles in state legislatures and in courts and to struggles between powerful business and political interests. Integration can involve the consolidation of parts of organizations, cooperation between different water services, and negotiations between different levels of government.

One of the most common scenarios is the financial situation, where there is a proposal for integration or cooperation, such as for the long-term rental of water or for a buyout of one system by another. Another situation is the one where risk or drought or pollution can be shared by cooperation. Another is the economy of scale scenario, where costs can be lowered by consolidation through integration or by joint development of projects, with contracts for cost sharing. Another is the political situation where integration becomes a bone of contention between competing parties. Finally, there is the environmental scenario where one group demands integration (and control) through regional or areawide integration. In this latter case the power of a large water organization could be diffused through such integration, and there would naturally be great resistance to such moves.

A comprehensive reference discussing the details of integration possibilities is the report of the urban water panel of the National Water Commission (1). The commission studied the management of urban water organizations in detail, and its observations are included in the commission's final report and in detail in the reference cited. There are not many such generalized references on the subject of integration of urban water management, however, since the possibilities for it are so site specific. McPherson's writing dominates the literature of the 1960s and 1970s, and in the 1980s it seems that the political realities have taken over and the few discussions that do appear in the literature are of realistic and case-by-case examples where actual improvements have occurred. In other words, managers know that the nitty-gritty of making integration and cooperation work is the most difficult part, and we do not see many theoretical papers; all of the work to be done is in the trenches.

Figure 2.1 shows the three general parts of the urban water system as they fit together. There are many linkages, but the obvious organizing concept is the hydrological cycle. The figure is itself an argument for integration of the management of the three urban water services together, since the rain that creates stormwater can also overload the wastewater system through inflow and infiltration, as well as serving as a source of drinking water supply. If the management authority can be unified, then the same information can be used to make coordinated decisions, and the resulting opportunities for improvements in management effectiveness are obvious. The main reasons that more progress toward integration has not been made are that the water systems were built incrementally and financial and political considerations have the effect of inhibiting cooperation.

The concept of the unified urban water system as used here originated from the work of the Urban Water Resources Research Program of the American Society of Civil Engineers (ASCE). This program was directed by Murray McPherson when it was evolving in the later 1960s and through the 1970s. A good comprehensive reference for the integrated urban water system as conceived by McPherson's program would be the

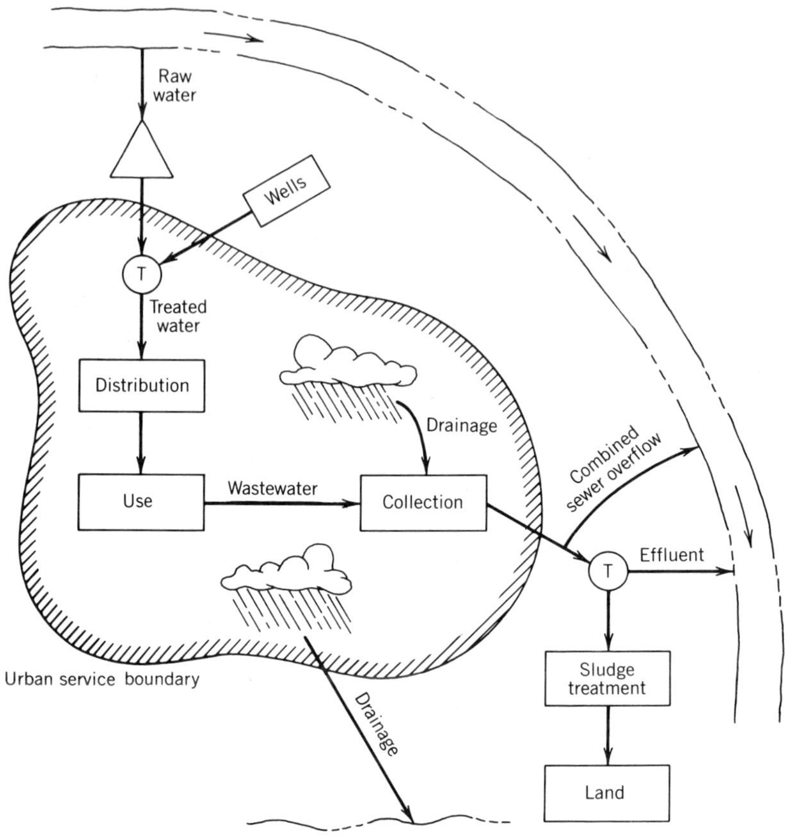

FIGURE 2.1 Components of the urban water system.

document "Prospects for Metropolitan Water Management"(2). Of course, knowledgeable engineers and water managers have been aware of the benefits of regionalization for a much longer time, and, as stated earlier, much of the real work has been done in the trenches.

As seen in Figure 2.1, the water systems of the city include all the interfaces with the hydrological cycle: tapping it for water supply, discharging treated wastewater to the receiving waters, and handling the storm and flood waters as they pass through the city. For this reason, we speak of the "urban water system" as consisting of the three water programs together. Since the wastewater stream for one city becomes the next city's water supply, it makes sense to look at them together. Also, since the same floodwaters that must be controlled can be a beneficial source of supply, they should be considered as part of the system. There are other interfaces to the system. For example, the beauty of a city will be dependent to a certain extent on how well the waterways are managed within the boundaries of the urban area. Also, the protection of the soils in the city and the control of erosion are directly connected to how well the adverse consequences of the hyrologic effects of urbanization are controlled.

The concept of a single "urban water system" is not fully accepted in practice due to custom, and some would consider it as a rather abstract concept not very useful in managing real systems, especially with all of the political constraints. Some are accustomed, for example, to referring to the water supply system alone as the urban water system, and that custom will endure for a time in the future since water supply has always been the most important of the urban water systems. However, it seems a good idea to get management thinking pointed toward integrated water management, and I urge the adoption of the above concept of the integrated system. As stated before, this is a book to stimulate thinking about how management can be improved. The ideas will not appeal to everyone, but they are intended to suggest how the overall management of water in urban areas can be improved by considering the integration possibilities.

The separate viewpoints for different water services are largely due to fragmented thinking brought about by historical, organizational, and political factors. These are very important constraints inhibiting the improvement of urban water service. Two initial examples will illustrate this. Later in this chapter some more detailed examples will be discussed.

First, consider the problem of providing an assured metropolitan water supply for all the municipalities in a region. It is the political factors that inhibit the cooperation and sharing of supplies that can solve this kind of problem. If the municipalities would only agree to get together and provide interconnections and guarantees for risk sharing, the overall cost of water could be reduced and the level of security increased. Some U.S. cities have made great progress in cooperating in their metropolitan water supply efforts; others have not. These are discussed later in the chapter.

As a second example, if one city's wastewater becomes another's water supply, why don't we have better wastewater treatment? Again, it is the social, political, and financial forces that cause the need for regulation and the resulting adversarial process that raises costs and causes conflict. By this I mean that even if individual citizens in an upstream community would by conscience prefer to send their neighbors downstream clean water that it might not work. The political process might well corrupt their desires and prevent them from acting according to conscience. This is not always the case, but it appears to be the usual case to this observer. As a result, it has been the exception rather than the rule that cooperation works in wastewater management. Areawide integration has worked in wastewater treatment since charges can be levied that compensate for different levels of waste load treated.

From the ecological point of view the city can be viewed as an organism that superimposes human behavior on the natural environment.

In fact, the concept of "urban ecology" has evolved in the field of biology to describe this point of view. The water system is the most important part of that urban ecological system, both from the points of view of survival, including basic human functions such as waste disposal, and from that of the higher human aspirations, such as for a clean and beautiful environment, one that is conducive to "communing with nature."

From a practical point of view, meeting the water part of these requirements means that certain services must be provided. These services are those that deal with the water part of the total urban services requirement. In studying the spectrum of services supplied in cities, there are several ways to classify them, but the one I use was introduced a few years ago to classify the "urban system" in Denver: The urban system = socioeconomic + socioenvironmental + linkage subsystems. This classification was developed in an attempt to apply urban systems engineering to urban drainage and to solid waste together. The effort to apply systems engineering was part of the general trend in the early 1970s to try to make space technology work to solve civilian problems. This particular project was called "Project REUSE," for "renewing the environment through urban systems engineering." Both urban drainage and solid waste management would be part of the "socioenvironmental subsystem." The other subsystems, socioeconomic and linkages, place the total system into perspective. All of the water systems would fit in the same socioenvironmental subsystem.

The socioenvironmental subsystem corresponds to the public works management categories and includes the following elements; the urban water system, land-use control, air quality management, solid waste, streets and traffic control, airport management, and public buildings. There are related aspects of infrastructure, especially transportation services and telecommunications, that would fit into the linkage category. These are expected to grow in importance in the future. The socioeconomic and the linkage subsystems are the ones that relate more directly to human activity such as business, education, travel, and communications.

Another way to classify these systems, and one that appeals to me more, is to view the physical infrastructure as the support structure of the city, and to let the social and economic activities be part of the social system that is superimposed onto the infrastructure. Viewed in that way, the infrastructure is seen in its proper role: to provide the support so that the socioeconomic activities can take place.

It seems imperative to this observer that if in the face of limited resources and rising urban problems we are to continue to maintain and improve the quality of life in the future, management will have to move toward more cooperation. There are plenty of examples where divided forces and political opposition has worked against the public interest, and the opportunity to cooperate through the sharing of water management responsibilities is one of the natural areas to develop cooperation. This can work toward better service to the public in advanced urban areas, and in some parts of the world where water services are practically nonexistent, it can be a life or death matter. Said another way, we cannot afford not to cooperate in the solution of urban problems. The remainder of this chapter will discuss some detailed aspects of integration and cooperation and provide some examples of where they apply.

As discussed earlier, there are two principal reasons to consider managing urban water systems in an "integrated" way. The first is the undeniable fact of the natural interconnection or connectivity of the systems through the hydrologic cycle, and the resulting observation that nature demands an integrated viewpoint. The second is the possibility to get real benefits from integrated management through the concept of synergism; meaning that by integrated management we can get more benefits from the system than by independent action. These two reasons for integration are the subjects of the next two sections.

CONNECTIVITY OF URBAN WATER SYSTEMS

It is basically the hydrologic cycle itself that connects the urban water system together. This is what led McPherson to call for a "metropolitan water inventory and balance," meaning that we need to have a water balance sheet and to understand what is happening to our water, quantity and quality, as it flows through the urban area. This is an expansion of the concept of the "water audit" of supply systems and of the infiltration and inflow survey of wastewater systems, and represents a concept that will become increasingly important in the years to come. The key will be the information system, including data collection, which will be required to monitor and report on the events in real time. Without a valid information system there is really no hope for any real long-term and large-scale improvements in management. The information system has to be effectively designed and managed since managers have no patience with useless information, and the systems will die if they are not used since there will be no financial support for them. Chapter 8 is devoted to the discussion of how to develop and use information systems to improve management.

The connectivity can be described best by a diagram. Figure 2.2 shows some of the ways that the subsystems are linked. The basic input, precipitation, becomes surface runoff from both the pervious and the impervious surfaces. It occurs more rapidly and in greater volume once a surface has been paved. Paving removes the possibility of biological and physical filtering of the urban runoff and prevents the local groundwater table from being recharged.

The diagram shows the impervious runoff going to a small surface

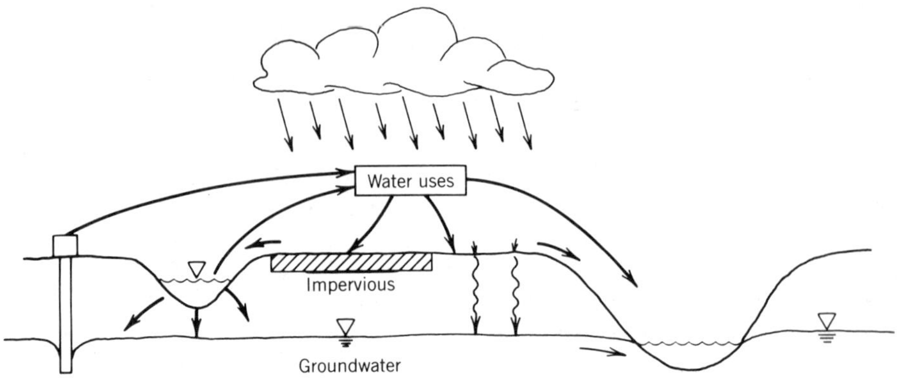

FIGURE 2.2 Some connections between urban water subsystems.

stream that infiltrates to the groundwater table (an "influent" stream). This will carry contaminants from the surface washoff to the groundwater. A well is shown pumping from the groundwater to the water users, along with additional water supplies from the same surface stream. The water is discharged to the larger stream after use and treatment with the stream also receiving inflows from the groundwater table. The rainwater is prevented from infiltrating by the impervious surface but is able to infiltrate through the pervious portion of the city's soils.

The two principal effects that result from the natural human processes shown in Figure 2.2 are rather obvious: a worsening of flood peaks and a deterioration of water quality. These effects are discussed in more detail in Chapter 5, which deals with stormwater, and they illustrate the importance of proper management of the stormwater subsystem in maintaining an integrated approach to overall urban water management. It may well turn out that controlling these effects will be the major challenges to urban water management in the years ahead.

Another less obvious feature of the hydrologic cycle is the influence of urban areas on the climate (or microclimate) that surrounds cities. This makes the cycle complete in that the city sends heat and contaminants to the atmosphere that are returned in precipitation. Thus, in nature the urban water system is integrated whether or not we manage it that way.

The hydrologic effects of urbanization illustrate why it is so important that urban water management be integrated. At the present time it is normally the stormwater program that is responsible for monitoring these effects, but stormwater programs are the most neglected and least understood of all the urban water services. There has been some movement toward making stormwater quality a responsibility of wastewater managers, and in the future we must do a better job of stormwater management. This is one of the legacies of McPherson's research: an understanding among a few engineers and planners of the importance of stormwater in the big picture of urban water management.

INTEGRATION, COOPERATION, AND SYNERGISM

The opportunities to take advantage of the connectivity of the urban water system follow naturally from the identification of the connections. The ones that seem most important are: the possibilities for improvements through areawide integration; cooperative management and functional integration; the opportunities to use stormwater as a source of supply; the opportunities for water reuse; the opportunities to protect water supplies through good land-use management; and the possibility to improve water management through the general concept of the dual water systems.

Some large systems have built-in opportunities to take advantage of economies of scale. The systems of supercities such as New York, Sao Paulo, Tokyo, and others are probably so large that there are few opportunities for further improvements through integration and cooperation. However, there are many urbanizing areas, particularly those around large core cities, that have obvious opportunities for improvement. Even the supercities, however, can benefit from cooperative ventures. In the case of New York, for example, the city reaches into upstate New York for its supplies. It relies on the continued protection of its watersheds for the future quality of its raw water supplies, and it is certainly possible for it to enter into cooperative ventures relating to water supplies with its neighbors. It would be difficult to see any advantage, however, in further regional cooperation in distribution of water supplies since the city is already serving such a large area.

The use of cooperative management for improving the general level of service and cost-effectiveness is well recognized by those that favor regionalism, but there are many who do not, and regional cooperation has proven to be a tough objective to achieve in U.S. local government. Functional integration, also an obviously attractive alternative, has many detractors as well. Often the management initiatives that seem most attractive are not feasible from the political and human points of view. These improvements should continue to be held up as goals, however, with the long-term intentions to make improvements when they are possible.

Areawide Integration

Areawide integration involves the combination of services across an area, such as a large metropolitan area or a county. Examples include the wholesale water supply operation of the Metropolitan Water District of Southern California or the metropolitan sanitary districts found in cities such as Chicago, Milwaukee, Seattle, and Denver. The main arguments for areawide integration are economy of scale, better use of natural resources, and improved operating efficiency.

The recent literature has several examples of advances in areawide integration. Failures and frustrations are generally not reported. In some cases the differences between integration and cooperation are not distinct. Some examples of moves toward areawide integration will be discussed first, and the next section presents some examples of cooperative ventures.

One recent example of a move toward integration is the interest in Fairfax County, Virginia, to acquire the City of Falls Church water system (3). This would be an example that involved consolidation. Specifically, the Fairfax County Water Authority, a publicly owned system, would buy out the Falls Church system, another publicly owned system. There are not yet any journal publications on this activity since the studies are still underway at this writing, but if the effort is successful, it would involve an acquisition, complete with compensation, and the partial consolidation of services on an areawide basis. This case is typical of the scenario where water authority that has been created in a growing area becomes stronger over time, and ultimately begins to take over the smaller entities around it. In this case the authority was created in 1957, and there has been tremendous growth in its service area since then. Dulles International Airport, for example, is within the service area.

There are many ongoing examples of the creation of regional water authorities, such as the one created in the west coast area of Florida in 1974 (4). In this case the cities of St. Petersburg and Tampa and the counties of Hillborough, Pinellas, and Pasco have banded together to form the West Coast Regional Water Supply Authority. In this form the authority can join together with other local, state, and federal agencies to engage in joint planning and project development. It remains to be seen whether this innovation results in integration or cooperation, since the degree to which local water purveyors surrender autonomy is still unknown. There are numerous other examples in the United States of water and other service authorities that are in the formative stage, and there are many successful examples of water authorities working within limited mandates.

Sometimes the presence of a powerful central city water agency creates a de facto water authority. The Seattle Water Department represents such a case, and its recent experience in dealing with rate setting for its suburban customers represents a typical problem encountered by water authorities (5). Achieving success in solving this kind of problem can lead to goodwill and the avoidance of problems of noncooperation, which often result in hard feelings and expensive "go-it-alone" solutions.

There is often merit in considering the formation of entities for providing areawide services where there were none in the past. For example, Wubbena and Maxwell describe some countywide improvements that are taking place through the Public Water System Coordination Act of 1977 in the State of Washington (6). In this case we have an example that involves both integration and cooperation. The satellite system allows for a lead utility to take over the ownership of a smaller one or to provide partial assistance with services such as billing and administration. From the report (6) it is working fine so far, and offers some hope for the problems of small water systems.

Small water systems management and rural water problems are of concern all over the world. The development of regional authorities for rural water service is an attractive possibility for improving management in the developing countries where providing adequate water service is a matter of life and death. For a discussion of some experiences in Latin America, for example, the reader is referred to Donaldson (7). The effort involved and reported by Donaldson goes back to 1938, and even since 1961, has involved some $7 billion in international loans and national matching funds. In these cases, areawide integration is not just a management improvement; it is essential if there is to be any service at all.

Cooperative Management

It is only a small step from areawide integration to cooperation among agencies in a common area. Real and effective cooperation has often been the exception rather than the rule, however. The lack of cooperation, a very human problem, has been one of the stumbling blocks to water managers. I personally know of any number of cases where the technical people involved in a water problem will say, "outside the meeting rooms we get along great with our colleagues from the other water organization, but due to politics we can agree on nothing in the negotiations." The loser in this kind of problem is usually the public.

There has been quite a bit of recent attention given to a success story in the Washington metropolitan area where several water organizations got together and determined that if they cooperated they could reduce future water supply storage requirements and capital expenses considerably. At issue were problems that went back 20 years. They were resolved through a task force approach among federal, state, and local agencies, with the important feature that local leaders were involved at every step along the way (8).

Another example of emerging cooperation is in the California Bay area where a number of water organizations are studying ways to cooperate. Gilbert points out that there are some in this effort who immediately call for regionalization and consolidation, but that the most productive effort seems to be through working together in small steps, and maybe the combining of supply efforts, with the retention over management of distribution systems by the existing utilities (9). This conclusion is one that is often seen: look for ways to achieve the benefits in small steps without asking for too many sacrifices by existing authorities. This is no doubt a viable principle for negotiations of all kinds.

The theoretical basis for joint water ventures is sound, and Hardten makes a strong case for the benefits of developing joint water projects

(10). He makes the point that the difficulties in raising capital suggest that real benefits can be realized by working together to develop projects and by cost sharing. This can result in the sharing of resources and risk, in the elimination of duplication, and in economies of scale in operation and capital investment. The complete article offers other principles and is recommended reading for water managers looking for tips about working together.

An area of the country that is still looking for the formula to develop cooperation is the Tidewater region of Virginia. There, a rapidly growing city, Virginia Beach, has depended on a core city, Norfolk, for wholesale water, and now feels the need to develop an independent supply (11). In spite of years of effort and many studies, the area has not yet been successful in resoving its water supply difficulties. Virginia Beach's latest proposal for a new raw water supply has embroiled it in an interstate dispute with the State of North Carolina, as well as in difficulties with the neighboring communities. The proposed project would cost some $160 million in 1982 dollars, and result in a raw water supply of about 60 mgd (million gallons per day). This estimated cost is down from the previous favored proposal, which would have cost more than $300 million, and even the lower figure will impose an additional debt of some $650 per capita on the existing residents. The willingness of the community to incur so much debt reflects a certain degree of frustration with their inability to solve the problem through cooperative ventures, and is a good example of the "go-it-alone" approaches in some urban areas.

Functional Integration

Functional integration means to integrate the different water services into a single organization. It may be the most difficult way to integrate systems, and the benefits are not always so obvious. There will always be pressure for functional integration within systems, since on the face of it there seem to be opportunities for increased efficiency and cost savings. This may be true in some of the administrative areas especially, and in some but not all of the operational areas.

The act of simply combining systems stops short of real integration. With integration there should be some economies of scale and some savings that result from combining of activities at the subunit level. There are numerous examples of combined water and wastewater utilities where there really is little integration of management. There are even fewer examples where stormwater is effectively combined with water and wastewater management.

In organizational theory there are some good reasons to be cautious about functional integration, and some of them apply to water organizations. I had an unpleasant experience with it in a state government office a few years ago, and it taught me some lessons. One of the lessons is that the functional integration of organizations tends to obscure some functions that people identify with, and this causes confusion. For example, if we integrate water and wastewater treatment staffs, and if the state government has different departments to regulate water and wastewater, there will be confusion in the communication links, and problems would occur. As another example, if some employees have worked for years in water supply, and all of a sudden they find themselves in a combined organization, some may not adapt. They will need some training, and measures may even have to be taken to avoid serious problems that result from their unwillingness to be assimilated in the new organization. There is an additional discussion of organizing water programs in Chapter 6 for the reader with more interest in this subject.

In spite of the problems with functional integration, there are some real success stories. In my own community of Fort Collins, Colorado, for example, the water and wastewater utilities are combined under a single manager, and the level of integration at lower levels is evolving. The water and wastewater utility is combined with the electrical utility in a single department, but there is very little integration between the water and electrical operational groups. The water and wastewater treatment facilities are separated, each with its own laboratories. The supervisors of the treatment systems are different people, and many of the operational requirements of treatment are handled by different staffs. The complexities of water and wastewater treatment are sufficiently different that it makes sense to have individual managers for some parts of the systems. On the other hand, water distribution and wastewater collection offers an area of management where integration can occur, and they are combined in a single section. In this section, there is a different supervisor for wastewater and for water distribution, however. It seems that the integration of water and wastewater in Fort Collins has gone about as far as it can go at the present. There are few obvious advantages at this time of integrating stormwater into the utility, but some may develop in the future. The organizational chart for the Fort Collins water utility is presented in Chapter 6.

The Fort Collins' example of integration must be studied carefully to identify the real benefits of integration as a function of level integration. There seem to be benefits of combining water and wastewater, but the only real result of the integration of water with power is the reporting line of the water utilities, and this may change with a subsequent reorganization. There does not appear to be any integration of electricity and water at the subunit level.

Thus the subject of functional integration is not a simple one. Integrating organizations by function rather than separating them by service has many pitfalls and cannot be approached lightly. Many managers are quick to jump to the conclusion that "sure, water and wastewater ought to be managed together, since they are part of the same hydrological system," but the reality is not that simple.

Stormwater as a Source of Supply

It makes sense that stormwater should be considered as a source of supply. There are not many urban areas that do so, however, since stormwater is recognized to be of low quality and not too suitable for drinking water. Stormwater, or rainwater, would be the only source of supply in an arid area that relied on cisterns for supplies. Viewed in this way, the concept is rather ancient. Even though stormwater is often contaminated, not being used much for water supplies in U.S. cities, it makes a lot of sense to look for ways to protect it and use it for supplies in the future, even for industrial and public uses. Stormwater is really "ambiental water," meaning water that is around us in our overall quality of life in urban areas. A natural way to use it is to recharge aquifers with it. This application is increasing.

The better use of stormwater was one of the main focal points of the ASCE urban water research program. McPherson saw stormwater as the environmental water that freely passes through both the water supply and the wastewater management systems. Even though at the present time it is often ignored, the future will undoubtably see more attention given to stormwater as we struggle to improve our environment. Certainly, the concept of managing stormwater comes through in discussions of urban ecology and how cities really should be managed, as in the proceedings of the conference on the role of water in urban ecology held in the Netherlands in 1979 (12). They discussed several elements of the urban water subsystem, including artificial canals, "balancing lakes" (detention ponds), reservoirs, artificial lakes, river fronts, and harbors. The ambiental features of the urban landscape are all related to stormwater. The concept of balancing lakes is particularly interesting in that they can serve to control stormwater peaks and provide for recharge to the groundwater table. These viewpoints have been reinforced in the United States as some have come to call stormwater a "resource out of place." They will be discussed in more detail in the chapter on stormwater.

Water Reuse and Dual Water Systems

The concepts of dual water systems and water reuse go together. Dan Okun of the University of North Carolina has been writing about dual water systems for many years. As he originally presented the concept, pure water was portrayed as of higher value than water not treated to drinking water standards and was envisioned as carried in a separate system for potable use. Lower-quality water would be considered for appropriate uses, such as lawn sprinkling, industrial uses, or perhaps to flush toilets. As Okun presents the concept, even the lower-value water would be treated to minimum standards, and then the expected objections about cross-connection hazards would be answered. Okun's ideas go well beyond these simple concepts, of course, and different aspects of the possibilities of dual water systems are being considered. There seem to be three general branches to the concept. One is the concept of the dual system itself; another the concept of water reuse, where the reused water goes to nonpotable uses; and the third is the concept of using only pure water for sources and not trying to clean up polluted sources. The concept could be extended to consider stormwater as used water and subject to the same uses as reclaimed water.

For all intents and purposes the dual system itself is not very practical in cities that already have large infrastructures in place and where the buildings have been constructed to use the traditional systems. However, since infrastructures will have to be rebuilt in future years, it makes sense to begin considering how economies can be effected at the same time that basic needs, such as the need for pure drinking water, can be met. The message seems clear: while the concept of dual water systems may not seem practical to some, it may emerge in future societies as a viable solution to many problems we are beginning to see.

The concept of water reuse is with us already. Anyone who has been following the literature of recent years knows that there have been innumerable conferences and papers on the subject, mostly focusing on the technical and social aspects of the problem. There are very few examples today of direct potable reuse of water, one being the experiment in Southwest Africa which has received a good deal of attention. Heaton, who directed a project on water reuse for the American Water Works Association for several years, states that no one has yet produced a treated wastewater over a long period of time that is suitable for drinking (13). Heaton specifies three degrees of potable water reuse: direct reuse, planned indirect reuse, and groundwater recharge. The latter two are more or less in use today. Planned indirect reuse refers to the deliberate discharge of wastewater upstream of a water supply intake, and there are many examples of this all over the world; in fact we might say this is the rule rather than the exception. Later in the text Okun will be cited again with a principle he advocates as a "preference for pure rather than polluted sources of water." This means not to draw water from streams with wastewater but to catch the water from protected upland sources before it is mixed with wastewater.

Protecting Watersheds

The principal of protecting watersheds from contamination has a lot of merit. As Okun's principle, it is clear: the preference for pure rather than contaminated sources of water. But the concept also has merit for water supplies that are already being taken from watersheds with contamination. These watersheds can be cleaned up and the sources protected. This is why it is important not to forget about nonpoint pollution control. It will only be through the eventual solution of some of the most difficult nonpoint source problems that cleaner drinking water sources can be provided.

Combinations of Measures

In addition to its organizational innovations, Fort Collins has demonstrated some innovative approaches to water management, and I would like to cite a few that relate to integrating the urban water system. This is not to brag, but just to give some actual examples of things that can be done. There are many other examples in other cities, of course, that can be cited. The city's approach to water management seems to illustrate combinations of the principles discussed above.

The city's organization for urban water management is working well and reflects the emphasis on good management through functional integration. As the regional area is growing rapidly, the city is also pursuing avenues for regional cooperation and has recently signed an agreement with other water and wastewater districts about studying future cooperation.

Figure 2.3 is a diagram from one of the city's plans for the future (14). The diagram shows the basic character of the system whereby the city consumptively uses 2.3% of the river basin supply of some 410,000 acre-feet annually. This includes treated water consumptive uses and raw water consumptively used in the parks and on the golf courses. In effect, there is already somewhat of a dual water system in operation in the city, as the raw water used from ditches and ponds for watering is not treated before use. This does not, of course, meet Okun's original vision for a true piped dual water system. The city has also developed a water reuse plan that is now in operation. It delivers treated wastewater to a large power plant north of the city to be used as cooling water and in return stands to get water rights credits for a water supply project that is under development. So far, this arrangement for water reuse is going exceptionally well. The city is also studying options for using raw water that comes through the city in older irrigation ditches for additional use on park lands. This will bring greater benefits to the ratepayers who will then not have to pay for treating water to potable standards and then have it used for watering.

As in many other fields the state of practice sometimes runs ahead of the literature and the theoretical basis for actions. Fort Collins, like other progressive cities, is bubbling over with ideas for improving water management through integrative measures. By transferring ideas such as the above about the improvement of water management, other communities can benefit as well.

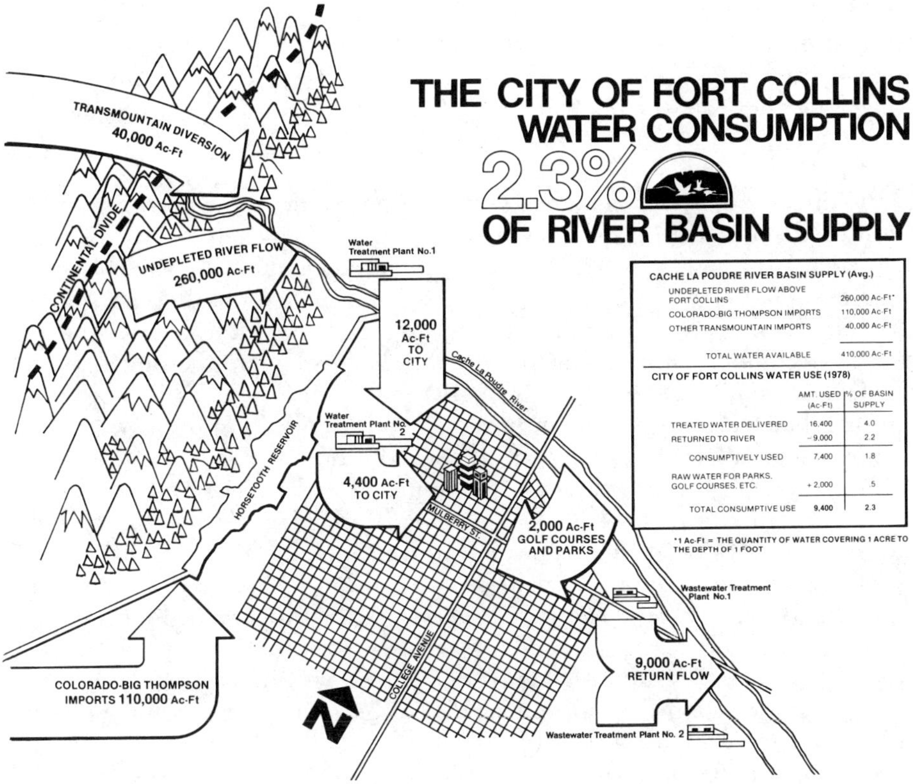

FIGURE 2.3 Diagram of Fort Collins water supply.

OPPORTUNITIES TO IMPROVE SERVICE

The examples of connectivity and synergism illustrate abundant opportunities for improving the management of urban water systems through technical measures such as using stormwater for recharge and other approaches that take advantage of the natural systems. The barriers to improving service are not usually the technical problems or the technical understanding, however, but the social and political problems. These have to be worked out on a case-by-case basis, of course, but the impact must be felt eventually through integration within a utility, such as in the Fort Collins example, or through some cooperative integrative approaches.

There are powerful political and bureaucratic obstacles to both functional and areawide integration of services, and even to cooperation. The advantages seem obvious, however, and the trend would seem to be in the direction of increasing integration and cooperation since the barriers tend to disappear over time. One obvious example of a barrier to functional integration is the fact that some managers may lose their jobs. A barrier to areawide integration is the fact that some governing boards may lose their authority. As the workers begin to retire and the boards change from oldtimers to new appointees, there are opportunities to overcome problems such as these.

There are many advantages to integration. Some of the most important will be discussed in the following paragraphs. First is the obvious management advantage that will accrue from the cooperation in data collection and information systems. The need for a balance sheet of the entire water system at one time is apparent, and by combining the subsystems, it might become possible to make more rapid steps toward the development of such a balance sheet. What it would consist of is discussed more fully in the chapter on decision support systems. In addition to developing a better balance sheet, there can be cooperation in monitoring for water quality changes and in the development of decision support systems.

The sharing of regional water supplies and facilities is another obvious example of a benefit that translates into reduced cost and increases reliability. The reason for the increased reliability—basically that the sharing of supplies reduces the risk of shortfalls and results in less need for contingency storage and redundant supplies—is discussed in the chapter on engineering essentials. The observations by Gilbert support the idea that cooperation in developing water supplies may be one of the most fertile areas to explore in terms of offering real advantages at the same time that it is feasible (9).

Environmental protection can be enhanced through the cooperation between the kinds of utilities and across areas. There is an old joke in the water business that the best way to get good wastewater treatment is to require that every community take its water supply just downstream from where it discharges its wastewater (see Figure 2.4). Even though this approach will not work in practice, there are obvious reporting advantages to such integration. If there is areawide cooperation, then any accident or mistake will certainly be reported since it may affect operations in the same organization, rather than in one that is in another community.

Another advantage to cooperation is the possibility of enhancing the total utilization of water resources. Stormwater really can become a "resource out of place" rather than just a nuisance, and wastewater can be recycled or used for groundwater recharge. There are signs of increasing interest in this kind of "total resource thinking" in many communities today.

Coordination in utility construction and repair is a big area for savings. The old story about the telephone company tearing up the street, then repaving it, just to be followed by the water department to dig it up again, need not happen, at least with the different water services. This

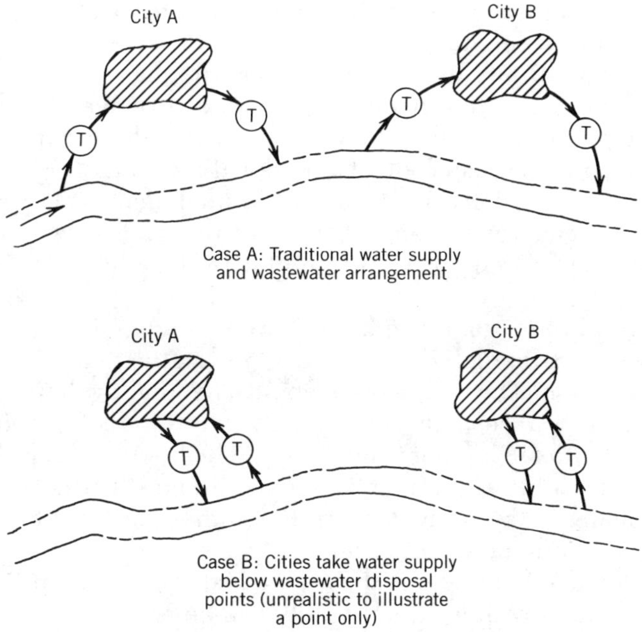

FIGURE 2.4 Two arrangements for water sources and wastewater disposal points.

depends on the level of cooperation and on the effectiveness of information systems.

Finally in the area of efficiency there are many areas for savings. These depend to a large extent on the organizational design, of course, but some of them should be quite significant. They result in economies of scale. Examples would be cooperation in training, laboratory services, computing, billing, legal, shop and maintenance, engineering and construction, and in executive management. There are also physical economies of scale in water supplies, treatment of both water and wastewater, and in management of the distribution and collection systems.

CONCLUSION

As this chapter began with the admonition that the integration of urban water service is not necessarily easy, it closes with the hopeful statement that if management can be improved to the extent that even partial integration becomes possible, then we will really see some steps forward. It seems that making progress on integration of urban water systems is one of the critical management issues we face today, but certainly not the easiest due to the many sociopolitical realities in water management. The path to improvement would seem to be in exercising the "art of the possible," and to carry out the improvements and management innovations that can be done within the space of control of the managers interested in improvement.

There are many fronts where progress can be made, however. They begin with the organization itself, where better management practices can be developed to recognize the connectivity of the urban water systems. The organization can look at functional integration and examine the avenues to regional cooperation. In the final analysis the measures discussed in this chapter are those that innovative managers would think of anyway as they search for ways to do business better in their organizations.

Going beyond the organization, there are many possibilities for areawide improvements through cooperation and even the ultimate measures: areawide integration. These cannot be approached lightly, however, since political measures are usually necessary. Even to develop cooperative ventures, the political element is necessary. The bottom line is that there do exist many hopeful water management strategies that need to be explored, and many of them involve bringing other parties into our plans through cooperation and negotiation.

This chapter has been about the water system as it is integrated in nature, and about how this natural reality can be recognized in management initiatives. The next three chapters examine the individual water services and illustrate the complexities they bring to managers even without integration.

REFERENCES

1. Victor A. Koelzer, and A. B. Bigler, *Policies and Organizations for Urban Water Management*, Water Resources Publications, Littleton, Colorado, 1975.
2. M. B. McPherson, *Prospects for Metropolitan Water Management*, ASCE Urban Water Resources Program, ASCE, New York, 1970.
3. County of Fairfax and Fairfax County Water Authority, Committee to consider the Falls Church and Fairfax City water systems, agenda notes, February 13, 1985.
4. Richard J. Hesse, A Regional Approach to Public Water Supply, in *Energy and Water Use Forecasting*, An AWWA Management Resource Book, Denver, 1980.
5. James W. Miller, Old Water–New Water: *JAWWA*, September 1984.
6. Robert L. Wubbena, and J. M. Maxwell, Managing Countywide Water Service with Satellite Systems, *JAWWA*, Denver, April 1983.
7. David Donaldson, Regional Authorities Support Small Water Systems in the Americas, *JAWWA*, Denver, June 1984.
8. Robert S. McGarry, Potomac River Basin Cooperation: A Success Story, in *Cooperation in Urban Water Management*, National Academy of Sciences Press, Washington, D. C., 1983.
9. Jerome B. Gilbert, Coordination of Major Independent Systems, in *Cooperation in Urban Water Management*, National Academy of Sciences Press, Washington, D. C., 1983.

10. Ronald D. Hardten, Developing Joint Water Projects, *JAWWA*, Denver, April 1984.
11. Neil S. Grigg, *Water Resources Planning*, McGraw-Hill, New York, 1985.
12. H. Hengeveld and C. De Vocht, eds., *Role of Water in Urban Ecology*, Elsevier Scientific, Amsterdam, 1982.
13. Richard D. Heaton, Potable Reuse: The U.S. Experience, in *Municipal Water Systems*, David Holtz and Scott Sebastian, eds., Indiana University Press, Bloomington, 1978.
14. Fort Collins Water Utilities Department, Water Utilities Progress and Projections: Report to the Citizens, March 3, 1980.

CHAPTER THREE

URBAN WATER SUPPLY SYSTEMS

SERVICE PROVIDED

Water supply is the most important of the urban water services and constitutes a primary need for people everywhere. Without an adequate water supply the result is chaos, disease, and a low standard of living. In some developing countries the breakdown of social systems is evident in the failure to take care of basic water supplies. Even in advanced economies there is increasing evidence of bad management of water supplies through pollution by toxic chemicals. We cannot stress the importance of urban water supply systems too much.

Even beyond the need for a safe supply of drinking water, people will go to great lengths to get water they enjoy, even to the extent of buying expensive imported bottled water. In recent years the U.S. market for bottled drinking water seems to be growing at a rate of about 15% per year, and it has already exceeded $1 billion annually. When people think their drinking water is not pure, they protest vigorously through the political process. So even though some consider drinking water to be a dull political subject, it is one that stirs powerful passions when things are not exactly right. Water supply is one of those issue areas that lies dormant until something goes wrong, and then everything can go wrong all at once.

The water supply for an urban area must meet needs across all economic sectors. This includes the needs for drinking and household uses, for industrial and commercial uses, for public uses, and for emergencies, primarily fire fighting. Figure 3.1, adapted from Mcpherson (1), illustrates these uses by showing a conceptual water balance for a typical urban area. To fully account for the water, a category of "unaccounted for water"

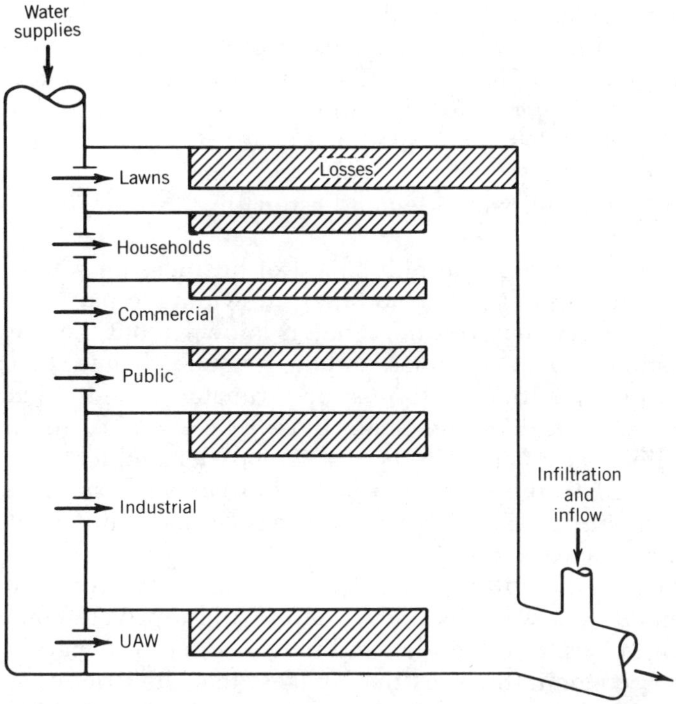

FIGURE 3.1 Metropolitan water balance showing unaccounted-for water.

(UAW) is included. This category is discussed in detail in Chapter 9, which is concerned with operations and maintenance, since one of the objectives of system operation and maintenance is to minimize losses. Since it is an important management concern, the American Water Works Association Research Foundation authorized a research project on the management of unaccounted for water in 1985, and one of the outputs of this project, expected in 1986, should be a manual with standardized definitions of UAW and recommended procedures for controlling it.

There have been a number of recent studies concerned with forecasting the demand for water. Most of these have resulted from the interest in water conservation or the recent and continuing worldwide experience with drought. The studies are not able to provide estimates of the actual demand for water in individual areas, however, because the variations are so extreme. Because of the need for information on water use, ASCE's urban water program published three technical reports in the 1970s covering household, industrial, and commercial water use (2–4). These reports provide some baseline information on water use, but they are just a beginning on a very complex subject. Our gap is not theoretical knowledge, but the lack of data. However, this problem is giving way to the efforts of individual utilities to collect and analyze water use data for their own management purposes. With the increasing availability of computerized data bases the data problem seems likely to ameliorate, and studies of local water management problems can increasingly rely on the availability of local data. The American Water Works Association (AWWA) and the National Association of Water companies both are publishing statistics that reveal national water use patterns, and the U.S. Geological Survey (USGS) has collected some water use data as well. Some guidelines for water demand estimating are provided later in this chapter.

The dependence of the public and of business on safe and reliable water supplies is undisputed, and they are willing to pay adequate prices for water supply; however, not much is known about the way water demand responds to rate changes, and the revenue changes that will result from rate manipulations still must be estimated in individual cases. By the mid-1980s considerable data is beginning to appear in the literature about willingness to pay and the elasticity of demand for water, however, and by the 1990s the science of pricing urban water supplies will have made great strides. This subject is discussed more in Chapter 6 under financial management topics.

In studying the willingness to pay for water, we are confronted with the dilemma that water is both a public and a private commodity. Its price can be set to reflect demand much as that of a market good, and there are economic theories that suggest that the pricing arrangement should be market oriented in that way; or the price can be set through public pricing policies that reflect the political judgments of the community and the government. There is at present an uneasy truce between these two approaches. There are a number of municipal objectives that fall outside the market mechanism, one being the general need for public uses of water, such as fire fighting or recreation in low-income areas and another being to sustain beautiful cities through lawn watering. The economic theories that advocate pure market approaches to the evaluation of desirable services seem too limited, as do those that propose no market disciplines at all in the process. A reconciliation between the two views of rate setting is in progress through the political process in individual communities. So far there has been little federal and state government influence in the rate-setting process for water supply, and this trend is expected to continue, except for the interest in water supply that is growing among state public utility commissions.

CONFIGURATION OF URBAN WATER SYSTEMS

The general configuration of the urban water supply systems is as shown on Figure 3.2. There is a three-way classification of the system components as acquisition, treatment, and delivery subsystems, and this classification generally helps later when we discuss costs and the data bases

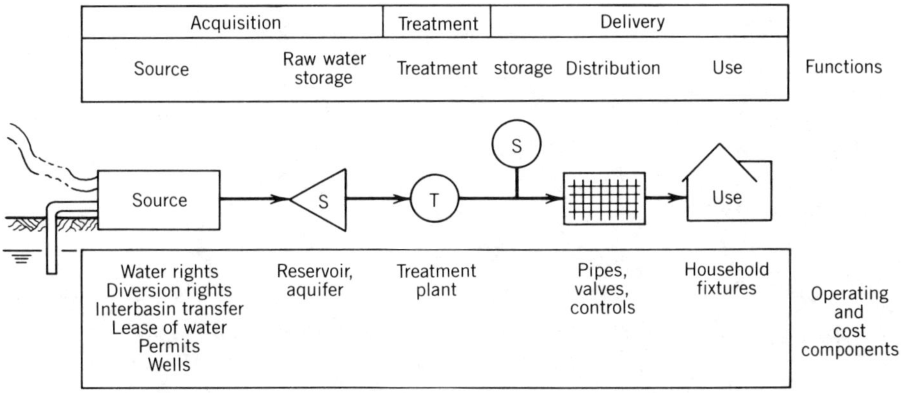

FIGURE 3.2 Configurations of urban water supply system.

needed for management. The system can also be classified in terms of service zone, and this becomes important in financial decision making about equity. Another classification system is the domestic, commercial, and industrial use classifications already mentioned. It is necessary to be able to break the system into the different components to gain information for management decisions, especially financial.

The subsystems of the urban water supply system divide further into the components shown in Table 3.1, which illustrates some of the choices available for the components of the subsystems. The treatment process classification is from an overview paper by Stacha in 1984 (5). As every designer knows, there are a myriad of choices available for the components at each point of the system, and the industrial engineer would view system design as the challenge to consider all the possible combinations and select the best one. From the practical standpoint, most water systems are fixed in the configuration of the components due to past decisions, the availability of certain supplies, or by the availability of financial resources. All over the world the configuration of systems varies. Okun has written several papers about water supply practices in different countries, and others are now contributing frequent articles in the *Journal of AWWA* about their experiences. The June 1984 edition, for example, has a photo essay by Okun about water supply practices in the People's Republic of China (6). The worldwide awareness of different approaches to water supply engineering is helping to transfer knowledge back and forth, and many innovations are resulting from it.

It is important that the best configuration of a system be selected. Designs are constantly changing, and the influence of new system configurations, new hardware, and automation will be felt dramatically in the next 20 years. In some areas the need is mostly for reliable sources of clean water. In other areas the need is for better management of existing supplies at lower cost. With the latter in mind, Chapter 10 is devoted to

TABLE 3.1 Choices of System Configurations

Source	Raw Water Storage	Treatment[a]	Treated Water Storage	Distribution	Use
Surface fresh water	Surface reservoir	Initial mixing	Reservoirs	Gravity pumped	Household
Fresh groundwater	Shallow aquifer	Flocculation	Tanks		Industrial
Brackish water	Deep aquifer	Sedimentation			Institutional
Seawater	Tanks, cisterns	Filtration			Public
Mixed waters	Artificial aquifers	Disinfection			
		Advanced techniques[b]			

[a]Classification of treatment processes is from (5).

[b]"Advanced techniques" refers to family of techniques available to treat against inorganic, radiological, and organic contaminants.

the emergence of automation as a force in urban water management. Also, with the attention currently being given to improving operations and maintenance, it is a safe bet that substantial advances will be made in solving problems such as leakage, toxic chemical control in supplies, and some of the other critical problems that face the industry.

Okun, of the University of North Carolina at Chapel Hill, has also given considerable thought to the subject of improving the security of water supplies and avoiding the problems relating to toxic pollution. He was an early advocate of the concept of the "dual water system," a concept that basically means the separation of the potable water system from the other uses, including household and industrial. While the concept of the dual water system has not been widely adopted yet, it certainly has merit, especially when we confront monitoring and treatment requirements for the many new trace metals and organic chemicals that are considered undesirable in public water supplies. While the concept of the dual water system has not found wide acceptance yet, like many innovations, it seems to creep in unnoticed, and the trend in many communities to go to more reuse is a verification of the validity of the concept, since a system that has industrial or irrigation uses of nonpotable recycled effluent, has, in effect, a dual water system.

Okun's concepts of a dual water system are based on an underlying philosophy that drinking water should enjoy a special status and be separated from other water supplies. He advocates this philosophy in other ways as well. A few years ago Okun described his perspective of the principles of water management in an ASCE article, and the principles apply directly to configuring of urban water supply, including providing for protection of supplies, so they are described here (7):

1. *Uniqueness of water projects.* Water projects defy generalization since the unique settings for implementation of projects are all different.
2. *Economies and efficiencies of scale.* There is merit in the concepts of consolidation and regionalization of water systems.
3. *Sound financing.* There must be provision for sound financial underpinning of water projects.
4. *Preference for pure rather than polluted sources of drinking water.* This is another way to view the objective Okun had in mind in proposing the dual water system—minimize the cost, complexity, and risk in treating potable water.
5. *Integration of water supply and pollution control services.* Okun had completed several years of study of the British experience with integrating services and is an advocate of integration. This subject was already discussed in Chapter 2 covering the possibilities for integrating water services.

Okun's principles also apply to wastewater and water quality management, but the central idea that is applicable to water system configuration is the preference for the pure rather than polluted source. What does it mean in a practical sense? It means that there are substantial benefits from developing clean upland water supply sources and then protecting them or from developing supplies from reliable underground sources and protecting them—a difficult task in the absence of assistance from regulatory agencies. The water supply that is most accessible may be a nearby stream, but one where the water has already been used a few times. With adequate treatment such a source may be perfectly safe, but generally inferior to the protected source.

The configuration of water supply systems has been changing over the years, with a number of factors influencing the changes. Some of them are growth and urbanization, safe drinking water legislation, new materials and processes, research into drinking water hazards, and changes resulting from financial pressures on all three levels of government. In the 1980s the changes will be affected more by the financial pressure on the industry and by the emphasis on improving operations, especially with the advent of more automation and computer control. There will also be continued emphasis on making water systems safer and improving health safeguards.

DIMENSIONS AND PERFORMANCE STANDARDS

Demand for Water

The first consideration in planning for the capacities and characteristics of a water supply system is to make sure there is enough water. The usual starting place to study this question is to refer to the "per capita demand," but planning to meet the volumetric needs is only the first step in meeting demands. Today, there is a need for more detailed information that can be used to improve planning, especially to consider water flow rates, the location of demands, and the quality of the supplies.

The concept of per capita demand does allow a gross first estimate of the needs of an urban area, but water demands vary so much due to type of land use, housing density, climate, and standard of living that the first estimates based on per capita needs must be regarded as only preliminary. Also, rapid changes in lifestyle today have made many of the old estimates obsolete.

Estimates that were used to plan many of today's systems were based on per capita needs in the range of 150 gallons per capita per day (gpcd), a figure that is based on annual average uses. In AWWA's 1981 report of water utility operating data the average per capita use is reported to be 167 gpcd, and there were some 1397 utilities reporting, so the 150 gpcd is still the correct range (8).

Typical of the recommended ranges was a list that appeared in a well-known reference text: domestic needs, 50 gpcd; commercial and industrial needs, 65 gpcd; public uses, 10 gpcd; and unaccounted for water, 25 gpcd (9). Domestic uses were given as 41% for toilet flushing, 37% for washing and bathing, 6% for kitchen use, 5% for drinking water, 4% for washing clothes, 3% to household cleansing, 3% to watering the garden, and 1% for washing the family car. These figures originated in a 1964 USGS report (10), and, as far as generalized estimates are concerned, they are still an acceptable place to start. There have been many changes in lifestyle, however, that affect these figures, and changes are still occurring. Changes in water demand reflect many of the changes in society, such as family size, income, women at work, the conservation ethic, and income distribution.

Household uses are usually expected to be somewhat constant, around 50 gpcd, depending on social and economic factors such as income and lifestyle, but sprinkling would be very sensitive to climate and to the size of lots. The category of sprinkling and lawn irrigation can be expanded to be a major water using factor in some climate zones. Also, the effects of metering and rates will influence the water devoted to optional uses such as sprinkling.

Commercial use, the second category shown on Figure 3.1, is very variable due to the extreme differences in the types of businesses and institutions. Table 3.2 provides a summary of the data from the study reported in Wolff and co-workers (2) that was conducted in the Baltimore metropolitan area. These data, although not intended to represent any kind of generalized approach, will at least provide some experience factors. My subsequent research into commercial water use patterns reveals a lack of much systematic data collection and analysis such as that reported by Wolff and co-workers (2). One additional article that provides some insight into the question is by McCuen and others (11).

Industrial uses are even more variable than commercial, and because industry's needs are so unique, many of them choose to develop their own sources. Tucker (3) reported in 1972 that 67% of the water used in some 224 metropolitan areas was self-supplied by industries, including manufacturing and mineral industries. This study was actually based on 1964 data. By 1980 the estimates, according to the U.S. Geological Survey, were a national industrial self-supplied use of 45 bgd (billion gallons per day) out of a national total public and industrial water use of 79 bgd, or 57% being industrial self-supplied (12). The above estimates used different data bases and should not be considered as precise, but they do indicate the substantial dependence of industry on self-supplied water. Neither of the estimates include thermoelectric cooling water. The estimates do not provide any help in estimating the water needs of individual industries, but such data is available from the Department of Commerce and the Environmental Protection Agency. The Department

TABLE 3.2 Summary of Commercial and Institutional Water Use

Type of Establishment or Institution	Selected Parameter	Annual Water Use (gpd/unit)		Maximum Day Water Use (gpd/unit)		Peak Hour Water Use (gpd/unit)		Hours of Peak Occurrence
		Expected	Design	Expected	Design	Expected	Design	
Primary and secondary schools								
Public elementary	gpd/student	5.38	8.67	9.68	13.0	49.1	52.4	1 p.m.– 2 p.m.
Public junior high	gpd/student	5.64	9.75					
Public senior high	gpd/student	6.63	12.2	19.6	25.2	121	127	4 p.m.– 5 p.m.
Private elementary	gpd/student	2.27	6.09	3.10	6.92	25.7	29.5	10 a.m.–11 a.m.
Private senior high	gpd/student	10.4	18.6	15.7	23.9	38.7	46.9	4 p.m.– 5 p.m.
Combined (grades 1–12)	gpd/student	8.49	18.7	16.8	27.0	51.3	61.5	5 p.m.– 6 p.m.
Colleges								
Students in residence	gpd/student	106	179	114	187	250	323	1 p.m.– 2 p.m.
Nonresident students	gpd/student	15.2		27.0		57.8		8 a.m.– 9 a.m.
Hospitals	gpd/bed	346	559	551	764	912	1120	11 a.m.–12 noon
Nursing homes and institutions	gpd/bed	133	209	146	222	424	500	4 p.m.– 5 p.m.
Apartments								
High rise	gpd/occupied unit	218	322	426	530	745	849	11 a.m.–12 noon
Garden type	gpd/occupied unit	213	315	272	374	671	773	8 a.m. & 6 p.m.

Hotels	gpd/ft²	0.256		0.294		0.433		6 p.m.– 7 p.m.
Motels	gpd/ft²	0.224	0.326	0.461	0.563	1.55	1.65	10 a.m.–11 a.m.
Office buildings								
General offices less than 10 yr old	gpd/ft²	0.093	0.164	0.173	0.244	0.521	0.592	12 noon– 1 p.m.
General offices more than 10 yr old	gpd/ft²	0.142	0.273					
Medical offices	gpd/ft²	0.618		1.66		4.97		11 a.m. & 4 p.m.
Department stores	gpd/ft² of total sales area	0.216	0.483	0.388	0.655	0.958	1.23	12 noon– 1 p.m.
Shopping centers	gpd/ft² of total sales area	0.160		0.232		0.412		2 p.m.– 3 p.m.
Car washes	gpd/ft²	4.78		10.3		31.5		12 p.m.– 2 p.m.
Service stations	gpd/ft² of garage and office space	0.251	0.485	0.590	0.824	4.89	5.12	12 a.m.– 1 p.m.
Laundries								
Commercial laundries and dry cleaners	gpd/ft²	0.253	0.639	0.326	0.712	1.57	1.96	2 p.m.– 3 p.m.
Laundromats	gpd/ft²	2.17	6.39					
Restaurants								
Drive-ins (parking only)	gpd/car space	109						
Drive-ins (seating and parking)	gpd/seat	40.6						

(Continued)

TABLE 3.2 *(Continued)*

Type of Establishment or Institution	Selected Parameter	Annual Water Use (gpd/unit)		Maximum Day Water Use (gpd/unit)		Peak Hour Water Use (gpd/unit)		Hours of Peak Occurrence
		Expected	Design	Expected	Design	Expected	Design	
Conventional restaurants	gpd/seat	24.2	55.2	83.4	114	167	198	2 p.m.– 3 p.m.
Clubs								
Golf clubs	gpd/membership	66.1						
Swimming clubs	gpd/membership	16.5						
Boating clubs	gpd/membership	10.5						
Churches	gpd/member	0.138		0.862		4.70		8 p.m.– 9 p.m.
Barber shops	gpd/chair	54.6	97.5	80.3	123	389	432	2 p.m.– 3 p.m.
Beauty salons	gpd/station	269	532	328	591	1,070	1,330	2 p.m.– 3 p.m.

Source: Wolff, Jerome B., F. P. Linaweaver, Jr., and J. C. Geyer, Commercial water use, ASCE Urban Water Resources Research Program TM 27, Dec. 1975 (2).

of Commerce had an industrial water management study effort underway for a number of years, and Kollar presented papers and reports of industrial water use that can be of help in making estimates. Typical of these papers is found in reference 13.

Water demand projections have a lower limit to be found in the studies that have been done about water needs in the developing countries. One report, by White and others who have worked in Africa, reports on the literature about minimum water supplies to sustain life (14). They report that the literature suggests that minimum water needs are in the range of 1.8–3.0 liters per person per day under the conditions of East Africa, which are not climatically that different from parts of the United States. They also show a U.S. Army graph that relates water needs for soldiers to the mean air temperature and to the daily workload, with the values ranging from about 3 to about 8 liters per day for temperatures in the range of about 75°F.

At the other extreme, there are reports of per capita water use exceeding 250 gpcd in parts of the United States, even for entire cities without heavy water using industries. This will be the case when substantial water is needed for lawn sprinkling, such as in cities like Denver where per capita use exceeded 200 gpcd for years. Phoenix is said to have one of the highest per capita demands in the country. In climates such as this the inside-house use would not be much greater than in humid climates, but the difference would be the lawn sprinkling. The water use in Denver has been studied extensively, usually with policy studies concerning metering in mind. In a recent study, for example, it was concluded that metering would reduce the water use about 20%. The data from a sample in the Denver area showed water use in metered homes to be about 453 gallon/day, and in flat-rate homes about 566 gallon/day, both figures based on 3-year data collection programs (15). The average occupancy was not given, but presumably it would be about three persons per home, and indicated per capita uses for the study area of about 150 and 190 gpcd for the metered and unmetered customers.

Thus, the general approach to planning for water supplies is to make estimates based on the population and other socioeconomic characteristics of an area. This will provide estimates of the annual average water needed, but there is a lot more to water supply planning than that. We also need to consider the variability of supply and demand. The variability of supply can refer to climatic variation, or wet and dry cycles, and to the possibility that supplies can be cut off for one reason or another. The former situation can be referred to as drought contingency planning, and the latter to emergency operations planning. Both are essential in water supply operations. Demand variations have to do with variations in human activities.

Water Supply and Storage Planning

Droughts are a condition of nature and must be anticipated. Some managers are young and have never seen a drought; others are older and remember the dust bowl days, which combined drought with other adverse conditions and with the effects of poor land-use management. In recent years, however, almost everyone has some experience with drought, whether it is with the northeastern drought of the 1960s, the late 1970s national drought, or the television accounts of starvation in sub-Saharan Africa as a result of drought-related desertification.

The oldest story of drought known to this author is:

> Now Joseph said to Pharaoh, ". . . seven years of great abundance are coming in all the land of Egypt; and after them seven years of famine will come, and all the abundance will be forgotten in the land of Egypt and the famine will ravage the land." (Genesis, 41:25–30, New American Standard Bible)

This shows that drought has always been known by humankind, and the point is that every water organization must do its own water contingency planning. There is no set formula for how to do it, but the important thing is that the management understands the risk.

A good way to approach the beginning of drought planning is to conduct a system simulation study that demonstrates the behavior of the system under conditions of stress. These might include the simulation of the worst drought of record to occur simultaneously with periods of high demand. What then would be management's options to deal with the situation? The *safe yield* of the system should be evaluated and judged for adequacy. Generally, safe yield is a concept meant to tell what the system, reservoir, or aquifer will deliver under the most adverse conditions to be expected. The concept has built-in problems since the worst conditions never reflect what really can happen, so probabilistic considerations have to be introduced.

In addition to raw water reserves to guard against drought and other contingencies, such as the loss of water through legal problems, there must be system reserves provided for more localized management purposes, such as demand variations, emergencies, breakdowns, and to regulate system behavior. The latter are mostly furnished through treated water system storage rather than through raw water storage, since the raw water still requires treatment before it becomes available to the system.

System storage requirements will usually be volumes adequate for a few days of use, depending on where the storage is located and other factors that might vary from system to system. Raw water storage might be as much as a year's water needs or more, depending on the statistical variability of the expected supplies.

Requirements for raw and treated water storage are set locally or by state boards, and not at the national level, although the standards and the policy statements of AWWA are very influential in the setting of local standards. As an example of the kind of standards that are encountered, the Texas Board of Health has regulations for drinking water systems that specify the storage volumes required (16). In the case of raw water the statement in the rules is that communities are required to have facilities with safe yields that can meet maximum daily demands during *extended* periods of critical hydrologic conditions and peak usages. *Extended* is emphasized since how long an extended period lasts is a question that must be answered by an analysis of the acceptable risk. This topic is covered in some detail in Chapter 6, since the analysis of risk is one of the requirements of planning.

In the case of treated water storage the Texas Health Department generally requires 200 gallons per connection with some variations for system size and type. Storage is required to ensure adequate chlorine contact time and to maintain pressure during all anticipated operating conditions, with a service pressure of at least 35 psi being maintained, and a minimum of 20 psi at all times. While these kinds of parameter requirements vary from state to state, it makes sense for each system manager to study the unique needs of the particular area, and to affirm or change the generalized parameters that might be required by the state. Many of the requirements are based on decisions made without much data, and it is wise to continually question them. The manager might find, for example, that considerable cost savings can accrue from changes in some of the system parameters, without violating any required levels of service.

Peak Rate Planning

In addition to quantity of water available we are also concerned with timing and the rate of delivery. Demands for water vary during the year and during the day, much as demand for electricity, telephone, and transportation. Concepts that have come to be accepted for planning, design, and operation are "peak day" and "peak hour." Actually, these are representative values with statistical connotations, and as information systems improve, more detailed information will start to become available.

Some data that is now somewhat dated showed peak-hour rates to fall generally in the range of 1.5–12.0 times the average hourly demand, and for peak-day rates to fall generally in the range of 1.2–4.0 times the average hourly rate where the average hour is the average over an entire year (17). Although there are newer figures available, they are still not appropriate to estimate these extreme values. They would be expected to be functions of the socioeconomic characteristics of the urban area and to be somewhat manageable, both through education programs and, in the future, through what is called "load management" in electrical utilities. In the meanwhile, the best course of action for any particular study is to collect local data.

In dealing with questions of peak use, the utility has the option of providing either the treatment capacity or the storage for treated water. The decision requires an optimization of the system after considering the advantages and disadvantages of the feasible combinations. Since the decision about how much treated water storage to provide also involves emergency planning and other distribution system considerations such as pressure control, the design decision cannot be made in a vacuum.

The general trade-off curve will look like Figure 3.3, which is based on a design period as shown in Figure 3.4. This covers strictly a hypothetical situation, a critical design period which should be selected with great care. The critical design period presumably would cover a period such as had been observed in the past for several successive days of high demand. With different levels of treatment available, the critical period would demand storage as shown in Figure 3.3. This curve has the expected shape of a trade-off curve, which enables the calculation of financial advantages of trading off storage against treatment. The curves themselves are not sufficient to make the decision since the cost information must also be analyzed. Also, the curve shown in Figure 3.3 cannot be the sole basis of the decision, again due to the other design factors involved in any distribution system.

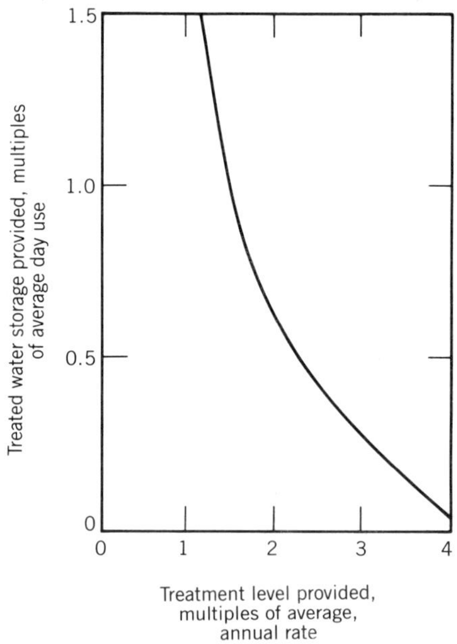

FIGURE 3.3 Trade-off curve of treated water storage versus treatment capacity.

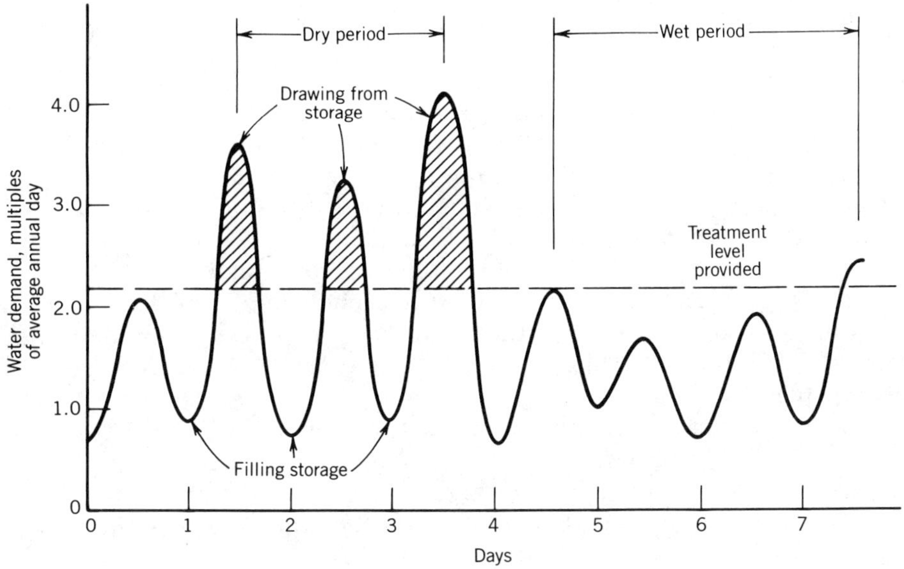

FIGURE 3.4 A typical time series of water demand over a period of several days.

In AWWA's operating report, the amount of distribution storage reported was 59,584 million gallons, corresponding to a population served of some 104 million. This corresponds to treated water storage per capita of 573 gallons, or 3.4 days of average annual use (8).

The dimensions of the system also must be adequate for distributing water at the desired rates and pressures during times of peak demand. The general capability of a distribution system to deliver water at certain rates will vary as the square root of the pressures provided; thus, pressures represent both a necessary condition for adequate distribution system capacity and a tool for distribution system management. At the present time, utilities in general are not managing distribution systems in this way, but with greater automation on the way, it is only a matter of time before the practice is widespread. There have been recent reports that the Welsh Water Authority, for example, is practicing distribution system management to control leakage at night. Leaks can also be expected to vary approximately as the square root of pressure, so this measure can be expected to save water. There are increasing possibilities for managing systems to control the distribution rates and pressures due to advancing technology. One possibility, for example, is the use of controllable check valves, such as reported by Lescovich (18). Another set of possibilities is computer control in general, the subject of Chapter 10.

Maintaining adequate water pressures and delivering the required rate of flow means having large enough pipes that are in good condition, as well as adequate provisions for supplying the necessary head, either through the gravity system or through pumping. Walski has recently published a book on managing distribution systems that deals with these subjects (19), and the subject is covered later in this book under operations and maintenance, since maintaining distribution system capacity is a matter of fixing leaks, preventing corrosion and postprecipitation, and maintaining the capacity of the pumps.

Maintaining pressures of 60–75 psig in most municipal distribution systems seems to work quite well, but of course, special provisions may be necessary for tall buildings and for emergency situations. It does not take that much pressure to service most household appliances, sometimes as little as 5–10 psig, but it is the losses that must be overcome on the way to the fixtures and the need for reserve capacity that require the higher distribution system pressures.

One consideration in maintaining distribution system capacity is the requirement of a minimum pipe size, ranging from 2 inches for minimum service locations to 6 inches and larger when fire flows are planned. In the case of the Texas Department of Health the minimum suggested pipe size is 8 inches for systems with more than 250 connections, with provisions for smaller minimum sizes for smaller systems.

Staging Construction

In providing the capacity of a water system, we must keep in mind that the systems are built in stages and the planning for making the system capacities adequate should consider the best strategy since additions must be made in rather large increments. The problem of finding the best way to stage construction is well known among economists and is referred to as the investment timing or staging problem. The concept is illustrated in Figure 3.5. The figure shows a system with demand growing over a

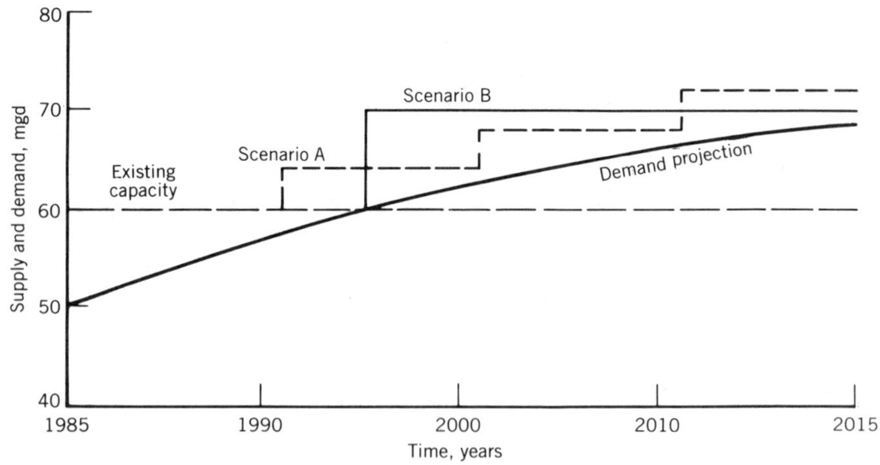

FIGURE 3.5 A staged plan for adding capacity to a water system.

20-year planning horizon from a current level of about 50 mgd to about 70 mgd in the 20-year period. Two alternative scenarios for additions are shown, one with three stages of additions and the other with one large addition. Each of these scenarios will have advantages and disadvantages depending on the uncertainty of the projection, inflation, the local indebtedness, the possibility of regional consolidation, and perhaps other factors.

System Quantities and Parameters

The system parameter requirements basically ensure that performance will be adequate. Performance is expected in the areas of flow quantity, quality, pressure, and reliability, with reliability being of great importance. System management must take all necessary measures to ensure reliability, and the need for this is usually reinforced when there are system failures since the public usually begins to notice the reliability of the water service when problems begin to occur, although they take service for granted when everything is working satisfactorily. Realibility is a responsibility of management and constitutes a constraint on operations; all of the great management performance in the world is to no avail if the system is not reliable. This explains some of the criticism water system managers are sometimes subjected to—they are too conservative—but the truth is that they are making double sure their systems are reliable.

It is always interesting to compare your own system with the others. Some parameters listed in AWWA's operating report (8), based on some 1400 systems, provide estimates of system quantities. These provide the following per capita figures:

Item	Quantity	Per Capita
Fire hydrants	1,663,636	0.01600
Water line miles	456,702	0.00440
Steel tanks	5,188	0.00005
Service connections	19,546,810	0.1880*
Metered	17,947,884	
Unmetered	1,598,926	

The statistic on service connections marked with an asterisk corresponds to 5.32 persons per service connection. This number no doubt reflects many apartments and multifamily connections. The area where performance is under the most scrutiny is the quality of the water. This is covered under the section about regulation of water systems, since the approach taken is to ensure the protection of water through regulating water utilities.

MANAGEMENT ORGANIZATIONS

The most common water supply management organizations today are city water departments, private water companies, and special-purpose districts. Although public systems predominate today, the early history of water utility management began with private initiatives, and it has only been in the last 100 years that the responsibility for water supply became predominately focused on public agencies.

AWWA's management manual refers to management structure in terms of ownership and control of water utilities. It reports that there are about 50,000 systems serving about 200 million persons (20). Since its latest revision was prepared in 1979, we assume that these figures should be updated with the appropriate population growth. Of these systems AWWA reports that 44% of the systems are publically owned and serve 80% of the population and 56% are investor owned and serve the remaining 20% of the population.

Investor ownership can involve all the forms of business organization, but the largest of the investor-owned water companies are corporate forms of public utilities such as the gas, electric, and telephone companies. The publicly owned companies, according to AWWA, are usually found in one of the following four forms: part of a city department, with mayor control; part of a city department, with the council-manager plan; a separate city department under a water board or water commission; or a separate utility district.

There are various sources to consult about the numbers of water organizations with the different forms. Perhaps the best current source regarding systems under federal control is EPA's Office of Drinking Water. It has established a reporting procedure for water organizations that distinguishes between "community water systems (CWS)," and "noncommunity water systems." Figure 3.6 shows the distribution of these two categories as of 1983, showing some 58,700 CWS, with 11,000 being served by surface water sources and 47,700 being served by groundwater sources. The 158,100 noncommunity systems are to serve transients and customers and are regulated for water quality by EPA's programs. (21).

The classification of system size by EPA is as shown in Table 3.3 and reflects the preponderance of small systems in the country. Figure 3.7 shows a statistical distribution of the sizes, with Figure 3.8 showing the cumulative distribution. It is clear from the figures that by the time you have accumulated 1000 systems you have covered over 60% of the U.S. population. These large systems would bear the names of the large U.S. population centers: New York, Chicago, Los Angeles, Philadelphia, Atlanta, Dallas, Denver, and Seattle, to mention a few. These large water supply companies contain centers of expertise about water supply that are very advanced in technical and managerial matters, but since there are so many small water systems with limited financial capacity, it is still

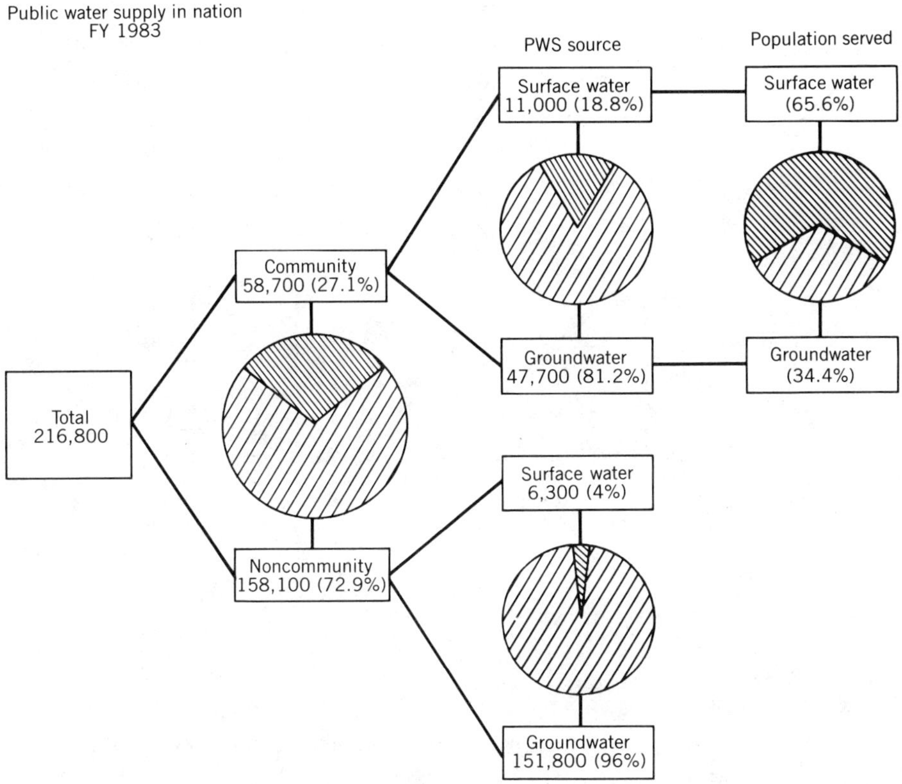

FIGURE 3.6 Diagram showing distribution of U. S. water supply systems.

TABLE 3.3 Size Distribution of U.S. Community Water Systems

Size	Population Served	No Systems	Percent National Population Served	Cumulative Population (%)
Very small	25–500	37,813	2.6	100.0
Small	501–3,300	19,915	8.7	97.4
Medium	3,301–10,000	3,943	10.4	88.7
Large	10,001–100,000	2,770	34.3	78.3
Very large	>100,000	277	44	44.0

Source: Environment Protection Agency, Office of Drinking Water, FY 1983 Status Report (21).

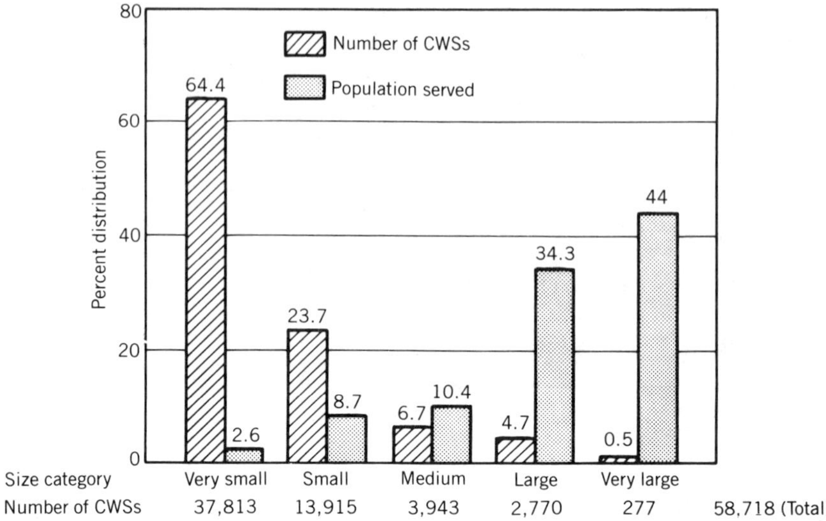

FIGURE 3.7 Size distribution of CWSs, fiscal year 1983.

difficult for the profession to hold on to managers and engineers that can provide the experience base needed to solve problems of obsolescence and increasing technological complexity.

The National Association of Water Companies (NAWC) lists financial and operation data for its membership, and the latest report lists some 86 member companies with revenues exceeding $1 million (22). How-

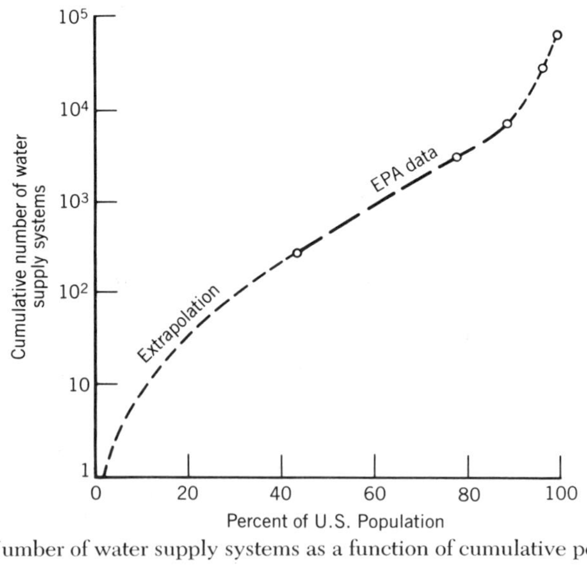

FIGURE 3.8 Number of water supply systems as a function of cumulative population served (15).

ever, not all investor-owned companies are members of NAWC, but this number provides a look at the general distribution of the private water company sizes. Revenues of $1 million would correspond roughly to a population served of 10,000–15,000 persons, according to my estimates.

AWWA also publishes water industry data, and its latest compilation shows 1247 utilities responding to its survey, with a population served of 104,043,324. However, this data cannot be plotted on Figure 3.8 since there may have been significant omissions due to companies not responding. Another report, from President Carter's intergovernmental Water Task Force, reported that there were 756 systems that served 50,000 or more persons (23). Although it is now somewhat dated, AWWA published a comprehensive summary of water supply industry characteristics in 1966, and that reference can be valuable for anyone who would like a starting point or historical perspective for studying the industry (24).

REGULATION AND LEGAL BASIS FOR SERVICE

There are two ways to approach legal questions about water supply service: What are water companies authorized to do? What are they required to do? The first question refers to the enabling legislation, which is generally not too specific, and the second to the regulatory legislation, which has received considerable attention.

For the most part there is not specific enabling legislation for water utilities except for the municipal codes, which are responding in general to state authority and local governments to undertake programs for the health and welfare of the people. As standards of living have risen, the definition of health and welfare has shifted and now the people are demanding services that were not foreseen in the U.S. Constitution. Because the rules of the game are still being written, there is so much political shifting around regarding water management in general.

The main driving force for regulation is the Safe Drinking Water Act, passed in 1974 as PL 93-523, with subsequent revisions (now PL 96-502). This act, in spite of its detractors, is the primary driving force behind much of the regulatory changes in the drinking water industry. Even if changes are made in federal, state, and local responsibilities under the act, the regulatory approaches involved, having taken place under considerable scrutiny, will most likely remain.

The Safe Drinking Water Act is the successor to years of incremental legislation and regulatory moves by the U.S. Public Health Service (PHS), beginning in 1914 with regulation of water on interstate carriers. At that time the PHS was part of the Treasury Department. There was a substantial revision of drinking water standards in 1962 and for years the 1962 regulations were the standard approach. As a result of increased concern about the safety of water supplies, culminating with scares about the cancer-causing effects (carcinogenicity) of water, the Safe Drinking Water Act was passed. It provides for primary and secondary drinking water standards that apply to all public water supply systems. The primary standards govern contaminants that threaten health and the secondary standards to those that threaten welfare. Obviously, *welfare* will be a term that requires considerable debate to define.

The National Interim Primary Drinking Water Regulations (NIPDWR) were first promulgated in 1975, and became effective in 1977. The National Secondary Drinking Water Regulations were promulgated in 1979, becoming effective in 1981. The secondary regulations cover contaminants that affect the aesthetic qualities of water, and are a response to the protection of "welfare" referred to earlier. An excellent overview of the development of both of these sets of regulations and their effect on the water utility industry is given by Alston (25).

Table 3.4 lists the contaminants that were included in the NIPDWR as they were issued and remain in effect, although revisions are anticipated over the years.

In addition to the above regulations there are standards for fluoride that depend on the temperature and standards for coliform bacteria that depend on the method of measurement. Trihalomethane (THM) regulations have also been promulgated that limit total THM, including chloroform, to 100 ppb. These rules apply to systems that serve more than 10,000 persons and add disinfectant during the treatment process.

The procedure used by EPA to revise the regulations is described by AWWA (25). EPA is approaching the development of the regulations in four phases. By October 1983 it was giving advance notice of rulemaking for phases II and III. Its intentions for staging the four phases are as follows: phase I, volatile synthetic organic chemicals (VOCs); phase II, synthetic organic chemicals, inorganic chemicals, and microbiological contaminants; phase III, radionuclides; and phase IV, disinfectant by-products.

The general approach to making these rules is first to publish an Advance Notice of Proposed Rulemaking (ANPRM), then propose recommended maximum contaminant levels (RMCLs), then declare maximum RMCLs and propose maximum contaminant levels (MCLs), and finally, declare the MCLs. This procedure is carefully worked out and apparently bureaucratically correct and requires considerable time, money, and scientific expertise to manage.

At the time of this writing the rulemaking is proceeding, in phases, as intended by EPA. The proposed rulemaking for phase I, volatile synthetic organic chemicals, has been issued (*Federal Register*, June 12, 1984), and the ANPRM has been issued for phases II and III (*Federal Register*, October 5, 1983). In this latter document, EPA spells out its approach to a three-tiered classfication system, where category 1 would be those contaminants that occur with sufficient frequency to warrant

TABLE 3.4 Contaminants and Maximum Contaminant Levels in the National Interim Primary Drinking Water Regulations

Contaminant	MCL	
Arsenic	0.05	mg/L
Barium	1.0	mg/L
Cadmium	0.01	mg/L
Chromium	0.05	mg/L
Lead	0.05	mg/L
Mercury	0.002	mg/L
Nitrate (as N)	10.0	mg/L
Selenium	0.01	mg/L
Silver	0.05	mg/L
Endrin	0.002	mg/L
Lindane	0.004	mg/L
Toxaphene	0.005	mg/L
2, 4-D	0.1	mg/L
2, 4, 5-TP (Solvex)	0.01	mg/L
Methoxychlor	0.01	mg/L
Alpha emitter		
Radium-226	5	pCi/L[a]
Radium-228	5	pCi/L
Gross Alpha Activity (excl radon and uranium)	15	pCi/L
Beta and photon emitters[b]		
Tritium	20	pCi/L
Strontium	8	pCi/L
Turbidity	1	TU[c]

[a]Picocuries per liter.
[b]Based on water intake of 2 L/day.
[c]Turbidity units based on monthly average.

national regulations, with consistent monitoring and reporting; category 2 would be those that justified only more flexible regulations to be administered by the states; and category 3 would be those contaminants that were of concern, but did not justify a regulatory approach. Examples of the three categories would be: category 1, coliforms, turbidity, and inorganic and organic chemicals that are of wide concern; category 2, natural radionuclides, certain pesticides, and some organics; and category 3, contaminants that require rapid determination of safe levels for short-term consumption after a spill or some other event.

These regulations will eventually be very extensive and technical. Managers will have to acquire some expertise in them just to deal with the public and with the regulatory agencies, and it is well not to leave all the regulatory matters to staff experts. Whereas a few years ago regulations covered only a few recognized constituents, mainly coliforms and turbidity, now the list of contaminants for reporting is beginning to contain many unfamiliar chemicals. For example, the VOCs proposed in phase I include trichloroethylene, tetrachloroethylene, carbon tetrachloride, vinly chloride, and several others. The RMCLs for the noncarcinogenics are proposed in the notice based on chronic toxicity data, and for the carcinogens they are set at zero. In the ANPRM for phases II and III, the contaminants discussed include, in addition to the standard ones such as coliforms and turbidity, now ones such as *Giardia lamblia* and *Legionella*. Inorganic chemicals included in the NIPDWR include familiar names such as arsenic, barium, cadmium, lead, and others, and another group is under consideration, including aluminum, molybdenum, asbestos, and copper, among others. Corrosion by-products are under scrutiny in the regulations since they can contain materials such as zinc, iron, and copper, and requirements have been set in the NIPDWR to report on the types of materials in distribution systems. Synthetic organic chemicals in the NIPDWR include endrin, lindane, toxaphene, and 2,4-D; others under consideration include household names such as 2,3,7,8-TCDD (Dioxin). Radionuclides include radium 226 and 228, and alpha and beta particles. The trihalomethane problem will be addressed in the phase dealing with disinfection by-products.

Obviously, there are many issues that managers of water systems will be dealing with in the future, and the above is just a sample of those that are under discussion relating to water quality. In addition, water utilities may come under more financial regulatory scrutiny than they have in the past. At the present time there is not much financial regulation of urban water supply systems, although some states have regulatory commissions that are beginning to get interested. Most of the water utilities that are regulated are small and rural, but a recent survey of 45 commissions that regulate water showed that nearly all are working to improve their regulation of water systems (26).

COST FUNCTIONS FOR WATER SUPPLY

There is great variation in the cost of water supply. In a recent discussion with the director of my city's water utility, I learned that the replacement value of the water supply facilities was about $560 per capita, the value of the city's water rights was about $500 per capita, and the annual operating and maintenance (O&M) cost was about $86 per capita, with the latter figure including some reserve for depreciation and debt service. Since this is a combined water and wastewater organization, he also had the comparable figures for wastewater: $500 per capita for replacement and operating cost of $50 per capita. There is some uncertainty about whether the replacement values quoted are high enough; indeed they may be depreciated "book values".

Using these figures to extrapolate would suggest that the national replacement value of water supply systems would exceed $130 billion, perhaps reaching twice that value, and the annual O&M, with some reserve for depreciation, would be on the order of $20 billion. While these figures have no real precise meanings, they do help to put the industries' economic dimensions in perspective.

It is difficult to obtain financial data that can be generalized. AWWA's survey of operating data does not, for example, provide the kind of financial ratios that allow the comparison of figures such as the above. The National Association of Water Companies financial data summary does provide some of the data for the private water companies, and a check of that data revealed that the annual revenues ran in the general range of $30–100 per capita, of which about half or a little less was for operating and maintenance. The depreciable plant was reported in the general range of about $125–425 per capita for the companies checked, but this represented some kind of book value that would vary considerably with the accounting practices used. One conclusion could be that it seems to take operating revenues of $50–100 per capita to sustain a system and that the general range of replacement values for water supply systems would be in the general range of about $500 per capita, all of the above figures being in 1985 dollars.

Robert Clark and others at EPA's Cincinnati laboratory have published a number of papers about the cost of water supply. In fact, a large percentage of recent published papers seem to have come from that research group. The most general paper presented by the group is "A Spatial Costing System for Drinking Water," where the authors utilize the utility functions of acquisition, treatment, delivery, and support services, and examine costs in the categories of labor power, chemicals, materials, and miscellaneous (27). Their approach allows the further classification of costs into subfunctions such as the division of delivery into pumps, storage tanks, transmission, and distribution.

Using this approach, costs can be allocated to cost centers within the service areas and reports can be generated as needed, such as details of cost by cost center, costs by function, cost comparison, and a summary of system information. The approach offers a method to study the cost, by function, to serve customers located anywhere in a system. This may be more than managers may want to know in some situations, but the information can be very helpful for growth-impact studies. Although financial management in most systems does not yet require information at this level of detail, all indications are that future requirements for management information are moving in the direction of more detail, particularly with the constant drive for more equity in water charges. Figure 3.9, taken from Clark and others (28) illustrates the kind of zonal function cost data that is now becoming available and that may be useful in the future.

Water costs fall into the two general areas of capital costs and operating

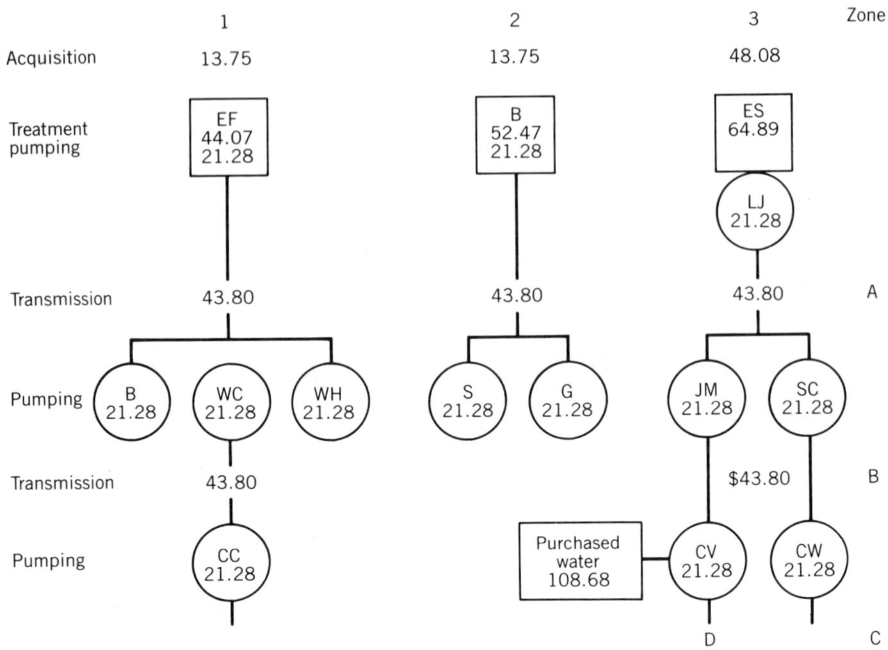

FIGURE 3.9 Allocation of capital and operating expenses to water system components for Dallas Water Utility.

and maintenance costs. These two categories may seem too simple for some financial managers, but the recognition of the distinction is critical to effective management.

Capital costs include the general categories of construction, debt service, and reserve for depreciation. Since debt financing is the form most favored for capital financing by water utilities, the capital budget usually involves some sort of debt financing. This subject is covered in more detail in Chapter 6. Operating costs include labor, materials and chemicals, energy, equipment and support costs, and require considerable detail to analyze. An analysis is made of the capital and O&M cost in 12 cities (28). As shown on Table 3.5, these vary considerably, especially due to variable interest charges. Five of the systems were analyzed to identify cost trends over a 10-year study period, and Table 3.5 presents this data. The five systems are Cincinnati, Dallas, Kansas City, Fairfax County, Virginia, and Elizabethtown Water Company.

The figures are in terms of revenue-producing water for the years 1965–1974. Total costs were increasing about 5% per year and labor costs nearly doubled during the period of analysis. The change in the consumer price index in the period seemed to be just under 5% with most of the inflation coming at the end of the period.

There are a number of interesting features of the data in Tables 3.5 and 3.6. First, we can see the distribution of the costs for the different components of O&M. As seen from Table 3.6, the ranges fall approximately as follows: support services, 26–32%; acquisition, 19–24%; treatment, 12–14%; power and pumping, 14–19%; and transmission and distribution, 15–21%. Total operating costs for the 10-year period ending in 1974 rose from a low of $204 to a high of $289 per million gallons. From Table 3.5 we can see that capital costs, meaning depreciation and interest, added from about $100 to about $125 per million gallons to the total (this must be deduced from the table by subtraction).

The report estimates that labor represents between 50 and 60% of total O&M costs. The relative share of labor's cost was rising during the study period. There was, however, a rising productivity of labor in terms of man-hours per million gallons; with the ratio declining from about 33 to about 28 over the study period. The level of confidence of this data would not be very high, but it would represent an annual productivity gain of about 1.7%, which would seem to be good for the water industry in a time of rising regulatory requirements, but the data was taken before the beginning of the Safe Drinking Water Act.

The analysis in the report suggests a breakdown of operating and capital costs for water, by function as shown in Table 3.7. The costs were also placed into regression equations, which can be used for rough calculations of the distribution of costs across the functions. The equations are as shown of Table 3.8. The r^2 value shows the level of confidence of the equations, reflecting the variability across the utilities studied. A value of 1.0 would represent, of course, a perfect fit, and the equations could be used with confidence. However, in the case of some functions, mainly treatment, the degree of confidence is low and the equations could only be used for the crudest of estimates.

To spot check some of these figures, I took the 1983 operating report of the local water utility (29) and compared some of the actual costs with those that might be predicted from the above data. This process was aided by additional data prepared for a rate study in Fort Collins that was conducted in 1984 and which provided data on capital and operating cost allocations. My analysis showed Fort Collins' revenues in 1983 to be about $493 for operations and maintenance and about $365 for capital-related expenses, each in terms of cost per million gallons produced. There are several reasons why this data is not entirely consistent with that reported by Clark and others (28), such as the use of water produced rather than revenue-producing water, and the fact that revenues rather than cost were used, so there can be some debt service beyond interest and depreciation. However, the figures still provide the kind of rapid comparison that is interesting to note, and the kind that can be done in most any utility with data on hand.

TABLE 3.5 Average Operating and Capital Costs for All Five Utilities Over the 10-year Study Period (1965–74) (28)

Item	Years									
	1	2	3	4	5	6	7	8	9	10
Operating Costs										
Support services										
$, in millions	1.126	1.198	1.474	1.560	1.837	2.031	2.268	2.437	2.705	3.127
% of total	26.0	26.4	29.7	30.2	31.6	32.4	31.6	31.5	31.7	31.1
$\$/10^6$ gal	55.29	54.60	62.51	61.89	71.66	76.19	79.35	83.49	91.72	89.98
Acquisition										
$, in millions	0.981	1.007	0.978	1.062	1.231	1.289	1.537	1.770	1.990	2.356
% of total	22.7	22.2	19.7	20.6	21.2	20.5	21.4	22.9	23.3	23.5
$\$/10^6$ gal	48.27	45.91	41.46	42.22	48.08	48.20	53.75	60.69	67.42	67.43
Treatment										
$, in millions	0.539	0.577	0.617	0.630	0.701	0.783	1.013	0.913	0.998	1.212
% of total	12.5	12.7	12.4	12.2	12.1	12.5	14.1	11.8	11.7	12.1
$\$/10^6$ gal	26.58	26.27	26.10	25.0	27.44	29.39	35.41	29.63	33.85	35.01
Power and pumping										
$, in millions	0.789	0.830	0.922	0.870	0.933	0.955	1.042	1.172	1.294	1.805
% of total	18.2	18.3	18.5	16.8	16.1	15.2	14.5	15.2	15.2	18.0
$\$/10^6$ gal	38.70	37.85	38.94	34.43	36.51	35.74	36.42	40.29	43.98	52.08

Transmission and distribution										
$, in millions	0.890	0.927	0.978	1.044	1.108	1.213	1.320	1.439	1.548	1.541
% of total	20.6	20.4	19.7	20.2	19.1	19.3	18.4	18.6	18.1	15.3
$\$/10^6$ gal	43.81	42.19	41.46	41.40	43.32	45.38	46.21	49.30	52.37	44.27
Total operating cost										
$, in millions	4.074	4.272	4.579	5.030	5.830	6.285	6.934	7.593	8.431	9.262
$\$/10^6$ gal	212.65	206.82	210.47	204.95	226.78	235.14	251.15	265.04	289.34	286.95
Capital Costs										
Depreciation ($, in millions)	1.241	1.296	1.430	1.547	1.604	1.661	1.693	1.828	1.904	2.145
Interest ($, in millions)	0.995	0.920	0.948	1.286	1.267	1.428	1.411	1.488	1.707	1.848
Total capital costs ($, in millions)	2.233	2.217	2.378	2.833	2.871	3.090	3.104	3.316	3.612	3.993
Total Operating and Capital Costs										
$, in millions	6.313	6.490	6.958	7.864	8.702	9.375	10.039	10.915	12.044	13.256
$\$/10^6$ gal	332.88	322.45	328.39	327.37	340.26	354.23	370.57	387.88	425.93	416.74

TABLE 3.6 Cost Analysis Summary for Latest Year of Record (1974) (28)

Utility	Revenue-producing water (10^6 gal/day)	Cost Categories ($/$10^6$ gal) Support Services	Acquisition	Treatment	Distribution	Interest	Total
Kansas City	26,855	$145	$ 15	$ 82	$138	$ 50	$430
Dallas	63,030	83	25	52	120	58	338
San Diego	47,192	96	277	28	106	7	514
New Haven	17,714	113	29	15	106	117	560[a]
Fairfax Co.	19,232	88	35	56	134	209	522
Phoenix	63,661	91	17	47	112	53	320
Kenton Co.	2,259	82	12	103	124	73	394
Orlando	12,522	110	42	22	135	85	394
Elizabeth	38,256	89	67	33	144	113	492[b]
Pueblo	6,793	99	38	84	232	164	617
Seattle	45,967	109	37	13	77	27	263
Cincinnati	38,104	85	17	36	139	18	295

[a]Includes $179 taxes.
[b]Includes $76 taxes.

TABLE 3.7 Utility Cost by Category

Item	Percent of Costs by Category				
	Support Services	Acquisition	Treatment	Power Pumping	Transmission Distribution
Operating cost	31	22	8	16	19
Capital cost	9.8	12.6	10.8	—	67.3

Using a rough increase in costs of 2.25 to represent the change in the consumer price index between 1974 and 1983, the figures in Table 3.8 would predict operating costs of about \$287(2.25) = \$645 and capital-related costs of about \$62(2.25) = \$140. These compare with observed costs in Fort Collins, Colorado, of \$493 for operating and \$365 for capital costs. The total costs of \$785 for predicted costs from Table 3.8 and \$858 for observed costs in Ft. Collins compare better than the separate capital and operating values, differing by less than 10%. This kind of estimating and comparing is very approximate, and while it would probably bother any financial officer not to be more precise, it seems informative to study the figures this way and to constantly strive to improve them.

The costs to operate small systems may vary considerably from large ones. Clark and his colleagues have published a report that describes the costs for several small water utilities (30), and in this report the portion of O&M costs devoted to labor are given as near 30%, which is lower than for the larger systems. There are several possible explanations: use of part-time workers, lower salaries, and shared use of employees with

TABLE 3.8 Relationship Between Annual Cost and Revenue-Producing Water (Corrected to 1974 dollars)

Item	Operating Costs[a] $C = aQ^b$			Capital costs[b] $C = aQ^b$		
	a	b	r^2	a	b	r^2
Support	141	0.95	0.94	3.61	1.06	0.76
Acqusition	2.1	1.23	0.64	2.4E-8	2.89	0.67
Treatment	4202	0.53	0.53	5.2	1.01	0.52
Transmission and distribution	358	0.82	0.91	25.7	1.06	0.89
Total	621	0.91	0.93	28.7	1.09	0.89

[a]Power and pumping costs have been allocated to other cost categories.
[b]C = annual cost in dollars, a = constant, b = rate of change, Q = revenue-producing water in 10^6 gal/year.

other municipal departments. This report also points out that O&M costs rose faster than capital costs in the study period, reflecting the importance of labor costs, and the fact that capital costs reflect costs of an earlier period. This approach, however, reflects certain shortcomings in the accounting practices of water utilities where the inflationary increases for capital are not fully considered. This shortcoming is partially remedied by the public utility's preference for using debt financing for new construction, but inhibits the charging of full costs for water and may exacerbate the public's perception that water costs less than it really does.

The capital costs for water distribution and treatment systems should be amenable to study and analysis to determine trends. Regarding the specific costs for individual components, however, it is really not possible to present the data available in any kind of brief form. EPA has published in four volumes a guide to estimating treatment costs which covers 99 unit processes (31). Curves for the 99 processes are given in the volumes 2 and 3 of the report and volume 4 is a computer program. The results of some sample calculations from the report may be helpful to give some idea of the general capabilities and results of the approach, so some of them are given here for illustration.

In the report the cost estimates are carefully prepared for a range of possible treatment plant sizes and configurations. An example of a configuration used would be as shown in Figure 3.10, which represents one that could meet the general requirements of the Safe Drinking Water Act. Figure 3.10 shows the basic stages that could meet the requirements of the act and modifications that would allow additional flexibility to the treatment arrangements.

The estimates shown in Table 3.9 are based on October 1978 costs and reflect the costs to build and operate the treatment facilities. They include estimates of costs for the following site-specific items, which are not part of the cost estimating curves: land, site work, engineering, contractor overhead and profit (ranging near 10%), legal, fiscal, administration, and interest during construction.

Cost estimating using the generalized procedures reflected in Table 3.9 requires a careful approach to the use of interest charges and corrections for inflation. For example, the annual charges shown are based on 20-year lives of the treatment facilities at 7% interest for the amortization schedules. Cost indexes to be considered include a number in addition to the consumer price index. Those considered and discussed by Gumerman and others (31) include the following:

Engineering News-Record (ENR) Construction Cost Index
Engineering News-Record (ENR) Building Cost Index
Engineering News-Record (ENR) Skilled Labor Index
Bureau of Labor Statistics General Purpose Machinery and Equipment Index

Bureau of Labor Statistics Concrete Ingredients Index
Bureau of Labor Statistics Steel Mill Products
Bureau of Labor Statistics Valves and Fittings
Bureau of Labor Statistics Electrical Equipment and Machinery Index

These are discussed in more detail in Appendix 1 on cost indexes.

All in all the manager must compile his own experience base on costs, and generalized estimates for the different parts of a system are very handy information to have. In 1985, for example, conventional water treatment capital costs seem to be in the range of $0.40–0.50 per gallon per day, and treated water storage might be in the range of $0.15–0.20 per gallon. This means that treatment capital costs for a 10-mgd plant would be in the very general range of $4,000,000–5,000,000, and that a 1-million-gallon storage tank would be in the range of $150,000–200,000. The astute manager will be watching bids on these items and updating his own data base constantly. These very general numbers will be subject to many variations, not the least of which will be the economy of scale factor.

The research by Clark and his colleagues has produced several other estimates of water supply costs, including an update on the water treatment plant estimating process. Another one where they studied the capital and operating costs of a large (260-mgd) and a small (20-mgd) utility resulted in estimates of $918 for the replacement value of the distribution system of the large utility with a corresponding annual maintenance of $2.6 million (both 1978 dollars). These figures translate to per capita estimates of $1224 and $3.47 for the population of 750,000 of the large area (32). Clark states in another publication that conservatively he estimates that the distribution system represents about 67% of the capital requirements of a water system (33).

In management planning it would be desirable to have a number of different ways to estimate costs: by function such as acquisition, treat-

TABLE 3.9 Representative Costs of Water Treatment Plants

Size	Population Served[a]	Total Project Cost[b]	Annual Cost[b]
700 gpm package	4,500	0.710	0.124
5 mgd conventional	20,000	2.364	0.396
40 mgd conventional	125,000	10.334	1.851
130 mgd conventional	575,000	26.050	4.399

[a]Population served calculated from assumed average demand of 150 gpcd and treatment capacity = 1.5 times average demand.

[b]Costs are in millions of October 1978 dollars.

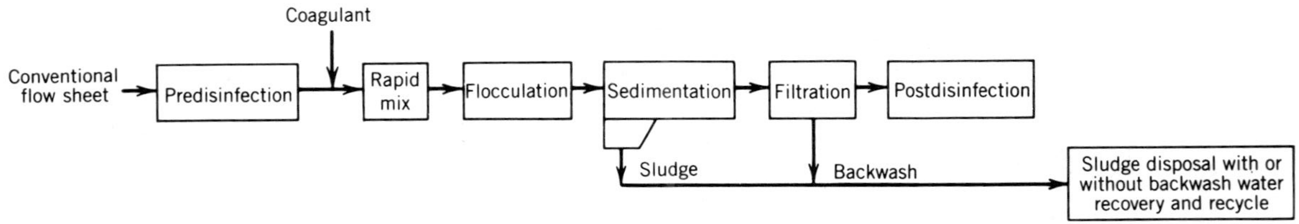

Treatment Step No.	① ② ③	④	⑤ ⑥ ⑦ Coagulant		⑧
Substance To Be Removed	Upper Limiting Raw Water Concentration	Maximum Contaminant Limit	Type	Dosage	Modification
Coliform bacteria	20,000 MPN/100 mL	1 per 100 mL	Usually alum	5–140 mg/L	None
Turbidity	1,000 TU	1 TU	Usually Alum	5–140 mg/L	None
Virus	1,000 pfu/gallon	None established (Turbidity 1 TU)	Usually alum	5–140 mg/L	None
Silver	0.17 mg/L	0.05 mg/L	Alum	30 mg/L @ pH = 6–8	None
		Modified conventional water filtration plants			
Coliform bacteria	Over 20,000 MPW/100 mL	1 per 100 mL	Usually alum	20–140 mg/L	Presedimentation, plus double coagulation, floc. & settling.
Turbidity	Over 1,000 TU	1 TU	Usually alum	20–140 mg/L	Presedimentation and/or double coagulation, floc. & settling.
Virus	Over 1,000 pfu/gallon	None established	Usually alum	20–140 mg/L	*Add:* Cl_2 to 1.0 mg/L free residual at point ⑦, polymer at points ② & ⑥, add pilot filter at point ②, add turbidity monitors to each filter effluent and point ⑦, use dual or mixed media rather than single media.
Arsenic, pentavalent, or trivalent	1.0 mg/L	0.05 mg/L	Alum or $Fe_2(SO_4)_3$ @ pH<7.5	20–30 mg/L	*Add:* Facilities for pH adjustment and for feeding ferric sulfate—Preoxidation required for trivalent form.

asbestos		None established	Alum	5–140 mg/L	*Add:* 2 stages to rapid mix, polymer feed, pilot filter, mixed media in filters, cont. turbidity recorders, adequate backwash, washwater recovery proper sludge disposal, and lab equipment for asbestos analysis.
Chromium, hexavalent	5.0 mg/L	0.05 mg/L	Ferrous sulfate	5–140 mg/L	No predisinfection; add $FeSO_4$ feeder and equipment for pH adjustment to 6.5 to 9.3 at point ④.
Cadmium	0.1 mg/L	0.010 mg/L	Alum or ferric sulfate	Varies	*Add:* facilities for pH adjustment at point ②, pH = 9 for alum, pH = 8 for $Fe_2(S)_4)_3$.
Fluoride (soft waters)	4–6 mg/L	1.4–2.4 mg/L	Alum	200–500 mg/L	High-capacity alum feed plus lime or caustic feed for final pH adjustment, provision for alum sludge disposal.
Lead	1.7 mg/L	0.5 mg/L	Alum & $Fe_2(SO_4)_3$	Varies	*Add:* ferric sulfate feeder at Point ②.
Manganese, inorganic	None	0.05 mg/L (Sec • std)	Alum & $KMnO_4$	5–140 mg/L 0.3–2.0 mg/L	*Add:* $KMnO_4$ feeder at Point ②.
Mercury, inorganic	0.006 mg/L	0.002 mg/L	Alum @ pH = 7 or $Fe_2(SO_4)_3$ @ pH 8 + powdered act. carbon (PAC)	Varies Varies 5–20 mg/L	*Add:* Feeders for ferric sulfate and PAC and provide for pH adjustment at Point ②.
Mercury, organic	0.02 mg/L	0.002 mg/L	Alum & PAC	1 mg/L PAC for each microgram Hg removed	*Add:* PAC feeder at Point ②.
Selenium, quadrivalent	0.05 mg/L	0.01 mg/L	Ferric sulfate	30 mg/L @ pH = 6.0	*Add:* ferric sulfate feeder and provide for pH adjustment at Point ②.
Silver	0.17 mg/L	0.05 mg/L	Ferric sulfate	30 mg/L @ pH = 6–8	*Add:* ferric sulfate feeder at Point ②.

FIGURE 3.10 Capabilities of conventional water filtration plants to meet maximum contaminant levels of the National Interim Primary Drinking Water Regulations.

ment, delivery, and support; by zonal characteristics such as elevation, distance, and land use; and by accounting categories such as labor, materials, and equipment. These are the kind of numbers that will be forthcoming for managers with the advent of better information systems such as those discussed in the chapter on decision support systems.

For now, it appears that the general cost for water in the United States would be on the order of $1.00 per thousand gallons, and higher, with a replacement value of systems in the range of $500–1000, with distribution systems representing about 67% of the capital value.

CAUSES AND RATE OF DEPRECIATION, DETERIORATION, AND OBSOLESCENCE

The components of the urban water supply system must be viewed separately to understand the factors that cause it to age, wear out, and require repair, replacement, and rehabilitation. This is one of the most important topics in water utility management today, and will determine costs and levels of service for years to come.

In most cases the acquisition system is a long-term asset, parts of it, such as water rights, being essentially perpetual property rights if properly maintained and used. Depending on service conditions, storage facilities may also last a long time if they are given the proper care. However, there are a number of factors causing obsolescence in storage facilities, over a long period of time, and there are also short-term hazards such as dramatic siltation of reservoirs, that can occur as extreme events. In recent years, with attention focused on dam safety, reservoirs have been subjected to obsolescence through the realization that they were not, in fact, safe. Another set of problems is the loss of water rights through legal encroachment, competition for use, and loss of permits. Some suburban cities, dependent on large central cities for a continued water supply, may lose the supply when a contract expires.

The acquisition system needs to be examined in detail to locate the components at risk of loss or failure. They may be in the legal rights or permit, the collection facilities, the reservoir or wells, the transmission facilities, or any of the support facilities.

The treatment facilities are essentially like an industrial process. They consist of land, concrete, buildings, and machinery, including instrumentation. The process components themselves are the most susceptible to wearing out and needing attention, of course, and this is where a good maintenance management system, discussed in Chapter 9, comes in handy.

The distribution system is the most vulnerable to unknown problems since it is, for the most part, hidden from view and constructed of very dissimilar materials installed at different times. Good record keeping is essential for proper management of this system, but utilities are just beginning to appreciate the need for it. Causes of problems in distribution systems include breaks, leaks, corrosion, postprecipitation, and general malfunctioning. These are discussed in some detail in Chapter 9.

DATA BASES FOR MANAGEMENT

Almost every topic discussed so far about water supply systems points to the importance of data and information to aid in management. What is needed is a careful approach to data base design. The approach to data base design can either be by function or by other classification systems such as the organizational approach. The general aspects of data base design are discussed in Chapter 8, which covers the use of data bases and decision support systems, and how they are used to improve management in urban water systems in general. Some of the information parameters needed specifically for the water supply system are given below in Table 3.10:

It will be recognized that these parameters of the management information system cover the same topics that have been discussed in the chapter so far. Management needs to know what is going on, and the

TABLE 3.10 Information Systems for Water Utility Management

Acquisition
Water balance, including flows and levels
Raw water quality
Water exchanges, sales, purchase agreements
Facility inventory and condition
Maintenance records
Treatment
Facility inventory and condition
Operational records
Reporting records
Maintenance records
Delivery
System flow, pressure and quality data
System inventory and condition
Maintenance records
Support
Financial information
Engineering records
Construction records
Administrative records

management information system informs the manager when he needs to know. There is a unity in the information items that are of interest to management, with the same needs in the planning, design, and management stages. A properly designed information system can go a long way toward improving management, and the design of information systems is one of the important needs in water utilities today.

IMPORTANT INTEREST GROUPS

Certainly, there are many interest groups concerned with the health, environmental, or business aspects of water supply. Since water supply is mostly a local function, the interest groups take on special importance for making the industry effective. The main interest group continues to be the American Water Works Association, now located in Denver after moving from New York in the early 1970s. AWWA has some 35,000 members and has been quite active in recent years in developing new publications, member services, and research to improve the water supply industry.

AWWA organized a research foundation in the 1970s, and it has been increasingly active in developing and managing research of interest to the water industry. Called the AWWA Research Foundation (AWWARF), it is an autonomous organization located in Denver with AWWA.

There are a number of other associations and organizations that affect the course of the water supply industry in the United States. Two that should be mentioned here are the National Association of Water Companies, which represents the private water companies, and the American Public Works Association, which has a number of management improvement programs for water supply utilities. A number of other organizations attract the participation of engineers working in water supply, such as the International Water Supply Association (IWSA), the American Society of Civil Engineers, and the Inter-American Association of Sanitary Engineers (AIDIS).

There is an important aspect of water management represented by the plumbing industry, and the American Society of Plumbing Engineers (ASPE) should also be recognized. There are several useful publications that come via the plumbing industry as well.

IMPORTANT ISSUES

There continue to be many issues affecting the water supply industry, especially scientific, financial, and regulatory issues. The scientific issues arise as research needs, and the financial and regulatory questions arise as "issues."

AWWA listed some of the most important research questions attended to in 1985 in the February 1985 issue of *Mainstream* (34). These are listed here to illustrate some of the questions that are currently plaguing the industry: bacterial regrowth in distribution systems, unaccounted for water, disinfectant residuals, plastic pipe practices, economics of internal corrosion control practices, scale formation and inhibitor chemistry, small system technology, case studies of disinfection practices, substitution of granular activated carbon for sand in filters, constant and declining rate filters, control of aluminum precipitate, dosage control, filter to waste procedure, survey of public attitudes and concerns, and sludge disposal practices.

One listing of the most important "issues" is from a recent survey by Malcolm Pirnie Engineering Company, which utilized the Delphi method to assess expert opinion to develop a list of the top 24 water issues. Although not all of them relate just to water supply, they are listed to illustrate the range of issues. They seem to relate more to water supply than to other water services. This is the list as reported in *Mainstream* (35) that utilized the Pirnie report as the primary source. The issues are given in order of current ranking of importance:

Inadequate rates
Insufficient health effects research
Small water systems
Capital funding
Organic contamination
Inadequate operations and maintenance
Groundwater quality
Convincing public to proceed without federal or state assistance
Disinfection side effects
Inability to raise rates
Groundwater supplies
Too much "politics"
Surface water quality
State and federal ability to implement programs
Lack of public confidence
Water revenue diverted to nonwater uses
Lack of competent leadership
Water rights issues
Inadequate treatment technology
Corrosion by-products
Disposal of water plant wastes
Natural emergencies
Mandated "overtreatment"
Sabotage

There are, of course, many other ways to approach these issues and others. For example, there is no mention about the need to develop better water plant instrumentation, yet the advent of better management solutions to operational problems depends on the instrumentation. There is also no mention of the wearing out of plants and equipment built during the urbanization of the 1960s. There would be many other such issues, but presumably the scientific approach of the study reported above considered them in developing this list of issues.

REFERENCES

1. M. B. McPherson, *Urban Water Balances Considering Conservation*, ASCE Urban Water Resources Research Program, New York, n.d.
2. Jerome B. Wolff, F. P. Linaweaver, and J. C. Geyer, Commercial Water Use, ASCE Urban Water Resources Research Program, TM 27, December 1975.
3. L. Scott Tucker, J. Millan, and W. W. Burt, Metropolitan Industrial Water Use, American Society of Civil Engineers, Urban Water Resources Research Program, TM 16, May 1972.
4. M. B. McPherson, Household Water Use, ASCE Urban Water Resources Research Program TM 28, New York, January 1976.
5. John H. Stacha and F. W. Pontius, An Overview of Water Treatment Practices in the United States, *JAWWA*, October 1984.
6. D. A. Okun, Water Supply in the People's Republic of China, *JAWWA*, June 1984.
7. Daniel A. Okun, Principles for Water Quality Management, *J. Environ. Eng. Div., ASCE*, December 1977.
8. American Water Works Association, 1981 Water Utility Operating Data, Denver, 1981.
9. Gordon M. Fair, J. G. Geyer, and D. A. Okun, *Water and Wastewater Engineering*, Vol 1, Wiley, New York, 1966.
10. C. N. Durfor and E. Becker, Public Water Supplies of the 100 Largest Cities in the U.S., 1962, USGS Water Supply Paper 1812, 1964.
11. Richard H. McCuen, R. C. Sutherland, and J. R. Kim, Forecasting Urban Water Use: Commercial Establishments, in *Energy and Water Use Forecasting*, AWWA Management Resources Book, Denver, 1980.
12. Wayne B. Solley, E. B. Chase, and W. B. Mann IV, Estimated Use of Water in the U.S. in 1980, *US Geological Survey*, Circular 1001, 1983.
13. K. L. Kollar and P. MacAuley, Water Requirements for Industrial Development, in *Energy and Water Use Forecasting*, AWWA Management Resource Book, Denver, 1980.
14. Gilbert White, D. J. Bradley, and A. U. White, *Drawers of Water, Domestic Water Use in East Africa*, University of Chicago Press, Chicago, 1972.
15. Brown and Caldwell Engineers, Effect of Water Meters on Water Use, Denver, March 1984.
16. Texas Water Utilities Association, Manual of Water Utility Operations, Austin, 1979.
17. John W. Clark, W. Viessman, and M. J. Hammer, *Water Supply and Pollution Control*, International Textbook Company, Scranton, 1971.
18. Joseph E. Lescovich, Check Valves with Brains, *JAWWA*, November 1983.
19. Thomas M. Walski, *Analysis of Water Distribution Systems*, Van Nostrand Reinhold, New York, 1984.

20. American Water Works Association, Water Utility Management, AWWA No M5, Denver 1980.

21. Environmental Protection Agency, Office of Drinking Water, FY 1983 Status Report, The National Public Water System Program, Washington 1984.

22. National Association of Water Companies, Financial and Operating Data, 1983, Washington.

23. President's Intergovernmental Water Policy Task Force, Urban Water Systems, June 1980.

24. AWWA, staff report, The Water Utility Industry in the U.S. *JAWWA*, July 1966.

25. Marc R. Alston, Revision of USEPA Drinking Water Regulations, Opflow, December 1984 and January 1985.

26. National Regulatory Research Institute, Commission Regulation of Small Water Utilities: Some Issues and Solutions, Columbus, Ohio, May 1983.

27. Robert M. Clark, V. F. Fender, and J. I. Gillean, A Spatial Costing System for Drinking Water, *JAWWA*, January 1982.

28. Robert M. Clark, J. I. Gillean, and W. K. Adams, The Cost of Water Supply and Water Utility Management, Vol. 1, EPA 600/5-77-015a, Cincinnati, November 1977.

29. City of Fort Collins, Colorado, Water Utility Operating Report, 1983.

30. Richard G. Stevie, R. M. Clark, J. Q. Adams, and J. I. Gillean, Managing Small Water Systems: A Cost Study, EPA-600/2-79-147a, September 1979.

31. Robert C Gumerman, R. L. Culp, and S. P. Hansen, Estimating Water Treatment Costs, EPA-600/2-79-162a, August 1979.

32. Robert M. Clark, C. L. Stafford, M. G. Langle, and J. A. Goodrich, Determinants and Options for Water Distribution System Management: A Cost Evaluation, EPA-600/S2-82-033, August 1982.

33. Robert M. Clark, *Maintenance, Repair and Replacement in Drinking Water Systems, in Cooperation in Urban Water Management*, National Academy Press, Washington, DC, 1983.

34. AWWA Mainstream, AWWA Seeks Proposals for 15 New Water Research Topics, *Mainstream*, February 1985.

35. AWWA Mainstream, Rates, Health Research Top List of National Water Problems, *Mainstream*, November 1984.

CHAPTER FOUR

WASTEWATER MANAGEMENT SYSTEMS

SERVICE PROVIDED

After water supply, wastewater is the next most important urban water service. The service provided, as normally viewed by the public, is getting rid of an unwanted commodity, wastewater. In this sense wastewater service is similar to stormwater management since both exist to get rid of something that is undesirable rather than to produce something that is necessary, which is the case with the water supply service. This simple but important distinction provides some insight into the reason wastewater has proven to be more difficult to finance and management than water supply. The distinction also explains the need for extra work to develop other management approaches, such as the needs to educate the decision makers about wastewater and to build a positive public image for the wastewater organization.

The term *wastewater* is coming into general acceptance to replace the older term *sewage*, but it may introduce some confusion unless we are clear about the difference between dry-weather wastewater (sanitary sewage), wet-weather wastewater (combined sewage), and stormwater (storm sewage). This chapter deals primarily with the first category, that is, the dry-weather problem, and the next chapter deals with stormwater and combined sewer management. However, it is not really possible to separate completely the discussions of these categories since each of the wastewater problems has some of the characteristics of the others. For example, dry-weather systems behave differently in wet weather even if they are not combined sewers; storm sewers also exhibit water quality problems; and combined sewers act like separate sewers in dry weather. More precise definitions of the categories are given in the next section.

Wastewater has not always been regarded as a high-priority public service since the health aspects of managing wastewater were not well understood until the end of the 19th century. Wastewater was often dumped out the door with little thought until public health and environmental laws began to be strengthened. The strengthening of these laws for wastewater has only become important in recent years, and the form of future regulatory programs is still evolving.

The linkage of good wastewater management to the protection of public health is probably the most important justification for regulatory programs. The existence and awareness of health hazards that relate to wastewater are increasing all the time with the advance of the industrial society and with population growth and urbanization. The problems are generally in the categories of microbes and chemicals, including toxics and radionuclides.

Microbial agents can be the causes of problems such as polio, hepatitis, cholera, typhoid, and amoebic dysentery. Chemical problems can include lead, mercury, arsenic, and organic chemicals of synthethic origin. Radioactivity can involve low- or high-level radioactive wastes from industrial or hospital uses. The list of health problems traceable to water origins includes everything from minor intestinal problems to major problems such as cancer. The current attempts to regulate and control toxics in drinking water were described in the last chapter, and this chapter includes a section describing the need to control toxics in the wastewater stream.

In spite of the public health implications, much of the wastewater part of the urban infrastructure is oriented toward pure convenience. The wastewater collection system, in particular, is appreciated mostly for this purpose. The main concern of most people is to get rid of waste material as quickly as possible and get it out of sight; so they are not concerned about the quality of the environment when they think about wastewater but about convenience. The environmental and health benefits of wastewater programs, based on acceptable water quality, are the result of good overall management of a wastewater program, including wastewater treatment.

Most citizens have no idea how complex wastewater management is, or that good wastewater management involves several sectors of the economy: local government's role in providing the collection treatment and management; the role of research institutes, business, and the private sector in providing the technology, innovations, and products; and the role of the federal and state governments as standard setters and regulators. As a result of these complexities, it seems fair to say that wastewater management places a real test on the effectiveness of the public administration process.

Modern city dwellers take wastewater service for granted. We don't realize that the improvements we enjoy are not very old. The invention of the flush toilet, for example, is a 19th century event, and models are still being improved. Since flushing toilets is now recognized to be one of the main water using activities in the home, there has been movement toward the development and use of water-saving toilets, which apparently work quite well. The 19th century discovery of the venting principle for sewer stacks in buildings made it possible to bring toilets inside buildings without the stack odors infiltrating the interior space. This recent development is not something most people are aware of at all.

In modern times the demand to take care of the increasingly complex products of our industrial society is rising. The litany of complex chemicals and waste products that could be imposed on wastewater systems is rather scary, and wastewater management in the future will increasingly have to be oriented to control these.

People are eager to get rid of wastewater but not so eager to pay to treat it for the benefit of the next downstream community, so government has had to take on a significant role in wastewater management, mostly as a regulator. This government role seems destined to continue, in spite of the fact that we now have well over a decade of experience with the Clean Water Act. For this reason the discussion of regulation takes on additional importance for the wastewater program. As a result, we can expect more scrutiny from higher levels of government about wastewater program effectiveness than either water supply or stormwater management. This government oversight has proven to be an important part of wastewater and water quality management.

CONFIGURATION OF SYSTEMS

The general configuration of the urban wastewater system is as shown in Figure 4.1. The wastewater system begins where the urban water supply system leaves off. The water use block in Figure 4.1 is the interface between water supply and wastewater, and it refers to the same uses discussed in the previous chapter on water supplies: domestic, industrial, commercial, and public. The collection system receives the wastewater from the residuals of these uses, and this origin of the wastewater system, with such widely varying characteristics, reflects the diversity of human social and economic activity.

There is some confusion about terminology in wastewater management, and some of the basic terms need defining. *Wastewater* and *sewage* are two synonomous terms that refer to the wasted waters that occur in the wastewater system. The term *sewerage* refers not to the wastewater but to the system of pipes, pumping stations, and other facilities necessary to handle the wastewater. This term seems not to be used much anymore. *Sanitary sewage* refers to regular wastewater, usually of the municipal kind. This is sometimes referred to with the more descriptive term *foul*

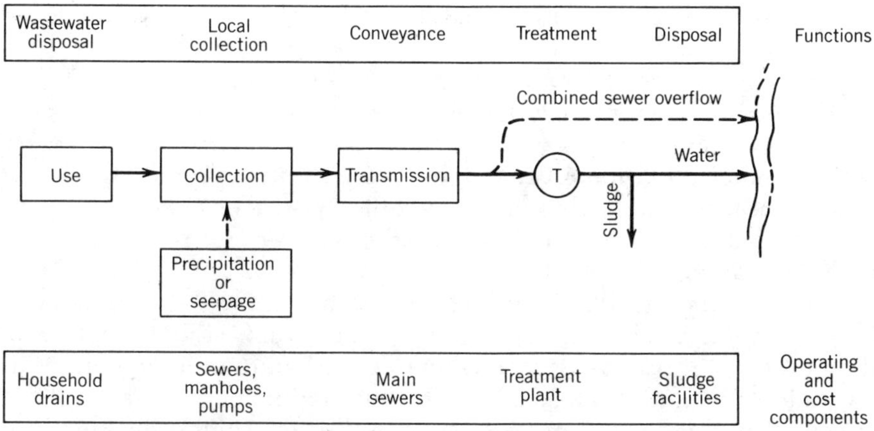

FIGURE 4.1 Configuration of urban wastewater management system.

sewage in England. *Combined sewers* transport both sanitary sewage and storm drainage. Sometimes this translates to "mixed wastewater" in other languages, such as German. *Separate sewers* transport only sanitary sewage, except for the inevitable quantities of infiltration and inflow that occurs. A *lift station* is a pumping station that overcomes the lack of a gravity route for wastewater. A *force main* is a section of sewer that is under pressure, and some call these *pressure sewers,* although the pressure sewer can be a different design for an entire sewer system where the terrain prohibits the gravity design. A *main or trunk sewer* is one of the principal sewers of a system, one that collects the *lateral sewers,* and an *interceptor sewer* is one that intercepts main sewers and carries the wastewater to a treatment plant. An *outfall sewer* is one that conveys wastewater to a point of disposal. Treatment plants have come to be classified as *primary, secondary, or tertiary,* depending on the degree of treatment, but other terms have come into use now, such as *advanced wastewater treatment (AWT).*

The collection system also is burdened with infiltration from the groundwater table and from runoff that gets into the sewers either legitimately, in the case of combined sewers, or illegitimately in the case of illegal connections from house drains and other sources. The collection system, like the distribution system for water supply, is where much of the capital investment is located and where the maintenance and upkeep is both important and difficult. It consists of collection sewers, manholes, pumps, and other support facilities.

The transmission system conveys the wastewater from the collection system to the treatment facility, which may be located at a remote location. It is normally from this facility that overflows will occur. For example, if the treatment plant is located downstream on a river, such as is shown in Figure 4.1, then there may be several relief points to prevent overloads on the treatment plant. Unfortunately, this may result in tremendous water quality problems and has been a source of concern to water quality management programs for a number of years.

The treatment plant is a 20th century addition to the wastewater management system. And, since the passage of the Clean Water Act, there has been much attention given to the wastewater treatment in the search for affordable, maintainable, and workable approaches. This topic will be discussed in greater detail later in the chapter and is of concern in the later chapters on operations and management.

Like treatment, sludge handling is an important and difficult part of the wastewater management program. No one wants sludge; there is no place to put it; it is expensive and complex to handle; the regulators are always looking; and it is a hazard to the environment. This topic will not be solved in this book, but there seem to be some lights at the end of the tunnel as we look at sludge more realistically than we did a few years ago when the objective was just to get rid of it. By this I mean that the emphasis on resource recovery and the increased awareness of the environmental hazards of sludge dumping are causing improved management programs to evolve. We will discuss the configuration of systems more under the section on cost estimating where we show some of the thinking about configuring systems to search for the most cost-effective approach.

DIMENSIONS AND PERFORMANCE STANDARDS

The basic objectives of the wastewater system are to have it convey wastes rapidly and neatly away from the points of origin and to treat and dispose of it in compliance with all regulations and without harming the environment. It is presumed that if discharges comply with the regulations, the environment will be protected. Most operators and system managers are in favor of protecting the environment, but they respond to the regulations, not to general goals like "protect the environment." This will be obvious to most, but it is an important principle of environmental management. To environmentally oriented persons it seems that everyone should be in favor of a clean environment, and it is important to see that the only way to have one is to have the appropriate enforceable rules and regulations.

To meet the goals of service and operations, there have arisen a number of norms of practice and standards that are usually built into wastewater systems. These include minimum velocity, size and slope for sewers, and many other design standards that determine reliability and other service parameters. The joint design manuals of the American Society of Civil Engineers and the Water Pollution Control Federation (WPCF) are a good starting place for design studies, as is the Texas manual (1,2).

The performance of a collection system is relatively easy to assess: either it takes the wastewater away or it does not. At least that would be the viewpoint of the customer. From the viewpoint of the system manager the assessment of performance would include the success of operations and maintenance strategies.

Performance standards for treatment and disposal facilities are more stringent and detailed, of course, since they involve meeting the spirit and letter of the law. In effect, the discharge permit requirements must be met. The nature of these will be discussed in the section on regulatory programs. There is an element of probability in the discharge requirements since the stream conditions when the discharges take place are random. For this reason the concept of the minimum design flow for the discharge has been established, and it has become common practice to take a "7–*Q*–10" as a standard for the design low flow. This refers to the level of the flow during the drought that occurs with a duration of 7 days once in 10 years.

Since the wastewater system is related to the water supply system, the demands on the system are related to the amount of water that is used and needs to be thrown out from the place of use. If we adopt the approximate figure used in Chapter 3 that the average per capita use of water is about 150 gpcd, then it would follow that much of this should show up in the wastewater system. The domestic component of the 150 gpcd was anywhere from 50 gpcd up, and this should show up in the system without any diversionary losses. The other uses may or may not show up. It is customary to use round numbers such as 100 gpcd for planning, and to look at individual cases such as communities with industrial loads for more detailed planning. The figure of 100 gpcd would be expected to include some allowance for infiltration and inflow, and this might not be tolerated by a financing agency such as the Environmental Protection Agency which might finance, say, 70 gpcd.

The wastewater system will be subjected to peaking factors just as the water supply system will. These will differ from the water supply peaking records and should be determined from local experiences. One factor that will be important to explain the differences between water supply and wastewater peaking factors will be the extent of self-supplied industrial water that is wasted into public sewers.

In handling wastewater, the constituents can be classified as physical, chemical, and biological elements. For the most part they are expressed in terms of milligrams per liter (mg/L) rather than the older term of parts per million (ppm), which is almost the same, depending on temperature. The parameters of wastewater are given in terms of the classification above. The biological intensity or strength of the wastewater is mostly expressed by the biochemical oxygen demand (BOD). The physical parameters of the wastewater include solids content, temperature, and other parameters that might be of interest such as the floatables. Some of the solids will be organic in nature and not purely inorganic. The chemical content of the wastewater will include organic and inorganic chemicals, and increasingly, all kinds of trace metals and inorganics. The nutrient content of the wastewater, principally nitrogen and phosphorus, will also be under the chemical classification.

Industrial wastes may contain a number of undesirable elements that simply cannot be handled in the wastewater treatment plant. These would have to be handled before the wastewater arrives at the plant through pretreatment. Examples would be acids, high temperatures, toxic chemicals, and oils and greases.

TOXIC SUBSTANCES IN WASTEWATER SYSTEMS

The handling of toxic materials in the waste stream promises to be one of the main challenges facing managers during the remainder of this century. In general the handling of the hazardous wastes from industries that manufacture or use chemicals is the cause of the problem, but there are other potential causes, including agriculture. Since industrial waste management is generally an urban problem, it is the responsibility of the wastewater management organization to make sure that problems relating to toxic wastes are not ignored. The extent of rural problems due to hazardous industrial wastes is also increasing.

Figure 4.2 shows the flow of hazardous materials in the environment, including through treatment facilities (3). The flow of the materials toward central treatment facilities is from the industries to the treatment plants. The wastes can also be treated by the industries themselves and then routed to the central treatment facilities. This preliminary treatment is becoming an important part of the trend toward controlling and financing solutions to the hazardous waste problem.

The approach to controlling toxics through preliminary treatment begins with the local sewer use or industrial waste control ordinance, which provides rules for how industries can discharge to the public sewer. There are two recent WPCF manuals of practice that describe the pretreatment program and the ordinance, as well as a number of government reports (4,5).

The need for an industrial waste control program is not limited to the need to control toxic substances. Wastewater management systems also need protection from problems caused from fire and explosion hazards, corrosive materials, potential obstructions, highly concentrated wastes, and excessive heat (4).

The local programs that seek to control toxics are still evolving since the national research knowledge base concerning what is toxic and what is not is still being developed. The status of the EPA development of a

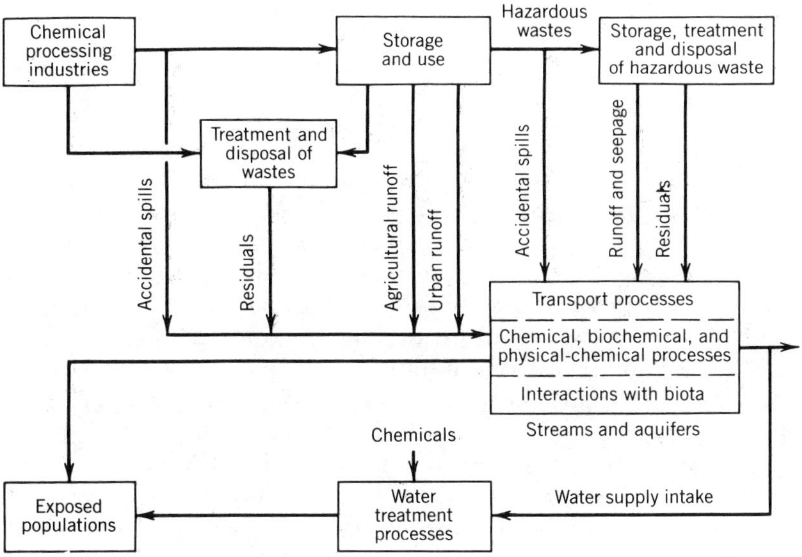

FIGURE 4.2 Flows of toxic materials in the water resource.

list of priority pollutants is described by a WPCF task force (5). A priority pollutant is one that has been identified by EPA as needing control in pretreatment standards. As of 1981, 129 had been identified and placed on the official list. The list will continue to be reviewed as new research is completed and new regulations are promulgated. The compounds on the list are not all that may be regulated, however, as there may be industries in local areas that produce compounds that are not on the list but which require control.

Some of the principal industries that may cause problems and that are candidates for control are as follows: electroplating, inorganic chemicals, leather tanning and finishing, petroleum refining, steam electric power plants, timber products, and nonferrous metals manufacturing. These generate waste streams ranging from the chemicals they produce to all kinds of acids, metals and suspended solids, grease, and oxygen-demanding wastes (5).

The need to pay attention to hazardous wastes and toxic substances will continue to increase, and this area of management challenge will continue to be a primary one for facility managers. Management issues identified for the future (5) are as follows: need for broader environmental monitoring, systematic identification of synthetic organic chemicals, need for aggressive follow-up of biological monitoring results to identify specific chemicals, need for more analysis of the kinds of fate of chemicals in particular areas, and the need for better disclosure of chemicals used in and discharged by the different industries.

MANAGEMENT STRUCTURES

Management of wastewater services has been centered for the most part in city departments. There are also a large number of special-purpose districts that offer wastewater collection and/or treatment. These are still municipal corporations, and the influence of the private sector that is felt in water supply is not present in wastewater management. There has been quite a bit of recent discussion about privatization possibilities under the emerging tax codes and regulations, especially since the federal government would like to ease out of the wastewater grant business. However, the privatization efforts are just beginning, and the jury is still out on the concept of private operation for wastewater treatment.

There are not as many statistics on the numbers and sizes of wastewater utilities as were given for water supply utilities in the last chapter. The character of the two kinds of services is different. EPA has extensive files of data on the U.S. wastewater facilities, and computer tapes of the statistics can be ordered from the agency. The statistics have been compiled in conjunction with the construction grants program.

There are on the order of 1000 communities in the United States with populations of 50,000 or greater, and most of these have their own wastewater utility. There are on the order of 3000 U.S. communities with populations of 10,000 or greater, and most of these can also be expected to be in the wastewater business. In its 1982 needs survey EPA listed a total of 32,511 "facilities" with there being different ways that a system or treatment plant could qualify as a "facility" (6). There are on the order of 30,000–50,000 wastewater systems in the United States, but the definition of what constitutes a "system" remains to be well defined. There are a large number of small systems that may be regulated, there are no reliable figures on how many of these there may be, but the number could certainly exceed 100,000 depending on the definition. Since wastewater management is still largely carried out by cities and special districts, the 30,000 figure seems to be a valid order-of-magnitude estimate, but further studies, particularly of the number and types of small systems, need to be conducted. The management of small systems is a real challenge for the same reasons that small water supply systems present management challenges.

LEGAL AND REGULATORY BASIS

Although the government has not been involved in wastewater management for many years, the legal basis for wastewater service is clear. Like water supply it esentially stems from the general powers of government to engage in programs that help improve the health and welfare of the people. The real legal driving force of wastewater service today, however, is the Clean Water Act (CWA). Since it was passed in 1972, there has been a different national approach to wastewater management, and there is every indication that the general approach enacted in the Clean Water Act will be preserved in spite of the fiscal difficulties of government in general in the 1980s and frequent debates in Congress about the reauthorization of the act.

Prior to 1972 there was little serious national planning and involvement in wastewater management. There was some attention given to minor assistance with planning and some wastewater grants, beginning in the mid-1950s, but nothing on the scale we have today. During the 1960s the momentum was building for a new approach, and the Clean Water Act was born out of the same environmental emphasis that brought us the Clean Air Act, the National Environmental Policy Act, and other environmental initiatives. Although it was enacted under President Nixon (who vetoed it), it seemed to rise out of the spirit that was present in the 1960s that called for government involvement to solve many social problems. That spirit has ebbed today, and some of the attention given to current issues under the CWA, such as to fiscal reform and the control of toxic substances, are the result of the shifts in emphasis.

The act is quite complex and could make an entire book to study the provisions, what they were intended to do, what they say, and what the results have been to date. This discussion will summarize the key provisions and programs of the act and discuss briefly how it affects wastewater management on a practical basis. For a detailed look at the act and its background, the reader is encouraged to get the WPCF guidebook on the act and its amendments (7).

The first part of the act where the goals and policies are stated received considerable initial attention, some of it due to the confusion over the intentions of the act. There is one overriding objective, two stated goals, and eight to ten policies, depending on what is considered a "policy." Briefly, they are:

1. The objective of the act is to restore and maintain the chemical, physical, and biological integrity of the nation's waters.
2. The first goal is to prohibit the discharge of pollutants into the navigable waters by 1985 (zero-discharge goal).
3. The second goal, an interim one, was to provide sufficient water quality for fish, shellfish, wildlife, and recreation by mid-1983.
4. The first policy is to limit the discharge of toxic pollutants.
5. The next policy is to provide federal financial assistance for the construction of publicly owned waste treatment works (POTWs).
6. The next policy provides for the development of areawide wastewater treatment planning.
7. The next policy provides for a major research program to find ways to eliminate waste discharges.
8. The next policy recognizes the role of the states in managing land

and water resources, including the construction grants and permit programs enacted by the act.

9. The next policy is to try to influence foreign governments to also eliminate pollution.
10. The next policy assigns the administration of the act to the EPA administrator.
11. The next three policies provide for public participation, provide for the minimization of paperwork, and provide that the act not supercede any existing law that allocated quantities of water.

The act has been amended a number of times since 1972. The 1977 amendments dealt with definitions of treatment levels and clarified state responsibilities. The 1981 amendments sought to improve the construction grants program, and the 1982 proposed amendments were addressed primarily at industrial dischargers. Debate on the act continues since the economic consequences are so important.

The initial provisions of the act were based on some powerful political feelings, and the maneuvers that accompanied the passage of the act make an interesting story. In addition to the Clean Water Act (7), another accessible place that the provisions can be studied is in a pair of WPCF articles (8). The provisions of the act can be understood better by looking at some of this background. For example, the zero-discharge goal, the object of many jokes among engineers that Congress was obviously not realistic with such a goal, was apparently based on the beliefs that technology was approaching (but not completely reaching) the point where 100% removal was possible on a cost-effective basis, and that our society should not in general be discharging wastes to the environment but should be recycling them. Both of these assumptions were reasonable at the time. The technological point led to some experiments with advanced wastewater treatment (AWT) that did not work too well, and we have learned quite a bit since then. Taken in context, the zero-discharge goal was not as ridiculous as some implied at the time. There are a number of other observations such as this that help in understanding the act, and, of course, this is true with any legislation, as one often has to study "Congressional intent" to get the full understanding of the law.

The general provisions of the act included planning, permitting, construction, monitoring, reporting, and enforcement. All of these have evolved into operational programs today.

The Section 208 planning program was supposed to tie everything together, and it has so many provisions that it covers about eight pages of the act. Section 208 did not get funded right away, however, and has not been as effective as the act intended. There, has been, however, quite a bit of planning under the act's total provisions. One thing that Section 208 did accomplish was to focus attention on nonpoint sources.

The general working of the Clean Water Act and its related programs seems to focus on the following process. Studies of stream capacities and standards lead to waste load allocations that become part of the discharge permit [NPDES (National Pollutant Discharge Elimination System)]. To ensure compliance with the act, there are two kinds of monitoring: ambient monitoring and effluent monitoring. Dischargers are required to report the results of their operational performance, and there is an enforcement program to maintain the standards of the act.

The most important part of all this to the operating agency is the NPDES discharge permit (Section 402 of the act). Compliance with this permit is the main operational goal of the urban water manager, along with the periodic review and renegotiation of the permit provisions, of course. In effect, the permit is the contract between the operational agency and the regulator, which is usually the state government, with EPA overseeing the process. EPA controls the authority of the state government to operate the permit program through a memorandum of understanding wherein the conditions of the program are set. If, in EPA's opinion, the state is not carrying out the full provisions of the act, then it can withdraw the authority of the state and begin to operate the permit program itself.

Monitoring consists of activities by a number of different parties. The operating agency monitors its own effluent and reports periodically to the regulatory agency. Usually it is the regulatory agency that conducts the ambient monitoring program, consisting of a network of stations. These are used to detect trends, provide data for water quality models, and detect spills.

Reporting is serious business since if it not faithfully and accurately done the process cannot work. To make sure it is effective, there is usually aggressive follow-up by the regulatory agency to make sure reports are accurate and on time.

The enforcement process is a joint one between the state governments and EPA. The process varies from state to state, but the intent is the same—to insure compliance with the rules and regulations. Enforcement against municipalities is usually based on negotiated approaches to agreement rather than a high degree of adversarial conflict, whereas enforcement against industries that have strong legal staffs will take a different approach. The most serious part of the enforcement program is that part that deals with criminal offenders and toxic waste dumpers.

The provisions of the CWA must be implemented by the states, and this requires parallel state legislation that enables them to operate the provisions of the act, including the permit program, the enforcement program, and the others. One of the most important state activities has been to operate the construction grants program, which has put billions of dollars into local areas since 1972.

Local ordinances are required also, such as a sewer use ordinance and a system of charges, especially for fair charges to industrial and commercial users.

COSTS OF WASTEWATER SERVICE

In one sense the wastewater system is simpler than the water supply system due to the predominantly gravity-based collection system, and on the other hand it is more complex due to the different kinds of wastewater that have to be handled. From the standpoint of cost estimating, however, the approach is about the same, and the accuracy available for preliminary costs is also about the same. The increased complexity of wastewater treatment means, however, that the results of treatment may be much more variable.

Much of the cost data given for water supply in the last chapter was generated to determine how much it would cost to implement the Safe Drinking Water Act. There has been little generalized cost data for water supply published elsewhere. One reason for this, of course, has been the absence of any construction grants program for water supply. In contrast to the water supply system, wastewater costs have received considerable attention. The most current reports were used to prepare this section, but the data base continues to improve and one should access EPA's construction grants files to obtain the most current information when the need arises.

EPA has studied cost-estimating procedures in its program to improve the needs studies. In the documentation accompanying the 1982 needs estimates, EPA describes levels of accuracy in terms of several different estimating procedures. Considered most accurate would be those from consulting engineering firms, either the final or preliminary estimates. For a collection system the sewer system evaluation report, with cost estimate, would be considered accurate. If a nearby project had been completed, it would be considered as a comparative basis for a valid estimate. Less accurate would be EPA's "rule of thumb" estimates, which would be based on curves and equations similar to those in this section. These kind of estimates can be considered as very approximate only, but they are useful in studying order-of-magnitude comparisons.

For the collection system an EPA report was used (9). This report analyzed the actual data from 777 collection systems with a construction cost of about $2.02 billion in 1979 costs. The data represented all ten EPA regions.

There are several aspects of the report that are interesting and potentially useful, and I will present those that would be most interesting to managers. First, there are the highlights of the study which report and provide insight on trends and findings from the construction (Table 4.1). The findings will confirm, no doubt, the general observations of engineers

TABLE 4.1 Highlights of Findings from Construction Costs Study (9)

Results obtained from the analysis of the large data base gathered by this effort fall within the normal range to be expected. The types of projects and their geographic diversity tend to cause some unexpected variation. Some trends are also surprising, but most are explained when the number of data points is considered. The highlights presented below are for various items that have sufficient data to be considered accurate, but which may seem to be anomalies. Some items are verification of suspected but undocumented trends. The items discussed are by no means a complete list. It is anticipated each reader will find similar cases upon detailed study of the data presented. These highlights are not presented in any order of apparent importance or significance.

Reinforced concrete was the most commonly used pipe, followed closely by vitrified clay. At diameters less than 16 inches, the latter was more prevalent.

ABS (acrylonitrile-butadiene-styrene) was the type least commonly used.

Relatively small differences in costs were noted for pipes buried less than 8 feet deep compared with those buried between 8 and 15 feet deep.

Pipes buried greater than 15 feet deep were significantly more expensive than those in the 8–15-feet deep range.

The cost of pipe delivered to the construction site was more expensive at construction sites located within a Standard Metropolitan Statistical Area (SMSA) for diameters of 21 inches and less than for non-SMSA (rural) areas. This trend was reversed for larger diameters.

The cost of labor for gravity sewer construction within SMSAs was significantly more expensive than in non-SMSAs, as would be expected.

Average step 3 nonconstruction costs (engineering, administrative, legal, etc.) are approximately 10% higher in non-SMSA areas than in SMSAs. This is probably due to the fact that non-SMSA projects are of smaller magnitude and, thus, these fees and expenses are a greater percentage of the total construction costs.

Step 1 costs are nearly twice as high in SMSA areas than in non-SMSA areas. This is attributed to the probable greater number of alternatives that must be considered in an urban area for step 1 planning.

Step 2 costs are higher in non-SMSA areas as a percentage of the construction costs than in SMSAs. This is probably due to the smaller size projects in the more rural areas, and the normal experience of engineering fees being a higher percentage for the smaller projects.

The unit construction costs for manholes were found to be less expensive for those 8–15 feet deep than for those less than 8 feet deep within SMSAs, and nationally.

Manholes 72 inches in diameter were more expensive in non-SMSAs than in SMSAs, probably because a unit of that size is a special situation in the more rural areas.

TABLE 4.1 ***(Continued)***

Bare-in-place construction costs for force mains smaller than 21 inches in diameter are significantly less expensive than equivalent size gravity sewers. This is explained by the fact that less attention to grade is required for the former, making construction faster with the use of less skilled labor.

Relatively few projects for only sanitary sewer construction were available for analysis. Most sewer projects were included as part of collection and treatment projects, at least for step 1 grants.

The bid items for cast-iron pipe averaged less than 500 feet each, based upon dividing total length in the data base by the number of individual bid items. This result indicates that many of these items were for special situations such as road or stream crossings or short reaches of shallow burial. For instance, of 85 bid items for 8-inch cast-iron pipe, the average length per bid item was 390 feet. For the same diameter reinforced concrete pipe and vitrified clay pipe, the average lengths were 1906 feet and 4457 feet, respectively.

Type of pipe was not specified for approximately 25% of the bid items. Undoubtedly these were cases where the contractor had a choice of materials and in the bid proposal did not specify the type to be used.

The depth of burial of sewer pipe was, on the average, greater in cities than in non-SMSA areas.

and managers in the field, especially on such items as the popularity of reinforced concrete pipe for sizes greater than 16 inches and clay pipe below that value.

Table 4.2 presents the national data for a number of different pipe sizes and depths. The same data is plotted in Figure 4.3. The data is interesting since it allows the study of the variation of cost factors. Of course, the actual costs in a particular location are apt to be very different from these. A term used in the report requires definition: *bare-in-place construction cost.* This refers to the cost just to install the pipe, including material and labor, but not including the cost of appurtenances such as manholes or pumping stations. The costs on Table 4.1 and Figure 4.2 include all of these, as well as the pipe costs.

Figures 4.4 and 4.5 present data for force mains and pump stations, two common parts of the collection system. All of the data is normalized to the first quarter of 1979 and for Kansas City, Missouri. For a discussion of the use of cost indexes to revise the data see Appendix 1.

A special cost index for urban sewer systems has been developed (Table 4.3). The costs will not be up to date, but the table does show the principal cities involved and the kind of variation that could be expected. Using these indexes and the previous figures, a person could make a ball-park estimate of the cost of collection sewer systems that were built in a particular city in 1979.

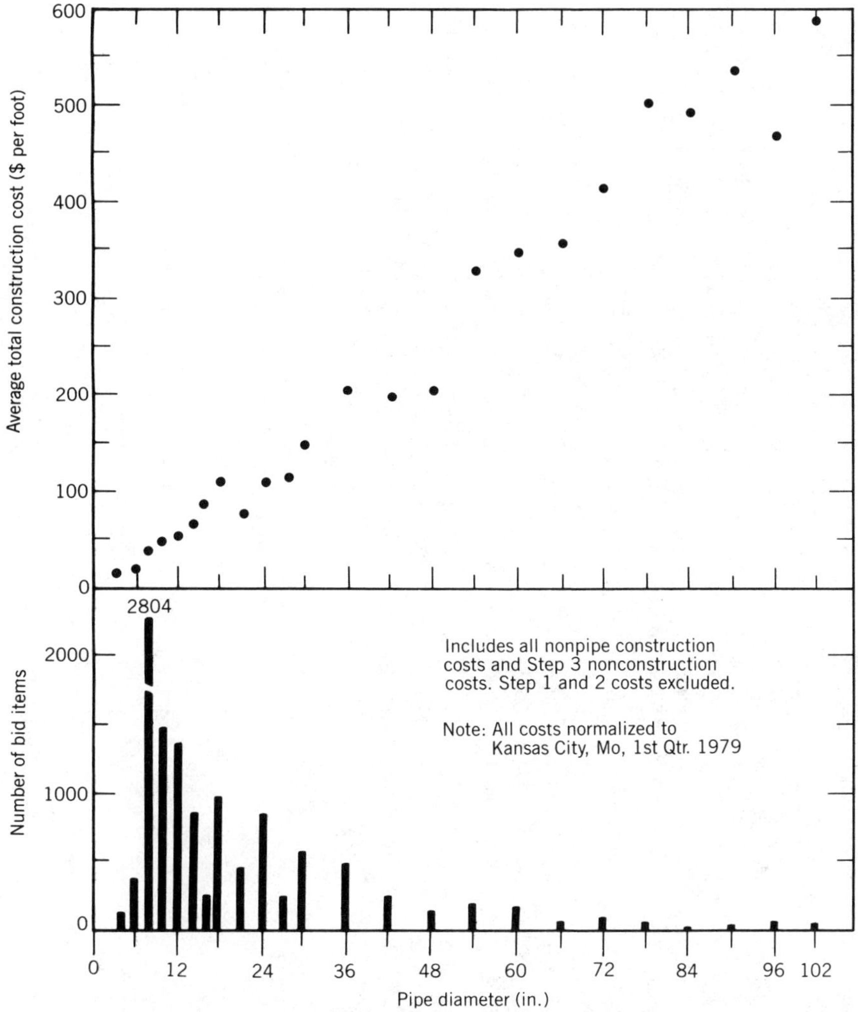

FIGURE 4.3 Total construction costs for gravity sewers (9).

Wastewater treatment is more complex than collection systems to approach from the standpoint of cost estimating. The report utilized here is (10). Again the data are dated and cannot be used for anything but the grossest of estimates, but they provide insight on the processes and the relative magnitudes involved.

The data are presented in terms of wastewater treatment and sludge handling. Wastewater treatment is viewed as a logical progression of unit processes, from preliminary treatment to final disinfection. These are shown in Table 4.4 from the report. Figure 4.6 shows the progression through the treatment processes with combinations of the unit processes

TABLE 4.2 Gravity Sewer Total Construction Costs[a] ($/ft) Pipe Type, Burial Depth, and SMSA Classification by Diameter[b] (9)

By Pipe Type Diameter	PVC	Asbestos Cement	Vitrified Clay	Cast Iron	Reinforced Concrete	Ductile Iron	ABS	Other/ Unknown
04	15.36 (18)	10.01 (6)	12.37 (16)	10.15 (5)	11.86 (4)	18.11 (4)	17.26 (2)	15.36 (61)
06	13.96 (43)	18.79 (13)	17.37 (119)	24.38 (15)	13.54 (22)	25.32 (22)	24.46 (2)	21.12 (160)
08	38.29 (392)	38.17 (378)	29.24 (784)	66.53 (85)	21.44 (49)	42.16 (252)	49.53 (17)	32.75 (847)
10	44.44 (252)	49.70 (229)	33.71 (421)	99.34 (52)	31.49 (23)	44.46 (99)	40.27 (2)	36.49 (421)
12	55.73 (169)	60.64 (174)	41.59 (420)	57.85 (49)	62.54 (56)	61.45 (94)	73.92 (5)	42.64 (381)
15	50.01 (82)	120.65 (88)	53.03 (284)	120.01 (4)	74.08 (149)	73.26 (7)	110.10 (1)	61.79 (218)
16	—	88.77 (40)	92.08 (14)	93.84 (18)	69.81 (12)	68.17 (112)	—	67.93 (24)
18	37.66 (1)	63.65 (132)	68.03 (208)	89.11 (19)	161.57 (299)	82.80 (109)	—	95.17 (206)
21	78.12 (3)	98.19 (45)	69.14 (104)	163.29 (1)	83.74 (212)	105.92 (4)	—	60.39 (71)
24	82.54 (3)	82.13 (50)	88.66 (120)	188.08(28)	116.13 (400)	107.26 (97)	44.60 (1)	92.39 (122)
27	—	34.48 (4)	116.56 (22)	—	117.52 (133)	—	—	106.83 (51)
30	—	32.16 (3)	135.47 (31)	114.96 (2)	168.60 (357)	173.36 (48)	—	115.81 (114)
36	—	—	154.74 (18)	109.61 (1)	225.90 (358)	212.61 (24)	—	147.24 (79)
42	—	—	79.47 (3)	—	181.06 (187)	550.52 (11)	—	202.81 (29)
48	—	—	132.95 (4)	—	209.63 (96)	—	—	232.54 (27)
54	—	—	—	—	313.59 (152)	—	—	411.14 (21)
60	—	—	—	—	349.47 (104)	—	—	361.31 (42)
66	—	—	—	—	360.64 (39)	—	—	321.52 (7)
72	—	—	—	—	416.46 (71)	—	—	373.87 (13)
78	—	—	—	—	415.44 (36)	—	—	737.73 (5)
84	—	—	—	—	559.90 (6)	—	—	385.55 (8)
90	—	—	—	—	444.64 (3)	—	—	632.95 (1)
96	—	—	—	—	369.76 (1)	—	—	462.21 (27)
102	—	—	—	—	583.58 (1)	—	—	—
Total No. of Bid Items	963	1162	2568	279	2752	883	30	2935

By Pipe Depth Diameter	<8 ft	8–15 ft	>15 ft	By Classification SMSA	Non-SMSA	National Average
04	13.58 (2)	12.53 (3)	—	16.51 (35)	13.41 (81)	14.37 (116)
06	15.29 (26)	17.99 (65)	23.94 (10)	22.31 (169)	17.40 (209)	19.60 (378)
08	33.28 (323)	36.97 (1233)	45.57 (441)	44.16 (1211)	30.81 (1493)	36.92 (2804)
10	31.18 (224)	48.83 (673)	47.25 (231)	54.69 (714)	36.79 (785)	44.95 (1499)
12	39.58 (190)	45.38 (553)	55.94 (234)	60.28 (656)	43.69 (695)	51.65 (1351)
15	46.27 (127)	58.33 (359)	63.55 (130)	73.18 (420)	57.88 (413)	64.73 (833)
16	61.86 (45)	65.52 (83)	78.93 (38)	70.68 (121)	78.71 (99)	74.97 (220)
18	63.31 (150)	63.11 (413)	82.15 (176	141.93 (504)	65.32 (475)	106.92 (979)
21	70.14 (57)	89.44 (173)	86.17 (101)	80.46 (101)	72.47 (289)	76.24 (440)
24	83.83 (95)	89.34 (316)	110.57 (200)	115.78 (379)	100.57 (446)	109.67 (825)
27	136.43 (16)	99.75 (58)	*c* (54)	115.76 (134)	103.97 (76)	111.52 (210)
30	120.01 (61)	117.13 (191)	*c* (170)	162.28 (344)	125.02 (219)	151.44 (563)
36	154.13 (24)	143.79 (156)	*c* (159)	216.42 (316)	168.84 (165)	204.37 (481)
42	170.64 (11)	194.58 (63)	*c* (77)	164.78 (138)	247.48 (92)	198.81 (230)
48	160.01 (4)	183.62 (20)	246.29 (27)	205.04 (92)	225.85 (35)	210.09 (127)
54	213.56 (8)	286.51 (44)	383.77 (89)	335.74 (148)	270.89 (25)	326.38 (173)
60	652.87 (1)	383.42 (33)	384.11 (65)	348.52 (96)	357.35 (50)	348.11 (146)
66	—	224.85 (1)	314.22 (10)	352.90 (39)	341.90 (7)	353.23 (46)
72	—	272.91 (6)	313.08 (26)	426.02 (65)	372.57 (19)	415.65 (84)
78	—	—	—	496.38 (41)	—	496.38 (41)
84	—	—	—	481.33 (14)	—	481.33 (14)
90	—	—	—	531.59 (4)	—	531.59 (4)
96	—	—	435.23 (25)	466.25 (28)	—	466.25 (28)
102	—	—	—	583.58 (1)	—	583.58 (1)
Total No. of Bid Items	1364	4445	2238	5970	5623	11593

[a]Includes all non-pipe construction costs and Step 3 nonconstruction costs. Step 1 and 2 costs excluded. [b]All costs normalized to Kansas City, Mo, October 1, 1979. ( )Number of bid items in sample. [c]Numbers not legible in source document; suggest using interpolated values.

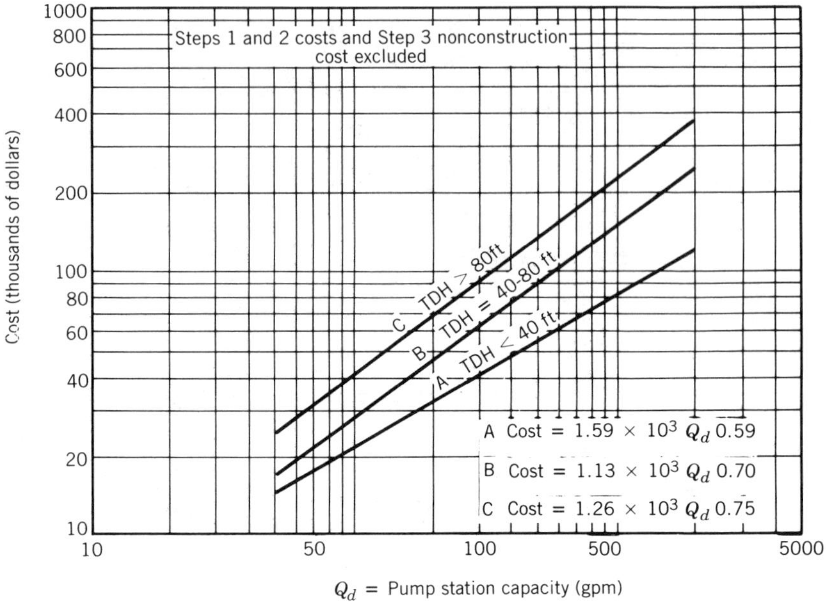

FIGURE 4.4 Total construction cost versus design capacity for pump stations (9).

that would fit together. The figure shows the levels of results that would be expected, as for example on the outer ring where treatment would achieve BOD of 5 mg/L and suspended solids (SS) of 3 mg/L, as well as the other parameters listed.

The sludge treatment facilities are presented in a similar method, as shown in Table 4.5 and Figure 4.7. The use of such approaches to make estimates of the costs of sludge handling is open to even more question than for wastewater treatment, but still the data provides some idea of the options available and their cost.

The reference presents cost equations for the various unit processes (10). These are presented in the form of equations with numerical values that can be adjusted for different circumstances. The basic equations are for amortized capital cost, fixed operation and maintenance costs, and flow-variable operation and maintenance cost. The variables are set up as shown in the Table 4.6.

where BMH is base man-hour requirement, from Table 4.7.

Variable operation and maintenance cost (cents/1000 gallons) =

The equations are

Total amortized cost (cents/1000 gallons)

$$= [\ (BCC)\left(\frac{STP}{177.5}\right) + (LR)\ (ULC)\ ]\left[\frac{100 + SIF}{100}\right]\left[\frac{1}{3650\ Q}\right] CRF,$$

where BCC is base capital cost (from table) and LR is land requirement in acres (from table).

$$\text{Fixed operation and maintenance cost (cents/1000 gallons)} = (\text{BMH})\ (\text{mhr}) \left(\frac{1}{3650\ Q}\right),$$

where BMC is base material cost in dollars per year from the table.

Table 4.7 presents the estimating parameters for the variables in the above equations. The estimates are provided as functions of the design capacity. These equations are useful for the planner to have some idea

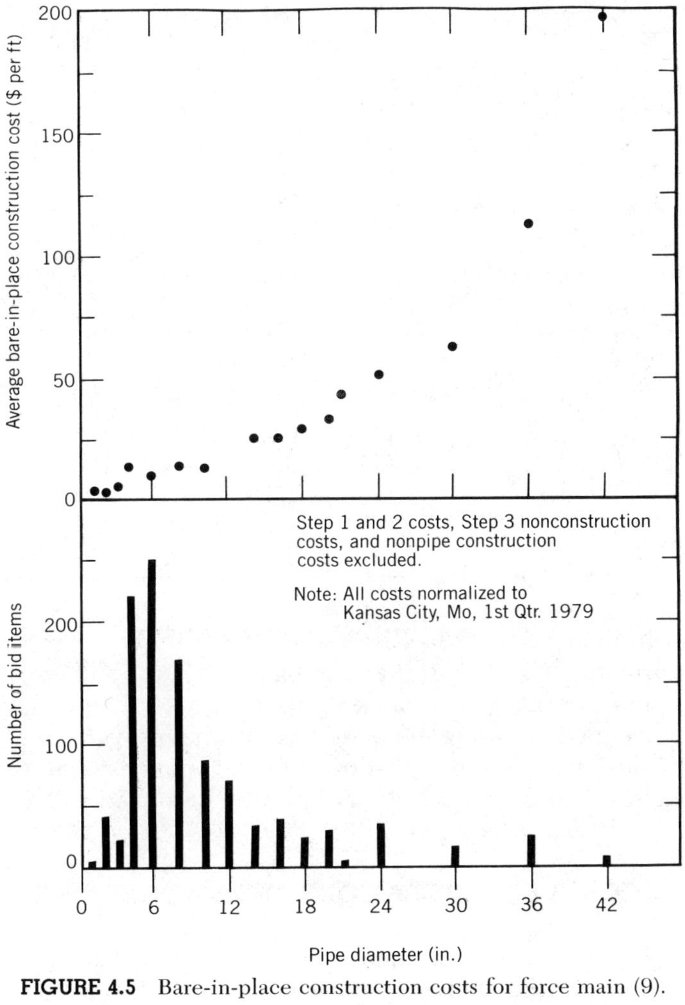

FIGURE 4.5 Bare-in-place construction costs for force main (9).

TABLE 4.3 Complete Urban Sewer System Cost Index (CUSS) From EPA (9)

City		1979				1980	
Code	Name	1st Qtr	2nd Qtr	3rd Qtr	4th Qtr	1st Qtr	2nd Qtr
101	Atlanta, GA	127	130	136	137	145	147
102	Baltimore, MD	153	158	160	161	165	168
103	Birmingham, AL	122	131	136	137	142	142
104	Boston, MA	175	179	180	192	191	191
105	Charlotte, NC	92	98	103	106	111	111
106	Chicago, IL	187	195	207	209	213	213
107	Cincinnati, OH	178	180	186	186	192	192
108	Cleveland, OH	184	187	195	199	205	205
109	Dallas, TX	113	114	121	123	129	129
110	Denver, CO	132	135	151	155	159	160
111	Detroit, MI	184	189	190	190	198	199
112	Houston, TX	129	130	133	135	140	142
113	Kansas City, MO	170	183	185	187	191	191
114	Los Angeles, CA	185	188	203	212	214	214
115	Miami, FL	136	137	140	144	152	153
116	Milwaukee, WI	170	177	187	190	195	195
117	Minneapolis, MN	154	156	167	167	171	173
118	New Orleans, LA	158	159	162	164	168	167
119	New York, NY	210	211	215	222	225	229
120	Philadelphia, PA	192	193	199	200	203	205
121	Pittsburgh, PA	172	172	176	180	185	191
122	St. Louis, MO	192	200	205	206	213	213
123	San Francisco, CA	176	177	195	197	201	208
124	Seattle, WA	185	188	191	191	194	200
125	Trenton, NJ	177	180	185	187	188	188

of the preliminary cost of facilities and should be used with only that rough information in mind. If they are incorrectly used, the results could be worse than having no estimates at all, of course. The equations were fit to data using curve-fitting methods and may not provide valid estimates for low Q designs. The reader should beware when using these kinds of approaches, but they can be very useful when used properly.

COSTS FOR OPERATIONS AND MAINTENANCE

Estimates for O&M are important in the same sense that a person would consider the miles per gallon for an automobile before buying it. The operational costs are likely to amount to more than the purchase cost

TABLE 4.4 Wastewater Treatment Unit Processes (10)

AA. Preliminary Treatment Influent: Raw wastewater	D. Filtration Influent: Effluent from A-2, B-2, B-3, C-2, C-3, C-4, C-5, F-1 or F-2, G-1, G-2, G-3, G-4, H, J, K
AB. Raw Wastewater Pumping Influent: Effluent from AA	E. Activated Carbon Influent: Effluent from D
A. Primary Sedimentation Influent: Effluent from AA or AB A-1 Conventional A-2 Two-Stage Lime Addition A-3 Single Stage Lime Addition A-4 Alum Addition A-5 $FeCl_3$ Addition	F. Two-Stage Tertiary Lime Treatment F-1 Influent: Effluent from B-1 F-2 Influent: Effluent from C-1
B. Trickling Filter B-1 Influent: Effluent from A-1 B-2 Influent: Effluent from A-3 B-3 Influent: Effluent from A-4 or A-5	G. Biological Nitrification G-1 Influent: Effluent from C-6 G-2 Influent: Effluent from B-1 G-3 Influent: Effluent from A-3, A-4 or A-5 G-4 Influent: Effluent from A-2, C-7, or C-8
C. Activated Sludge C-1 Conventional Influent: Effluent from A-1 C-2 Conventional Influent: Effluent from A-3 C-3 Conventional Influent: Effluent from A-4 or A-5 C-4 Alum Addition Influent: Effluent from A-1 C-5 $FeCl_3$ Addition Influent: Effluent from A-1 C-6 High Rate Influent: Effluent from A-1 C-7 High Rate & Alum Addition Influent: Effluent from A-1 C-8 High Rate & $FeCl_3$ Addition Influent: Effluent from A-1	H. Biological Denitrification Influent: Effluent From G-1, G-2, G-3 or G-4 I. Ion Exchanges Associated with A-2, B-2, B-3, C-2, C-3, C-4, C-5, F-1, or F-2 J. Breakpoint Chlorination Influent: Effluent from A-2, B-1, B-2, B-3, C-1, C-2, C-3, C-4, C-5, F-1 or F-2 K. Ammonia Stripping Influent: Effluent from F-1 or F-2 R. Disinfection Influent: Effluent from any treatment process

before the equipment is retired. Another reason that O&M costs are so important is that they come from local budgets rather than a federal grant, and the importance of maintaining an adequate supply is crucial to the effective operation of the system.

The reference for the following discussion and data is another EPA report (11). The costs that are reported generally consist of pure O&M

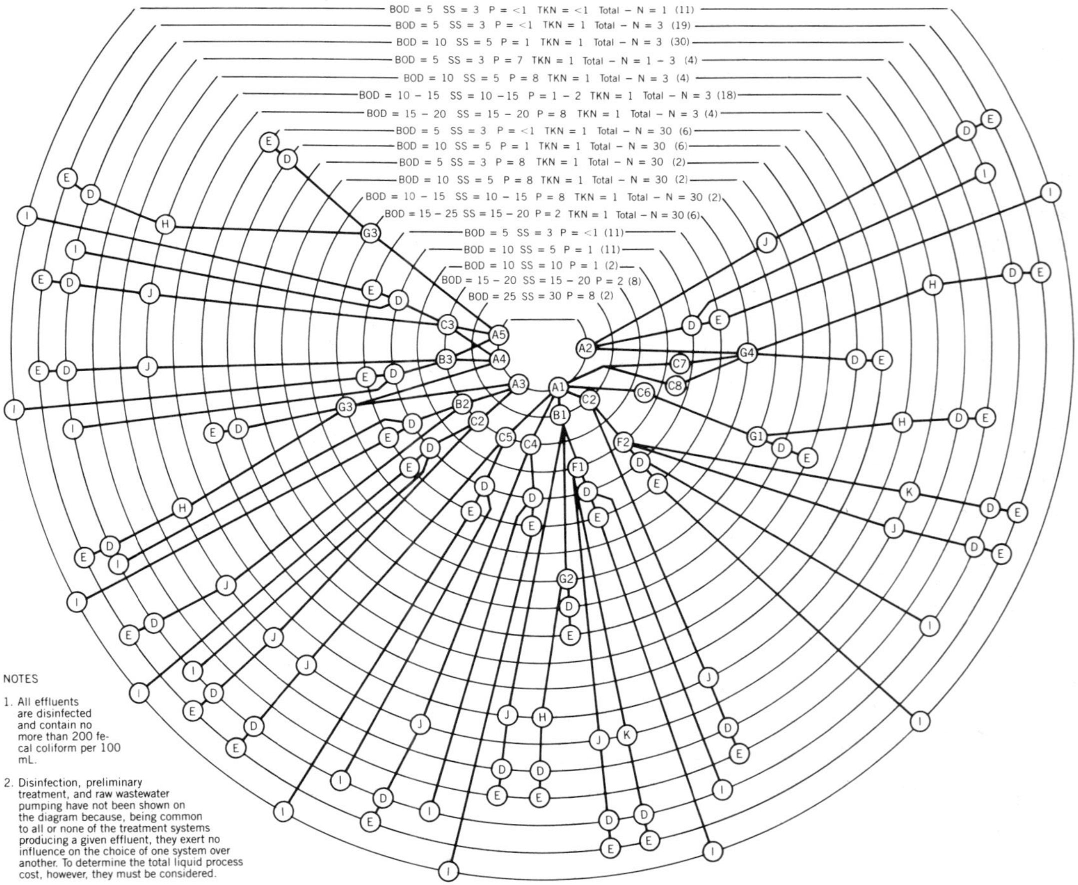

NOTES

1. All effluents are disinfected and contain no more than 200 fecal coliform per 100 mL.

2. Disinfection, preliminary treatment, and raw wastewater pumping have not been shown on the diagram because, being common to all or none of the treatment systems producing a given effluent, they exert no influence on the choice of one system over another. To determine the total liquid process cost, however, they must be considered.

FIGURE 4.6 Combining unit processes for wastewater treatment systems (10).

TABLE 4.5 Sludge Handling Unit Processes Description (10)

L.	Anaerobic Digestion		
	L-1	Sludge Influent:	Generated from A-1 + B-1, C-1 or C-6
	L-2	Sludge Influent:	Generated from A-1 + C-4, or C-5, or C-7, or C-8, A-4 + B-3 or C-3, A-5 + B-3 or C-3
M.	Heat Treatment		
	M-1	Sludge Influent:	Generated from A-1 + B-1, C-1 or C-6
	M-2	Sludge Influent:	Generated from A-1 + C-4 or C-5, or C-7, or C-8, A-4 + B-3 or C-3, A-5 + B-3 or C-3
N.	Air Drying		
	N-1	Sludge Influent:	Effluent Sludge from L-1
	N-2	Sludge Influent:	Effluent from L-2
O.	Dewatering		
	O-1	Sludge Influent:	Generated from A-1 + B-1, C-1 or C-6
	O-2	Sludge Influent:	Generated from A-1 + C-4 or C-5, or C-7, or C-8, A-4 + B-3 or C-3, A-5 + B-3 or C-3
	O-3	Sludge Influent:	Generated from A-2
	O-4	Sludge Influent:	Generated from A-3 + B-2 or C-2
	O-5	Sludge Influent:	Effluent Sludge from L-1
	O-6	Sludge Influent:	Effluent Sludge from L-2
	O-7	Sludge Influent:	Generated from F-1 or F-2
	O-8	Sludge Influent:	Effluent Sludge from M-1
	O-9	Sludge Influent:	Effluent Sludge from M-2
P.	Incineration		
	P-1	Influent Sludge:	Effluent Sludge from O-1
	P-2	Influent Sludge:	Effluent Sludge from O-2
	P-3	Influent Sludge:	Effluent Sludge from O-3
	P-4	Influent Sludge:	Effluent Sludge from O-4
	P-5[a]	Influent Sludge:	Effluent Sludge from O-7 + O-1
	P-6	Influent Sludge:	Effluent Sludge from O-8
	P-7	Influent Sludge:	Effluent Sludge from O-9
Q.	Recalcination (includes chemical storage & feeding)		
	Q-1	Sludge Influent:	Effluent Sludge from O-3
	Q-2	Sludge Influent:	Effluent Sludge from O-4
	Q-3	Sludge Influent:	Effluent Sludge from O-7

[a]Use pathway from O-1 to P-5 only when F-1 or F-2 is included in the complete system.

and do not include any debt service or even major repairs. They are adjusted to third-quarter 1977 costs and include data from 348 plants and 155 sewer systems, all with capacities greater than 1 mgd. This is a sizable sample since the 1980 needs today reported that there were some 13,200 wastewater treatment plants in the United States, with some 1900 having capacities greater than 1 mgd.

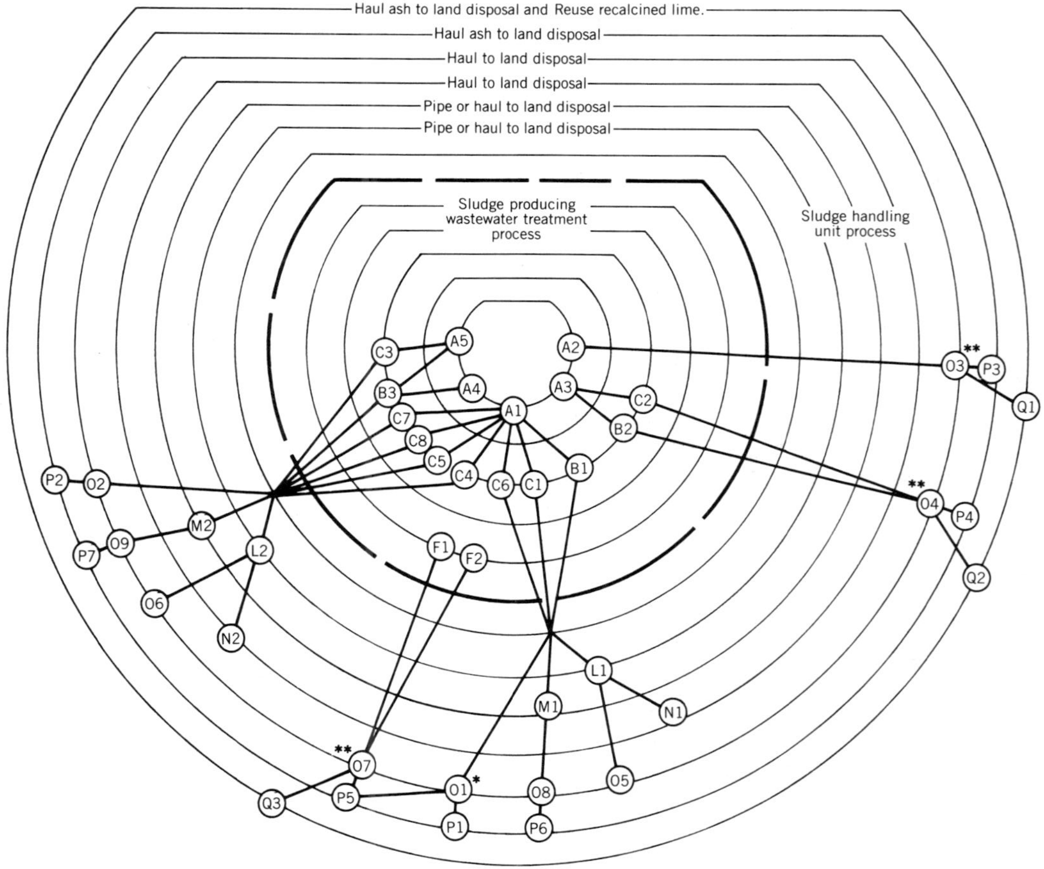

FIGURE 4.7 Combining unit processes for sludge handling systems (10).

TABLE 4.6 Variables for Cost-Estimating Equations

Variable	Units	Symbol	Value Used to Determine Cost Curves
Plant capacity	mgd	Q	1, 5, 20, 100
Amortization period	years	n	20
Interest rate	%	i	5⅝%
Service and interest factor	%	SIF[a]	27
Labor rate	$/man-hour	mhr	5
Land cost	$/acre	ULC	2000
Wholesale price index, industrial commodities	—	WPI[b]	120
National average wastewater treatment plant cost index	—	STP[b]	177.5
Capital recovery factor		CRF	

[a]The SIF includes allowance for engineering, contingencies, and interest during construction.
[b]Values given for the WPI and STP are for February 1973. Values for April 1975 had risen to WPI = 169.7 and STP = 232.5, and for 1975, i = 5⅞%.

Some of the general findings are interesting and helpful in estimating O&M costs. Obviously, if the load on the plant is well below the design load, then the costs are relatively higher, and the converse is true; if the plant is overloaded, the costs will be relatively lower. The values observed for activated sludge plants were: $192 per million gallons treated at the design load, $176 when overloaded, $198 for loads at 70–90% of design, $315 at 50–70% of design, and $436 at less than 50% of design. The best situation is a plant operating near the design value. Personnel, power, and chemicals account for around 80% of the total O&M cost, varying with parameters such as plant sophistication, where more automation has usually been installed, resulting in a need for fewer personnel, or with plant type. The AWT plant, as would be expected, would use relatively more power and chemicals. For sludge handling, incineration was the most expensive approach, while land spreading was the least expensive. Mixed sewer systems had lower collection system O&M costs than pure separate sewers. Treatment plant sizes of about 85 mgd seemed to be about the optimum.

All of these general findings must be taken with a grain of salt when it comes to specific applications, of course, and the best approach is still to look at site-specific data before reaching conclusions. Some of the details of the data will help in demonstrating the ranges involved.

TABLE 4-7. Flow Variable Cost Elements (10)

Unit Process	Base Capital Cost (BCC)	Land Requirement (LR)	Base Manhours (BMH)	Base Materials Cost (BMC)
AA	$32{,}331Q^{0.61}$	0	$1{,}379.2 + 143.1Q$	$860.6 + 247.7Q$
AB	$163{,}612Q^{0.62}$	0	$738.2 + 39.9Q$	$\frac{Q}{0.000885 + 0.000023Q}$
A1	$139{,}753 + 17{,}341.2Q$	$0.23 + 0.088Q$	$1{,}852.8Q^{0.42}$	$1{,}158.4Q^{0.62}$
A2	$307{,}785 + 33{,}538.6Q$	$0.16 + 0.18Q$	$4{,}259.3Q^{0.41}$	$2{,}956.2Q^{0.66}$
A3	$198{,}801 + 19{,}934.9Q$	$0.68 + 0.11Q$	$3{,}260.8 + 161.1Q$	$1{,}694.4Q^{0.65}$
A4	$241{,}226 + 33{,}921.4Q$	$0.26 + 0.16Q$	$2{,}783.4Q^{0.47}$	$\frac{Q}{0.0000662 + 0.00000036Q}$
A5	$269{,}563 + 33{,}561.5Q$	$0.26 + 0.16Q$	$2{,}805.5Q^{0.43}$	$2{,}982.5 + 14{,}255.3Q$
B1	$232{,}882 + 84{,}335Q$	$1.20Q^{0.81}$	$2{,}558.4Q^{0.51}$	$4{,}097.3 + 902.0Q$
B2	$241{,}083 + 63{,}200.5Q$	$0.79Q^{0.84}$	$2{,}500.8Q^{0.48}$	$3{,}525.8 + 895.8Q$
B3	$241{,}083 + 63{,}200.5Q$	$0.79Q^{0.84}$	$2{,}500.8Q^{0.48}$	$3{,}525.8 + 895.8Q$
C1	$359{,}744 + 84{,}786.7Q$	$0.76Q^{0.80}$	$4{,}574.8Q^{0.45}$	$10{,}499.7Q^{0.73}$
C2	$349{,}156 + 67{,}047.4Q$	$0.46 + 0.32Q$	$6{,}228.4 + 303.5Q$	$10{,}233.9Q^{0.73}$
C3	$349{,}156 + 67{,}047.4Q$	$0.46 + 0.32Q$	$6{,}228.4 + 303.5Q$	$10{,}233.9Q^{0.73}$
C4	$395{,}978 + 89{,}419.7Q$	$0.78Q^{0.81}$	$4{,}834.7Q^{0.47}$	$184{,}641.1 + 15{,}301.8Q$
C5	$411{,}240 + 89{,}839.2Q$	$0.78Q^{0.81}$	$5{,}093.2Q^{0.47}$	$18{,}720.2 + 14{,}714.7Q$
C6	$348{,}852 + 53{,}752.3Q$	$0.50Q^{0.84}$	$4{,}292.9Q^{0.44}$	$10{,}336.4Q^{0.73}$
C7	$395{,}350 + 59{,}528.6Q$	$0.78Q^{0.81}$	$6{,}959.8 + 360.6Q$	$18{,}466.2 + 15{,}301.5Q$
C8	$413{,}737 + 59{,}627.4Q$	$0.50Q^{0.84}$	$4{,}898.9Q^{0.45}$	$20{,}047.7 + 14{,}697.0Q$
D	$231{,}495.0Q^{0.66}$	$0.024 + 0.028Q$	$\frac{Q}{0.00068 + 0.00058Q}$	$16{,}491.9Q^{0.68}$
E	$629{,}840 + 90{,}719.4Q$	$0.024 + 0.028Q$	$1{,}600.0Q$	$\frac{Q}{0.00011245 + 0.00000014Q}$
F1	$327{,}175 + 33{,}438.9Q$	$0.16 + 0.18Q$	$2{,}981.1Q^{0.46}$	$2{,}027.6Q^{0.65}$
F2	$327{,}175.0 + 33{,}438.9Q$	$0.16 + 0.18Q$	$2{,}981.1Q^{0.46}$	$2{,}027.6Q^{0.65}$
G1	$210{,}055 + 59{,}204.6Q$	$0.50Q^{0.84}$	$3{,}503.5 + 192.4Q$	$8{,}756.5Q^{0.75}$
G2	$203{,}714 + 56{,}924.2Q$	$0.50Q^{0.84}$	$3{,}360.1 + 183.9Q$	$8{,}756.5Q^{0.75}$
G3	$209{,}599 + 65{,}633.4Q$	$0.44 + 0.24Q$	$3{,}820.6 + 226.0Q$	$8{,}756.5Q^{0.75}$

G4	$210{,}055 + 59{,}204.6Q$	$0.50Q^{0.84}$	$3{,}503.5 + 192.4Q$	$8{,}756.5Q^{0.75}$
H	$155{,}767 + 37{,}290.7Q$	$0.49 + 0.16Q$	$2{,}031.1Q^{0.42}$	$-3{,}559.4 + 8{,}110.1Q$
I	$163{,}270Q^{0.88}$	$-0.17 + 0.021Q$	$3{,}746.2Q^{0.72}$	$1{,}5161.5Q^{0.86}$
J	$136{,}587Q^{0.52}$	$-0.081 + 0.047Q$	$3{,}0432.2Q^{0.41}$	$2{,}399.3 + 39{,}947.7Q$
K	$93{,}029.1Q^{0.89}$	$-0.016 + 0.04Q$	$3{,}385.6 + 660.2Q$	$2{,}103.0 + 3{,}490.0Q$
R	$62{,}270.5 + 5{,}127.1Q$	$0.21 + 0.018Q$	$462.6Q^{0.60}$	$-1{,}748.7 + 2{,}739.3Q$
L1	$111{,}168 + 23{,}450.4Q$	$0.40 + 0.09Q$	$923.39 + 108.1Q$	$1{,}384.2 + 152.4Q$
L2	$103{,}721 + 21{,}188.3Q$	$0.32 + 0.09Q$	$896.1 + 94.2Q$	$1{,}441.8 + 140.7Q$
M1	$92{,}686.9 + 12{,}701.3Q$	0	$2{,}246.7 + 313.9Q$	$2{,}810.5 + 1{,}025.8Q$
M2	$101{,}672Q^{0.55}$	0	$2{,}039.8 + 314.7Q$	$2{,}148.8 + 1{,}025.0Q$
N1	$-14{,}885.6 + 57{,}978.60$	$1.5 + 1.9Q$	$\dfrac{Q}{0.000518 + 0.00001161Q}$	$5.4 + 8222.7Q$
N2	$\dfrac{Q}{0.0000097 + 0.000000101Q}$	$\dfrac{Q}{0.287 + 0.0000972Q}$	$\dfrac{Q}{0.000490 + 0.000002Q}$	$\dfrac{Q}{0.001156 - 0.0000136Q}$
O1	$153{,}201 + 27{,}538.5Q$	$-0.026 + 0.021Q$	$2{,}264.1 + 488.9Q$	$6{,}231.8Q^{0.86}$
O2	$123{,}189Q^{0.71}$	$-0.026 + 0.016Q$	$2{,}510.8 + 514.9Q$	$8{,}059.1 + 3{,}513.6Q$
O3	$194{,}601.0 + 45{,}218.2Q$	$-0.046 + 0.021Q$	$2{,}624.6Q^{0.78}$	$8{,}681.0Q^{0.70}$
O4	$173{,}784 + 37{,}399.2Q$	$-0.043 + 0.018Q$	$2{,}391.8 + 920.04Q$	$6{,}875.3Q^{0.73}$
O5	$140{,}189 + 22{,}599.2Q$	$-0.04 + 0.018Q$	$1{,}899.5 + 385.5Q$	$6{,}061.6Q^{0.86}$
O6	$152{,}608 + 22{,}520.1Q$	$-0.026 + 0.017Q$	$2{,}339 + 405.2Q$	$5{,}978.1Q^{0.88}$
O7	$168{,}827 + 31{,}955.5Q$	$-0.043 + 0.016Q$	$3{,}352.5 + 880.9Q$	$7{,}524.1Q^{0.71}$
O8	$127{,}953 + 13{,}386.5Q$	$-0.023 + 0.015Q$	$1{,}916.3 + 482.0Q$	$4{,}058.8Q^{0.73}$
O9	$123{,}931 + 18{,}524.6Q$	$-0.026 + 0.016Q$	$1{,}966.1 + 507.7Q$	$4{,}481.9Q^{0.71}$
P1	$378{,}837Q^{0.54}$	$-0.084 + 0.015Q$	$1{,}280.6 + 509.2Q$	$7{,}136.4 + 728.8Q$
P2	$386{,}161Q^{0.57}$	$-0.084 + 0.012Q$	$1{,}521.8 + 504.7Q$	$9{,}938.7 + 933.7Q$
P3	$631{,}877 + 67{,}043.6Q$	$-0.084 + 0.011Q$	$1{,}281.8 + 735.4Q$	$2{,}039.1 + 1{,}974.8Q$
P4	$410{,}798Q^{0.58}$	0	$1{,}286.8 + 609.1Q$	$13{,}163.9 + 1{,}214.7Q$
P5	$286{,}299Q^{0.53}$	$-0.084 + 0.013Q$	$1{,}519.3 + 291.4Q$	$7{,}926.8 + 549.8Q$
P6	$390{,}808Q^{0.54}$	$-0.084 + 0.015Q$	$1{,}433.9 + 492.1Q$	$12{,}703.3 + 1{,}267.2Q$
P7	$634{,}606 + 65{,}598.6Q$	$-0.084 + 0.012Q$	$1{,}464.1 + 697.8Q$	$8{,}507.3 + 891.9Q$
Q1	$422{,}409Q^{0.58}$	$-0.17 + 0.021Q$	$3{,}048.8 + 550.9Q$	$4{,}264.7 + 14{,}698.8Q$
Q2	$331{,}428Q^{0.54}$	$-0.17 + 0.021Q$	$1{,}770.4 + 356.9Q$	$3{,}267.5 + 7{,}913.7Q$
Q3	$534{,}671Q^{0.59}$	$-0.17 + 0.021Q$	$1{,}904.0 + 911.1Q$	$-262.8 + 12{,}458.3Q$

The tables that follow show some of the specific results and provide a basis for order-of-magnitude estimates of O&M costs. Table 4.8 shows the distribution of various categories of expenditures for O&M and is the basis for the statement that personnel, power, and chemicals make up about 80% of the costs. The table refers to the EPA survey since the report also includes data from a survey by the Association of Metropolitan Sewerage Agencies (AMSA), but the results are similar, differing only by small percentages in most cases.

Table 4.9 shows the variation of costs with treatment level or type and with EPA region. As expected, the costs increase with complexity. Table 4.10 shows the average cost per million gallons treated, again varying predictably with complexity and size. Table 4.11 shows the variation of cost with mile of gravity sewer system. The term *mixed sewer* refers to systems that have both separate and combined sewers in a network. This table refers to "OM&R" costs, which are the same as O&M but utilizing a different term to emphasize the inclusion of the minor repair costs. Tables 4.12 through 4.15 show some potentially significant O&M relationships for costs with flow, staff size, service population, and total length of gravity sewer system. As the sample sizes and the results of the statistical variation tests are shown, the general validity of the data can be

TABLE 4.8 Average percent Distribution of Various Expenditures to Total Costs by Treatment Level for All Size Plants (11)

	Level of Treatment			
		Secondary		
Object of Expenditure Category	Primary	Trickling Filter	Activated Sludge	Advanced (AWT)
Personnel	59	58	52	46
Power[a]	(14)	(13)	(22)	(19)
Total utilities	15	16	24	23
Chemical disinfection[b]	(4)	(3)	(2)	(2)
Total chemicals	10	9	7	12
Equipment	3	5	4	6
Materials	5	6	6	3
Contractual	4	3	3	5
Other	4	3	4	5
Total	100	100	100	100
Number of plants surveyed	56	82	137	28

[a]Power costs are also included in total utility costs.

[b]Chemical disinfection (usually chlorine) costs are lso included in total chemical costs.

TABLE 4.9 EPA Survey Average Operating Cost for Varying Treatment Levels by EPA Regions (Dollars per Capita per year) (11)

		Secondary		
EPA Region	Primary	Trickling Filter	Activated Sludge	Advanced
I	$6.94 $n = 6$	$9.29 $n = 3$	$22.74 $n = 11$	$35.21 $n = 2$
II	$7.95 $n = 5$	$18.60 $n = 10$	$13.03 $n = 18$	$18.64 $n = 5$
III	$4.46 $n = 4$	$9.80 $n = 7$	$12.25 $n = 19$	$32.32 $n = 3$
IV	$8.40 $n = 10$	$10.16 $n = 14$	$18.21 $n = 26$	$16.15 $n = 1$
V	$8.90 $n = 10$	$7.70 $n = 6$	$13.86 $n = 29$	$16.41 $n = 11$
VI	$3.08 $n = 2$	$5.31 $n = 14$	$7.44 $n = 18$	$ $n =$ [a]
VII	$5.09 $n = 5$	$6.81 $n = 5$	$10.19 $n = 3$	$14.29 $n = 2$
VIII	$3.55 $n = 3$	$6.03 $n = 7$	$22.72[b] $n = 6$	$ $n =$ [a]
IX	$8.92 $n = 11$	$8.64 $n = 6$	$9.47 $n = 16$	$8.05 $n = 6$
X	$7.60 $n = 7$	$7.27 $n = 3$	$14.32 $n = 7$	$ $n =$ [a]
National Average	$7.40 $n = 63$	$9.27 $n = 75$	$14.02 $n = 153$	$17.81 $n = 30$

[a] Per capita operating costs not reported in these regions.

[b] Abnormally hgih due to inclusion of two mountain resort areas; when these two Colorado resort areas are excluded, the average per capita cost is $8.01.

assessed. Figure 4.8 shows a plot of the cost of secondary treatment, based on the activated sludge process, versus the flow. The scatter of the data is apparent. This figure is the basis for the statement that the optimum plant size for O&M costs is about 85 mgd; obviously, the conclusion is very tentative since the data base is very scanty.

Figures 4.9 and 4.10 present some cost data for small systems. Figure 4.9 is capital costs for extended aeration and overland flow systems that

TABLE 4.10 EPA Survey Average Cost per Million Gallons Treated (11)

Level of Treatment	1.0–5.0 mgd	5.1–20.0 mgd	>20.0 mgd	All Plants
Primary	\$176[a] $n = 40$	\$137 $n = 12$	\$ 47 $n = 4$	\$159 $n = 56$
Secondary				
Trickling filter	\$212 $n = 61$	\$162 $n = 17$	\$ 95 $n = 4$	\$196 $n = 82$
Activated sludge	\$316 $n = 95$	\$165 $n = 30$	\$149 $n = 12$	\$268 $n = 137$
Advanced (AWT)	\$454 $n = 22$	\$251 $n = 3$	\$136 $n = 3$	\$398 $n = 28$

[a]The values appearing in this table were determined from the following equation:

$$\text{Average cost per million gallons} = \frac{\text{total annual O\&M costs in dollars}}{\text{actual flow (mgd)} \times 365}$$

TABLE 4.11 EPA Survey OM&R Cost per Mile of Gravity Sewers for Various Types of Sewer Systems (Costs in Dollars Per Mile) (11)

		Component Costs[b]				
Type of System	Total Cost[a]	Personnel	Power	Materials[c]	Contractual	Other
Separate sewer Number in sample	2783 15	1289	201	388	491	414
Combined sewer Number in sample	3565 2	1861	58	640	[d]	1006
Mixed sewer Number in sample	1272 2	427	217	398	219	11
WWTP + separate sewer Number in sample	1618 81	839	231	246	136	166
WWTP + combined sewer Number in sample	2142 4	981	522	164	324	151
WWTP + mixed sewer Number in sample	1154 27	614	133	180	89	138

[a]The values appearing in this column were determined from the following equation:

$$\text{Average cost in dollars per mile} = \frac{\text{total annual OM\&R costs in dollars}}{\text{total length (miles) gravity sewers}}$$

[b]$$\text{Component costs per mile} = \frac{\text{respective component cost in dollars}}{\text{total length (miles) gravity sewers}}$$

[c]Chemcials, if any, are included as materials.
[d]No cost reported.

TABLE 4.12 Potentially Significant O&M Relationships Total O&M Costs versus Actual Flow (11)

Title	Equation[a]	Sample Size (n)	Correlation Coefficient (r)	F-Test Value
Primary treatment national	$TC = 4.53 \times 10^4 Q^{1.01}$	57	0.83	119.57
Trickling filter (TF), national	$TC = 6.02 \times 10^4 Q^{0.94}$	71	0.86	194.15
Activated sludge (AS), national	$TC = 8.25 \times 10^4 Q^{0.96}$	143	0.89	515.76
AWT, national	$TC = 6.85 \times 10^4 Q^{1.44}$	28	0.71	25.75
TF, region II	$TC = 7.58 \times 10^4 Q^{1.10}$	9	0.77	10.31
TF, region III	$TC = 6.14 \times 10^4 Q^{1.04}$	7	0.91	23.54
TF, region IV	$TC = 4.66 \times 10^4 Q^{1.27}$	13	0.70	10.32
TF, region V	$TC = 8.08 \times 10^4 Q^{0.70}$	5	0.95	25.13
TF, region VI	$TC = 3.99 \times 10^4 Q^{0.90}$	12	0.95	96.03
TF, region VIII	$TC = 2.51 \times 10^4 Q^{1.29}$	11	0.96	99.55
TF, region IX	$TC = 8.55 \times 10^4 Q^{0.95}$	7	0.91	23.16
AS, region II	$TC = 1.11 \times 10^5 Q^{0.82}$	15	0.95	112.19
AS, region III	$TC = 1.08 \times 10^5 Q^{0.87}$	16	0.87	44.65
AS, region IV	$TC = 7.27 \times 10^4 Q^{0.98}$	26	0.83	52.60
AS, region V	$TC = 1.04 \times 10^5 Q^{0.87}$	20	0.95	176.73
AS, region VI	$TC = 4.36 \times 10^4 Q^{1.14}$	18	0.93	109.73
AS, region X	$TC = 9.55 \times 10^4 Q^{0.80}$	6	0.98	100.82

[a]TC equals total O&M cost in dollars and Q equals the average daily flow in million gallons per day.

would be appropriate for flows in the 1–2 mgd range. Figure 4.10 presents the operation and maintenance costs from the same systems (12).

CAUSES AND RATE OF DETERIORATION, DEPRECIATION, AND OBSOLESCENCE

The same general factors that afflict water supply systems apply to wastewater, although in general the service conditions are more severe in wastewater systems. For many of the reasons cited earlier, they are likely to receive less maintenance attention. Chapter 9 presents references on the causes of maintenance problems and the suggested remedies.

Referring back to Figure 4.1, we see that the facilities that needed attention for repair, replacement, and rehabilitation are those of the collection, treatment, and disposal systems. Like water supply systems, much of the capital investment is in the pipelines. It is imperative that the sewage collection system remain in good order through proper inspection

TABLE 4.13 Potentially Significant O&M Relationships Total O&M Costs versus Staff Size (11)

Title	Equation[a]	Sample Size (n)	Correlation Coefficient (r)	F-Test Value
Primary treatment national	$TC = 2.06 \times 10^4 S^{1.10}$	56	0.85	145.31
Trickling filter (TF), national	$TC = 1.63 \times 10^4 S^{1.19}$	72	0.86	199.19
Activated sludge (AS), national	$TC = 1.85 \times 10^4 S^{1.19}$	140	0.91	693.14
AWT, national	$TC = 3.32 \times 10^4 S^{1.02}$	30	0.97	464.19
Primary region IV	$TC = 2.01 \times 10^4 S^{1.08}$	11	0.86	25.25
Primary region V	$TC = 2.20 \times 10^4 S^{0.90}$	8	0.86	17.20
Primary region IX	$TC = 3.79 \times 10^4 S^{1.08}$	11	0.76	12.27
Primary region X	$TC = 2.90 \times 10^4 S^{0.97}$	7	0.90	21.38
TF, region II	$TC = 3.35 \times 10^4 S^{1.02}$	9	0.93	47.31
TF, region III	$TC = 1.39 \times 10^4 S^{1.35}$	7	0.86	14.70
TF, region V	$TC = 3.41 \times 10^4 S^{0.83}$	5	0.99	125.76
TF, region VI	$TC = 1.67 \times 10^4 S^{1.16}$	12	0.90	43.53
TF, region VIII	$TC = 1.85 \times 10^4 S^{1.09}$	8	0.94	45.16
TF, region IX	$TC = 3.35 \times 10^4 S^{1.11}$	7	0.94	40.18
AS, region I	$TC = 6.20 \times 10^3 S^{1.73}$	11	0.85	22.67
AS, region II	$TC = 1.99 \times 10^4 S^{1.15}$	15	0.98	304.45

[a]TC = total cost; S = staff size.

and maintenance arrangements. Service conditions are severe, and it is difficult to attract and train qualified and motivated staff.

A manual (13) from the WPCF lists the problems that need attention in the preventive maintenance program. These illustrate the range of difficult problems needing attention: improper pipe bedding, failure due to live loads and earth movement, root growth damage, and faulty building sewer connections. Some system deficiencies that need attention other than preventive maintenance are listed as misaligned pipes, grade deviations, porous pipes, solids buildup, hydrogen sulfide gas generation, pipe deterioration, hydraulic overloads, and manhole problems (13). Another manual describes procedures for evaluating and rehabilitation of existing sewer systems (14).

There are similar guidance manuals published for the operation and maintenance of wastewater treatment plants. Two are published by the WPCF (15,16). This subject is covered further in Chapter 9.

We are at the point of learning how well modern wastewater systems will last. When the construction grants program was initiated, the thought was that all kinds of miraculous things could be done through technology,

TABLE 4.14 Potentially Significant O&M Relationships Sewers Systems: Total Cost versus Service Population (11)

Title	Equation[a]	Sample Size (n)	Correlation Coefficient (r)	F-Test Value
Separate sewer system, national	$TC = 0.27 \times SP^{1.35}$	17	0.74	17.71
WWTP + separate sewer system, national	$TC = 0.09 \times SP^{1.39}$	92	0.80	160.45
WWTP + mixed sewer system, national	$TC = 0.012 \times SP^{1.55}$	30	0.89	108.07
Separate sewer system, region VI	$TC = 26.69 \times SP^{0.87}$	6	0.98	113.40
WWTP + separate sewer system, region II	$TC = 0.22 \times SP^{1.30}$	21	0.92	105.07
WWTP + separate sewer system, region III	$TC = 0.14 \times SP^{1.31}$	11	0.85	23.59
WWTP + separate sewer system, region IV	$TC = 4.16 \times SP^{1.07}$	26	0.78	37.86
WWTP + separate sewer system, region V	$TC = 0.10 \times SP^{1.35}$	19	0.71	17.15
WWTP + separate sewer system, region VIII	$TC = 10.44 \times SP^{0.92}$	8	0.95	54.49
WWTP + mixed sewer system, region II	$TC = 0.025 \times SP^{1.52}$	9	0.93	48.15
WWTP + mixed sewer system, region V	$TC = 1.20 \times 10^{-3} \times SP^{1.76}$	16	0.84	32.72

[a]TC equals total OM&R cost in dollars and SP equals the number of people served by the sewer system.

TABLE 4.15 Potentially Significant O&M Relationships Sewer Systems: Total Cost versus Total Length of Gravity Sewers (11)

Title	Equation[a]	Sample Size (n)	Correlation Coefficient (r)	F-Test Value
Total cost versus total length of gravity sewers, national	$TC = 56.35 \times L^{1.72}$	132	0.75	172.26
Total cost of gravity sewers, region II	$TC = 31.14 \times L^{1.90}$	34	0.78	50.86
Total cost of gravity sewers, region V	$TC = 14.71 \times L^{1.89}$	34	0.75	42.22
Total cost of gravity sewers, region VI	$TC = 1.44 \times 10^3 \times L^{1.12}$	6	0.94	29.14
Total cost of gravity sewers, region VIII	$TC = 1.46 \times 10^3 \times L^{0.97}$	10	0.94	57.16

[a]TC equals total OM&R cost in dollars and L equals the total length of the gravity sewer system.

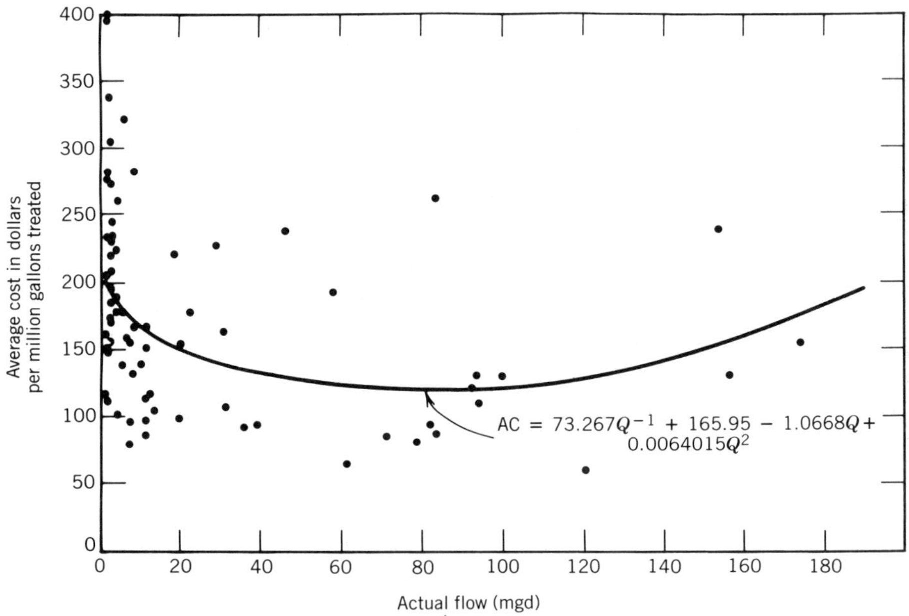

FIGURE 4.8 Average cost per million gallons treated versus actual flow (11).

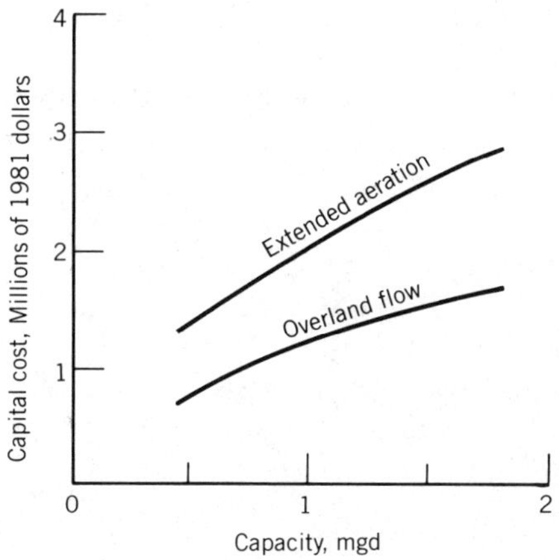

FIGURE 4.9 Capital cost of treatment facilities, first quarter 1979 (12).

and it can. However, the reality of operational difficulties, obsolescence, and the difficulty of making a wastewater system work well over the long haul makes the experience still an experimental one. There are some modern plants that have never worked the way they were designed. It will be interesting to see how well the technology endures and which processes turn out to be good ones over the long run. Until we gain this needed historical experience, it will be difficult to make too many observations about the service life and the true costs of wastewater facilities.

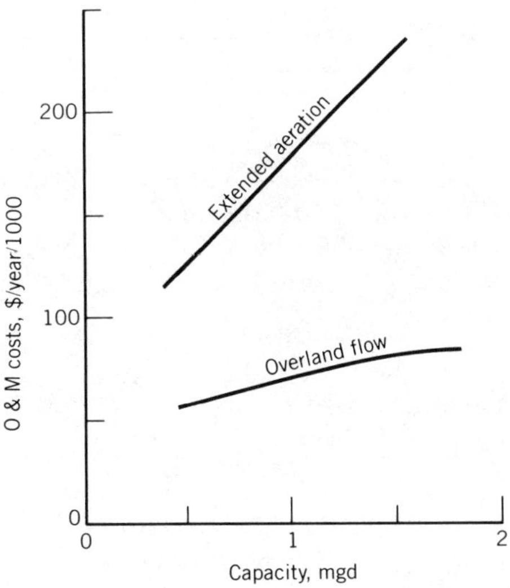

FIGURE 4.10 O & M cost of treatment facilities (12).

DATA BASES NEEDED FOR MANAGEMENT

The data base needed for management for wastewater is a blend between that for water supply and stormwater, as the wastewater system has some characteristics of both systems. Table 4.16 illustrates some of the data needed.

One of the most important uses of data for the management of wastewater systems is in detecting the presence or probability of toxic and other pollutant problems, especially those that deviate from normal conditions. For this reason the monitoring system is of critical importance. The integration of these data with the other for the management of systems is discussed further in Chapters 8 and 9.

PRINCIPAL INTEREST GROUPS

There are a number of special-interest groups that focus on wastewater rather than water management in general. Probably the largest is the Water Pollution Control Federation (WPCF), which represents a cross section of interests ranging from operators and wastewater managers, to design engineers, to government officials, to financiers and equipment manufacturers.

The Association of State and Interstate Water Pollution Control Administrators (ASIWPCA) represents the interests of the state water regulators and a few interstate management officials. During the last few years ASIWPCA has been especially active in its Washington office in

TABLE 4.16 Information Systems Needed for Wastewater Management

Collection
Wastewater loadings, quality, and quantity
Industrial waste information
Facility inventory and condition
Maintenance records
Treatment and Disposal
Facility inventory and condition
Operational records
Reporting records
Maintenance records
Sludge Disposal Facilities
Facility inventory and condition
Operational records

maintaining liaison with the EPA and with Congress, reflecting the growing importance of the role of state agencies in wastewater management. This trend will continue.

The Association of Metropolitan Sewerage Agencies (AMSA) represents some of the largest wastewater agencies. Although the number of agencies involved is not large, they represent some of the most important in the United States. The headquarters are in Washington.

The American Public Works Association (APWA), with headquarters in Chicago and a Washington representative, is also very important in wastewater management since some of the top wastewater administrators have chosen to participate actively in the association. APWA has also taken a strong international position in coordinating the activities of other municipal public works associations. The APWA research foundation has completed several projects related to wastewater and has also been active in the establishment of an instrument testing service.

IMPORTANT ISSUES

Like water supply the future for wastewater treatment is filled with questions about the role of government in setting regulations and providing support. Many of the most important regulatory issues have been settled during the 1970s and early 1980s, but for every one that was settled there seem to be at least two more.

The 1970s were concerned with implementation of the first version of the Clean Water Act. There were many new issues that dealt with conventional pollutants to consider, and much of the energy of the regulatory and environmental establishments went into them. The remainder of the 20th century seems destined to focus on different issues, some dealing with the questions of toxic materials in the environment. However, these will not be the only issues that arise.

Wastewater management issues for the immediate future seem to cluster in three categories: how to deal with toxic materials, how to "internalize" the costs of wastewater treatment, and how to deal with diffuse sources of pollution. These issues take on different focal points in some countries than in others. In the United States, for example, we have settled some of the basic institutional questions concerning environmental protection philosophies, and the major work ahead is implementation and the tackling of new problems, such as protection from toxics. In the developing countries, however, some basic problems still remain to be solved, and some of them are addressed at the most basic levels of human needs, like enough safe drinking water and protection of the primary sources of drinking water. In those countries they are still fighting disease and poverty that have already been dealt with in advanced countries.

The problem of dealing with toxic substances is no fad. It is an immensely complex technological and social problem. The extent of the problem ranges from the identification of sources to tracking the substances from creation through use through treatment to safe disposal or recovery. Of primary interest in wastewater management will be the technical and managerial issues of identification, measurement, treatment, and accountability. The external effects of bad management, such as groundwater contamination, must be prevented and mitigated by improved management. This is an issue area that will remain with us for a long time.

The internalization of the effects and the costs of managing wastes is a goal that has been sought in the Clean Water Act through the zero-discharge policy. While this has been a difficult goal to achieve, it is nonetheless the only viable one for the long term as far as toxic substances are concerned. To achieve it industries must deal with their waste products in a complete way, finding ways to recover and reuse the residuals and safe disposal places for those that are without any salvage value. During the years environmental engineering will be moving toward that general philosophy of waste management, and that the economic incentives created by government at all levels will be directed toward the same goal. In fact, the creation of incentives for the private sector to manage the environment will be a major task of federal and state environmental agencies in the 1980s and 1990s.

Dealing with diffuse sources of pollution refers to the "nonpoint source" problem in urban areas and to the problem of managing the pollutants that have been carelessly disposed of in the past, and which now threaten our groundwater quality. This problem concerns wastewater managers since how we manage tomorrow's wastewaters will depend on the extent to which we can control all threats to the environment. It is possible that in the future the responsibility for controlling all sources of contamination will be assigned to integrated agencies, with the task of managing wastewater being one of many responsibilities.

Through all of the emerging wastewater issues runs the problem of managing information. In fact, the information management problem seems to be more serious for wastewater than for water supply, since the extent to which the pollution problem can be controlled depends on the information available about pollutants, the environment, and on the effectiveness of treatment and monitoring programs. All of the indications are that the wastewater management problem will continue to grow in complexity, and that challenges to managers will continue to mount.

REFERENCES

1. Clayton H. Billings, Ed., *Manual of Wastewater Collection*, Texas Water Utilities Association, Austin, 1981.
2. American Society of Civil Engineers and Water Pollution Control Federation, Design

and Construction of Sanitary and Storm Sewers, New York, 1969; and Gravity Sanitary Sewer Design and Construction, New York, 1982.

3. News, Water Resources Research Institute of North Carolina, Monitoring for Toxic Substances in the Waters of North Carolina, March 1985.
4. Water Pollution Control Federation, Task Force on Preliminary Treatment, Preliminary Treatment for Wastewater Facilities, Manual OM-2, 1980.
5. Water Pollution Control Federation, Task Force on Industrial Wastewater Control, Industrial Wastewater Control Program for Public Agencies, Manual OM-4, 1982.
6. Environmental Protection Agency, The 1982 Needs Survey, EPA/43019-83-002, June 15, 1983.
7. Water Pollution Control Federation, The Clean Water Act with Amendments, Washington, DC, 1982.
8. Water Pollution Control Federation, Monitor, WPCF Roundtable Discussion—Congressional Staffers Take a Retrospective Look at PL 92-500, Washington, DC, August and September 1981.
9. Environmental Protection Agency, Construction Costs for Municipal Wastewater Conveyance Systems: 1973–1979, EPA-430/9-81-003, January 1981.
10. Environmental Protection Agency, A Guide to the Selection of Cost-Effective Wastewater Treatment Systems, EPA-430/9-75-002, July 1975.
11. Environmental Protection Agency, Analysis of Operations and Maintenance Costs for Municipal Wastewater Treatment Systems, EPA-430/9-77-015, February 1978.
12. Environmental Protection Agency, Costs of Municipal Wastewater Treatment Facilities for Small Communities, August 1982.
13. Water Pollution Control Federation, Operation and Maintenance of Wastewater Collection Systems, MOP 7, Washington, DC, 1985.
14. American Society of Civil Engineers and Water Pollution Control Federation, Existing Sewer System Evaluation and Rehabilitation, New York, 1983.
15. Water Pollution Control Federation, Plant Maintenance Program, Manual of Practice OM 3, Washington, DC, 1982.
16. Water Pollution Control Federation, Operation of Wastewater Treatment Plants, Manual of Practice 11, Washington, DC, 1976.

CHAPTER FIVE

STORMWATER AND COMBINED SEWER SYSTEMS

The water supply and wastewater systems are managed much differently than the third urban water subsystem, stormwater. Stormwater is the linking part of the urban water system; that is, it is at the same time a source of supply and a wastewater. We are not able to control when it arrives or in what amounts. It provides cleansing of streets and renewal for parched lawns and depleted aquifers. The stormwater management system really is a very important part of the urban water system, but it has historically received little attention. The reason for the neglect is that we have been busy attending to the more basic and immediate needs of water supply and waste management. The truth is that people can survive without taking care of stormwater, and they can even remain relatively free of disease, but the quality of the living environment will suffer. However, with growing urbanization, and with problems of non-point sources of water pollution, the real importance of stormwater management is becoming more apparent.

Stormwater management has been the victim of an "identity crisis." Prior to the mid-1960s the term was not really used very much, but the terms *drainage* and then *urban drainage* came into use. It was really into the 1970s before the term *stormwater management* appeared and started to stick. The identity crisis is that some use the term to refer to drainage only, and others only think of the water pollution control aspect. Really, stormwater management should be viewed as a multiple-purpose activity, including the water quantity and quality aspects.

In this chapter we focus on the separate stormwater system and the combined sewer system. The wet-weather design problems of combined sewer systems are similar to those of stormwater; the main difference is that the combined sewer system is subject to the "overflow" problem, which means that the stormwaters reaching the receiving waters are contaminated with full-strength wastewater as well as with stormwater runoff. Stormwater runoff itself, however, sometimes is higher in the level of contaminants than some full-strength wastewater. When speaking of a stormwater system, the term includes the flood control system as well. There is a difference in scale between stormwater and flood control systems, and this difference will be described later.

SERVICE PROVIDED BY STORMWATER SYSTEMS

Stormwater management systems provide two basic services: drainage of minor stormwaters and control of larger flood flows. These two basic services are distinguished from one another by the benefits provided: convenience, cleansing and safety from drainage of minor storm flows, and damage prevention and life safety from controlling flood flows. Combined sewer systems provide the same general drainage services, and they handle the dry-weather flows of regular wastewater systems as well. If so designed, stormwater programs can also provide benefits from three additional services: erosion and sedimentation control, control of the quality of urban runoff through best management practices and treatment of wet-weather flows, and the beautification of urban stream environments.

In evaluating the benefits of stormwater service, we think first of using the economic analysis techniques developed for other water programs. It is difficult, however, to apply standard economic evaluation techniques to stormwater management since many of the main benefits are intangible and cannot be quantified. Also, the benefits that can be quantified, mostly the prevention of urban flood damages, are not very visible to the public or to politicians, except right after flood events. The reliable estimation of potential urban flood damages is a technique that most engineers have little experience with, and there are only rare opportunities to improve our skills in this area. The other benefits are very much attached to the land and the convenience of everyday life and are not the kind that lend themselves to benefit–cost analysis.

Some agencies have resorted to unusual methods to decide on the best stormwater investments due to the difficulty in performing standard economic analysis. For example, the Los Angeles County Flood Control District was considering the use of voting behavior on flood control bond issues a few years ago to get an idea about public preferences about stormwater projects. Another agency had an implicit policy to make sure that funds spent on stormwater projects generally went to areas more or less in the proportion that the same areas were taxed for stormwater.

Even though the benefits are difficult to measure and to quantify, stormwater programs offer forward-looking water managers the oppor-

TABLE 5.1 List of Benefits and Costs of Drainage and Flood Control Systems (1)

Benefits	Costs
Tangible	Tangible
Reduced flood damage	Construction costs
Reduced risk to life	Land costs
Land value enhancement	Nonstructural program cost
Reduced traffic delays	Evacuation costs
Reduced business losses	Administration costs
Reduced cleanup losses	Insurance subsidy cost
Reduced relief costs	Higher reconstruction costs
Increased recreation opportunities	
Intangible	Intangible
Reduced inconvenience	Environmental and social costs
Increased security	
Reduced health hazards	
Improved aesthetics	

tunity to get "more bang for the buck" by introducing multiple-purpose water/land improvement programs into development projects. There is a recognized opportunity for "synergism" in project development when stormwater problems are corrected, and stormwater projects can serve the integrating function of bringing departments and agencies together for joint planning and efforts. The planning, development, and financing can all be shared, and the winner will be the public.

In spite of the difficulty in measuring stormwater benefits, there is an approach to making an economic evaluation. In a study of the economic performance of stormwater projects a few years ago, I listed all the benefits and costs in an ASCE paper (1). The list, which can be used to evaluate service levels as well as to carry out an economic analysis, is shown in Table 5.1.

Table 5.1 constitutes a rather complete listing of benefits and costs that can be used either as a description of the service provided by stormwater or as a starting point for an economic analysis.

There is also a service associated with the stormwater function that relates to water quality management. It is the cleansing function of the city, and the water quality protection that accompanies it. This is a wastewater management function. The idea is summed up by the caption that accompanied an EPA film on stormwater quality. It said, "When a city takes a bath, what do you do with the dirty water?" The services are included in the city taking a bath (the cleansing service) and in the water quality protection service (taking care of the dirty water).

CONFIGURATION OF STORMWATER MANAGEMENT SYSTEMS

The general configuration of the stormwater system is as shown in Figure 5.1. It shows existing features and some that may be emphasized in the future. The basic division is into the "minor" and "major" systems, corresponding to the drainage and flood control functions as previously described. These two systems generally rely on different program components for handling the demand.

The minor drainage system, which has also been called the "initial" system or the "convenience" system, relies on gutters, small ditches, small culverts and storm drains, detention ponds, and small channels to handle its load. It might be described as a "storm drainage system" in a community development program. The minor system is easily recognized since it is the visible part of the stormwater management system that people see operating and providing a service.

The major system is often not so visible since it has the purpose of draining larger floods and handling the rare events. It is like the emergency spillway on a dam. It might not be used for years, but it must be there when needed.

In combined sewer systems (CSS) the system configuration takes on a more complicated shape than minor drainage systems alone, since the elements of the system that deal with water quality are added in. Actually, a combined sewer system is a hybrid between the minor stormwater sys-

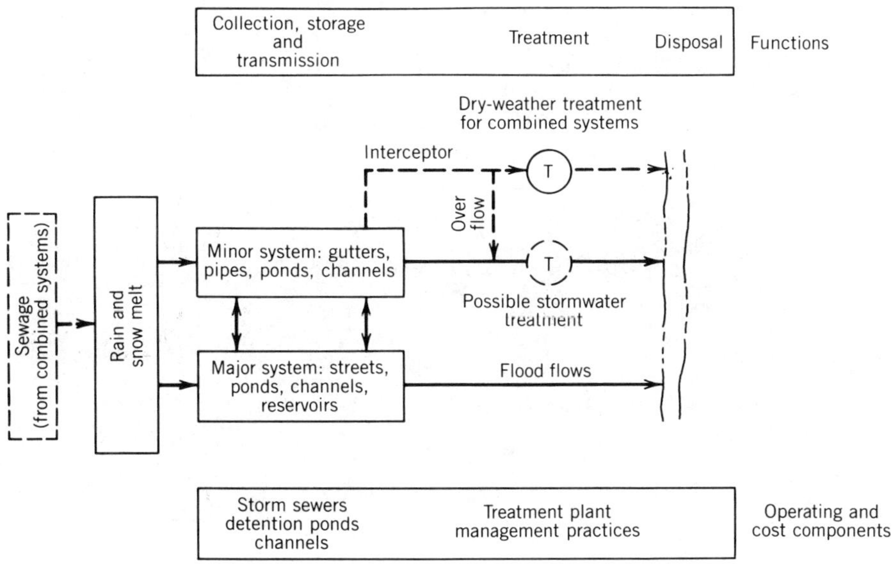

FIGURE 5.1 Configuration of urban stormwater system.

tem and the sanitary sewer system. Not much thought was given to this until recent years when it has become necessary to pay more attention to stream water quality, with the attendant concern over how to either separate combined sewer systems or treat the discharges.

In general the CSS is the same as the minor stormwater system except for the addition of the regulator and treatment facilities. Figure 5.1 shows how the CSS fits into the minor stormwater system, with it being part of the same collection system until it reaches the interceptor sewer that conveys the dry-weather flow to the treatment plant. If the capacity of the interceptor or the treatment plant is exceeded, then overflows will occur, and these are the source of severe water pollution problems. Figure 5.1 also shows where the treatment of separate storm sewage would occur if it became part of the system. For the most part treatment has not been incorporated in U.S. systems since it is too expensive. The Environmental Protection Agency has continued to study the problem, however, and it is by no means sure that we will never have requirements for treatment.

Figure 5.2 is a conceptual diagram that shows the storage, treatment, and overflow facilities for CSS. The figure shows the pollutant accumulation on the city and the contaminated runoff that may be stored, treated, or allowed to overflow. The figure is from a report that describes a useful planning model named "STORM" (2). It was developed to aid in planning storage and treatment options for combined sewer systems,

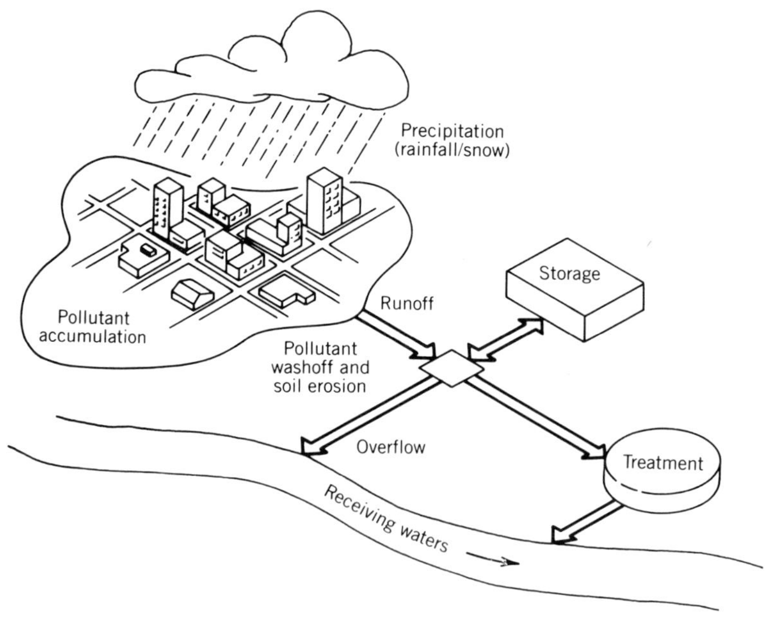

FIGURE 5.2 Conceptualized view of urban systems used in "STORM" (2).

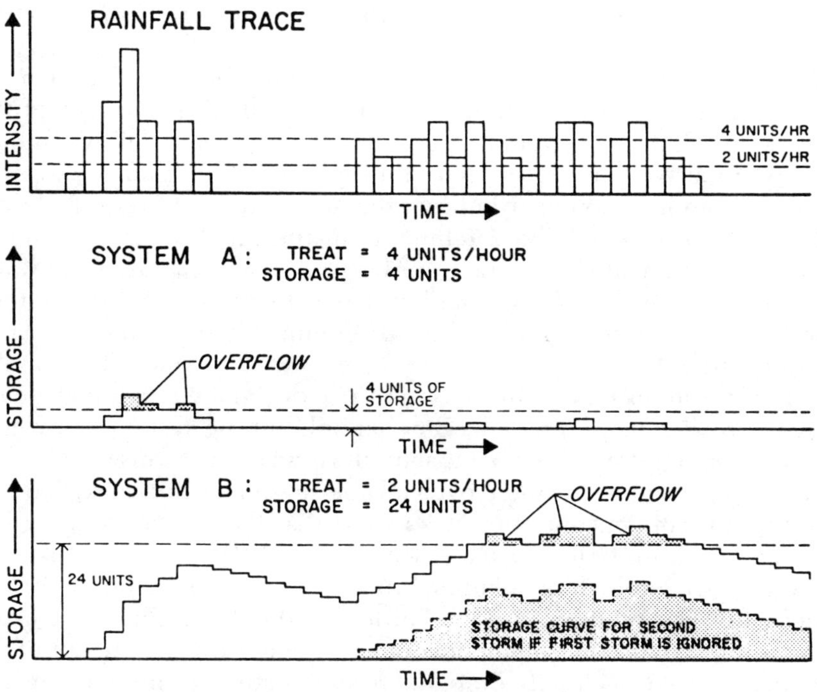

FIGURE 5.3 System response examples.

but not too many sophisticated storage–treatment systems have been implemented yet due to the high cost.

The design decisions concerning storage–treatment systems are shown in Figure 5.3, which illustrates the trade-offs between different wet-weather treatment rates and the amount of overflow that takes place (2). The figure demonstrates the use of detention storage in controlling the volume of CSS that reaches the receiving waters. In effect, both storage and treatment for CSS can be considered as measures to mitigate the pollution problem associated with overflows since if enough storage can be provided then the combined sewage can be treated after the storm has passed. Also, storage can serve to provide some treatment, such as sedimentation, if provision for adequate operation and maintenance can be made. Figure 5.3 shows the results of two competing policies: one that provides more treatment capacity and overflows as a result of a short burst of intense rain and the other that provides more storage capacity, but overflows for a longer rain that exceeds the storage capacity provided.

EPA's research office has published several comprehensive reports describing the management and treatment technologies available, a good example being the report by Lager and others (3). Unit processes such as physical and biological treatment and land treatment as well as storage and disinfection are described. Quite a bit is known about the management of combined sewer systems, and the problem now seems to be the cost.

The minor system may be sufficient to handle the runoff from small tributary areas even during periods of heavy rainfall. It consists of the regular storm drainage system that developers normally install when they develop property. This system will drain up to a certain range of catchment sizes before enough tributary area accumulates to require larger facilities. Typical would be catchments in the range of 10–500 acres, depending on rainfall characteristics. Above a certain size the stormwater management system no longer falls in the category of a minor system, regardless of how it is designed, since even minor storms can create large flow demands.

In larger systems part of the system can be designated for minor flows, such as a trickle channel for a larger channel, but the system as a whole must be designed for a greater capacity during larger storms. That is not true for small catchments where even a large storm may not cause flooding since the time of concentration is so short that the flow is gone before it has the opportunity to cause any trouble.

There is an interesting relationship between the size of watershed tributary to a certain drain and the "order" of the drain. The word *order* is used to designate the same concept as stream order, a concept of geology that is used to classify drainage basins. I studied this relationship for storm drains in two dissimilar cities, San Francisco and Englewood, Colorado, and found that there was a definite relationship between the order of the sewer and the tributary area of the watershed. There was also a relationship between the total length of pipe with a particular order and the order number. This finding could eventually be useful in inventory studies leading to maintenance planning.

Specifically, the classification shown in Table 5.2 was used to define sewer order (4). The statistics of number of sewer lines, average tributary area, and average length from the study areas are shown in Figures 5.4, 5.5, and 5.6. Although this site-specific data cannot be used as it is for

TABLE 5.2 Order Classification of Sewer Systems

Order	Category
1	House sewer
2	Local sewer
3	Collector sewer
4	Small main sewer
5	Medium main sewer
6	Large main sewer
7	Small interceptor sewer
8	Medium interceptor sewer
9	Large interceptor sewer

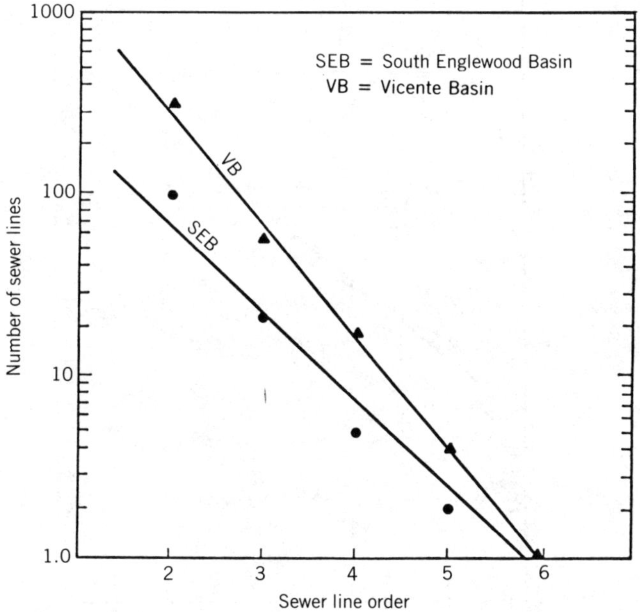

FIGURE 5.4 Number of sewer lines versus sewer line order.

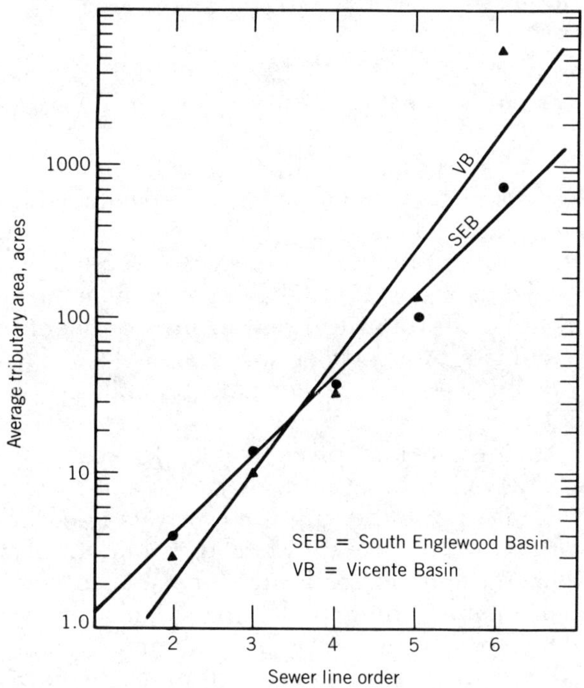

FIGURE 5.5 Average tributary area versus sewer line order (4).

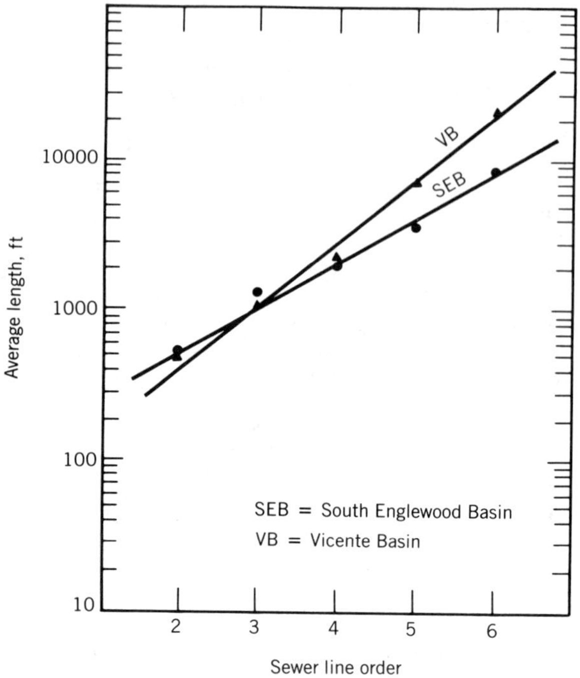

FIGURE 5.6 Average length versus sewer line order (4).

general planning purposes, the approach would help greatly in assembling data bases and estimates of the amount of storm sewerage in an urban area.

From the figures it can be seen that the order number of the sewers is basically dependent on tributary area. Watersheds with different climatic conditions and urbanization factors have not been studied, but it is obvious that the order number would also be dependent on these factors. The size and cost of sewers will be dependent on the order number, and there will be some definite relationships between urbanization characteristics, climate, and cost per unit of area of sewer facilities. This seems to provide an avenue to make some generalized estimates of infrastructure costs.

The major system is often not planned at all. In areas of frequent heavy rain it can be expected to be clear of obstacles and ready to pass flood flows as a result of experience with the previous flood flows. In many areas of the world, however, the statistical distributions of storms are not very well understood, and the floods hit at unexpected times, often when there is a lot of economic activity in the flood plain as well as settlement. Problems with heavy damage, loss of life and property, and social trauma can be avoided with good land-use control in the flood plain and good flood warning systems. This has been the thrust of the program in the

United States since the Flood Insurance Act of 1968. The planning of the major system also needs to consider street drainage, adequate channels, reservoirs, and outlets for the floodwaters. The planning and design of major drainage systems is probably the most neglected part of stormwater management today.

The configuration of major systems has to take into account both the physical facilities themselves and the arrangements for nonstructural flood plain management. In other words, with the U.S. adoption of the general philosophies of nonstructural approaches to flood plain management, the system for handling the floods has gone beyond the channels themselves into the management of land use in the entire widths of flood plains, and this fact must be recognized in developing complete management programs.

Specifically, the major system must include the *floodway* and the *flood fringe area,* two terms that have taken on programmatic meanings since all the activity of the 1970s in flood plain management. Figure 5.7 defines the basic concept of these areas, which taken together are called the flood hazard zone. This zone can be designated in different ways, depending on the return period of flood that is regulated. The 100-year flood has come to be widely accepted since the implementing regulations for the Flood Insurance Act were accepted. Since there is really no theoretical basis for accepting any other value, it is likely that the 100-year value will remain, at least for the foreseeable future.

The water quality subsystem is superimposed on top of the minor and major systems. There are really two water quality systems. One is the wash off of pollutants from the surface of the land and the other is the source of pollutants that comes from combined sewer overflow or the erosion of pollutants from the inside of sewers, both combined and separate. Figure 5.8 shows the water quality subsystem, with the different elements contributing to the problem related to land-use characteristics (5).

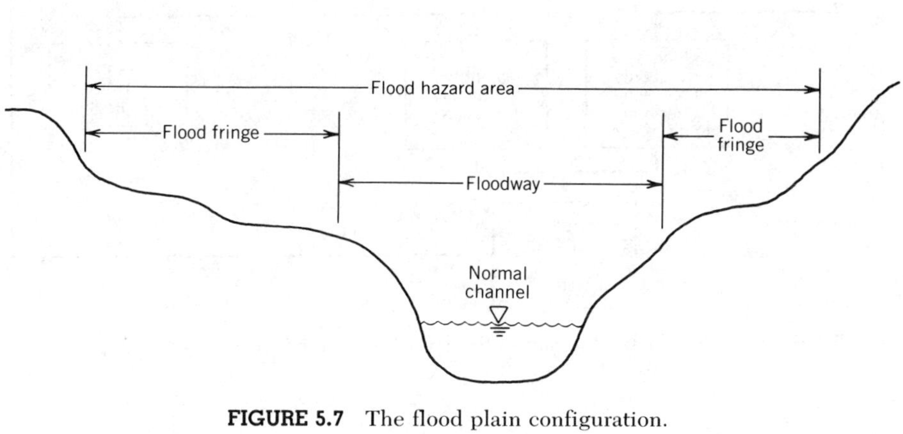

FIGURE 5.7 The flood plain configuration.

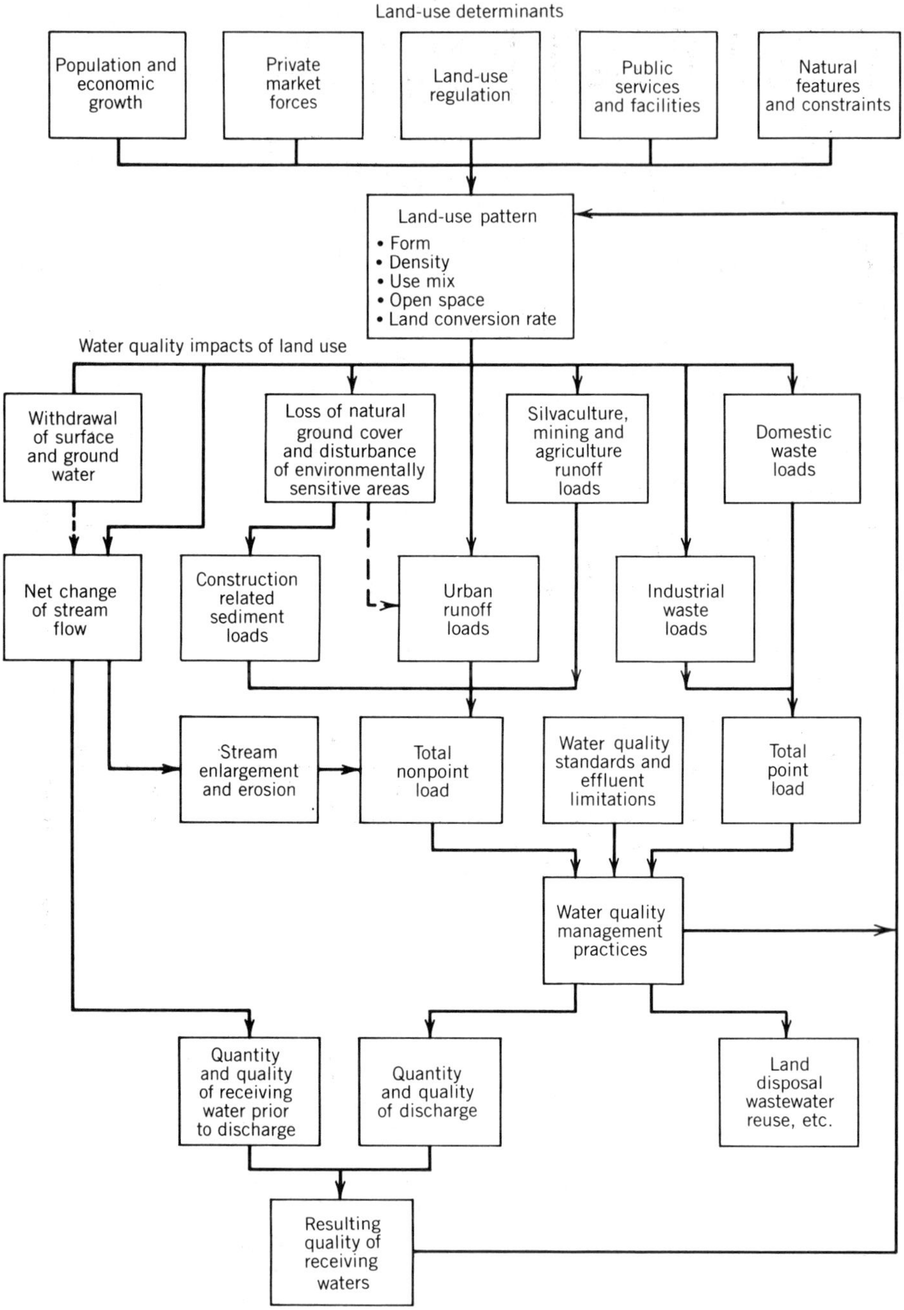

FIGURE 5.8 Schematic diagram of the land-use–water quality relationship (5).

DIMENSIONS AND PERFORMANCE STANDARDS

The standards that govern the design of stormwater systems are not as uniform as those for water supply and wastewater, mostly because there has been little federal activity in stormwater control (there has, however, been considerable federal research into the subject) and because the stormwater function has been left mostly to local governments. Moreover, there is no single professional association that is the recognized standard-setting authority for stormwater, such as the American Water Works Association has been for water supply. Several professional associations have been active in research and education, but not in stormwater standard-setting, and there does not appear to be any urgent need for such activity at the present time.

Common local standards revolve around the design storm return period, but there are some significant policy and technical questions that have not been well answered about the choice of these design parameters. The return period for design is important since it is the single most influential factor in system cost. For minor systems it varies in the range of 2–25 years with common recommendations being 2 years in residential areas and 5–25 years being recommended for commercial zones where less disruption due to flooding can be tolerated. The major systems are generally planned for the 100-year event since that was the requirement originally promulgated in the regulations that implemented the Flood Insurance Act, and many areas accepted that criteria so that they would qualify for various federal programs.

The design storm and the companion design return period are the answer of flood control planners to the need for what is called today "risk analysis." In other words the selection of a return period sets the level of risk that the system will fail or the capacity will be accepted. Some lay leaders are not acquainted with the concepts of risk, return period, and probability, but the concepts are essentially simple. Every year there is some exceedance probability P that a storm of a certain magnitude will be equaled or exceeded one or more times that year. For a storm of any given magnitude, given in depth of precipitation during a fixed storm duration, there will be a interval of time, called the recurrence interval or return period, which represents the *average* time interval during which the storm with the expected magnitude can be expected to be equaled or exceeded. This is usually calculated as the reciprocal of the exceedance probability.

$$T = \frac{1}{P}.$$

It is very important to recognize that this does not mean there will be a storm of the expected magnitude in every period of T years. For example, there may be no 10-year storm in the next two decades, but in the following 8 decades we may experience ten such storms, and then the average would be correct.

There is also a simple equation to relate the exceedance probability to the risk over a period of time. If J is the risk that a storm will occur one or more times in a period of n years, then

$$J = 1 - (1 - P)^n$$

As an example, using this equation allows the risk to be stated this way: If we design this system for a 20-year storm, there will be a risk of $J = 1 - (1 - 0.05)^{10} = 0.40$ that the capacity will be exceeded in the next 10 years.

Each stormwater management agency has a unique approach to setting design standards, but in recent years there seems to have been some trend toward uniformity, perhaps due to the increase in communication among agencies. Typical approaches are those such as were reported in the criteria manual adopted in the Denver region several years ago (6), which recommended the design criteria shown in Table 5.3.

None of the return-period-based criteria have been tested in general with any valid economic evaluation. Every project and situation is unique, and it is unlikely that any universal standard would ever be accepted. In the meanwhile all the intangible parameters and complexities, coupled with the relatively low stakes involved, prevent any close attention from the economists.

There is a recognized and genuine controversy among stormwater designers over whether the design storm approach to the assessment of risk is correct. The alternative is to examine actual or potentially actual storms that have occurred or could occur in the area, and through simulation modeling to examine all the possible consequences of the actual storms. McPherson wrote about the need to consider actual storms in his report on urban runoff planning, and he stated that the design storm or "synthetic storm" might be appropriate for the design of small facilities, but when

TABLE 5.3 Some Suggested Design Criteria (6)

	Design Storm	
Land Use	Minor	Major
Residential	2	100
High-value commercial	5	100
Public building area	5	100
Airports	2–5	100
Major airport terminals	5–10	100
High-value downtown area	5–10	100

complex systems are involved, particularly those involving storage and treatment, it is important to consider actual rainfall histories (7).

There is also a need for a design level to anticipate the really rare events, and the general concept of the "probable maximum flood" , caused by the "probable maximum precipitation," (PMP) has been widely accepted. A PMP is a storm that represents what the meteorologists consider to be the maximum that could occur under reasonable projections of the most severe climatic conditions. Few people have ever seen a rainfall approaching the PMP, however, and there is often skepticism over the validity of the concept. This writer believes that the concept is valid when properly used in the assessment of risk, and that when the consequences of a large flood can be severe enough, then it pays to consider the PMP.

In an attempt to address the maximum probable concept, the Corps of Engineers has adopted an approach called the "standard project storm," which is defined as "the most severe flood-producing rainfall depth-area-duration relationship and isohyetal pattern of any storm that is considered reasonably characteristic of the region in which the drainage basin is located, giving consideration to the runoff characteristics and existence of water regulation structures in the basin" (8). This provides an integrated picture of the worst conditions that could occur, considering watershed conditions as well as the meteorology.

MANAGEMENT STRUCTURE

Stormwater management systems are mostly operated by local governments. The typical situation is a drainage section located in the office of the city engineer. Smaller communities often locate the drainage service in the streets department since there are some common requirements for the maintenance and upkeep of streets and drainage systems. Since it has been traditionally difficult to get enough financial help to construct and maintain stormwater systems, there has been a trend toward the establishment of stormwater "utilities" in recent years. The idea is that stormwater is identified and managed as a separate service, and that the separate identification of the service will ensure that it gets the attention it needs.

Many county governments are into drainage and flood control services as a result of suburbanization and the connection with roads. In Michigan, for example, the counties have elected drainage commissioners. There has also been a move in recent years toward more special-purpose drainage and flood control districts, such as the county flood control districts in California and the Urban Drainage and Flood Control District serving the Denver region. Soil and water conservation districts have a mandate to engage in drainage also, but their activities in urban areas have been limited.

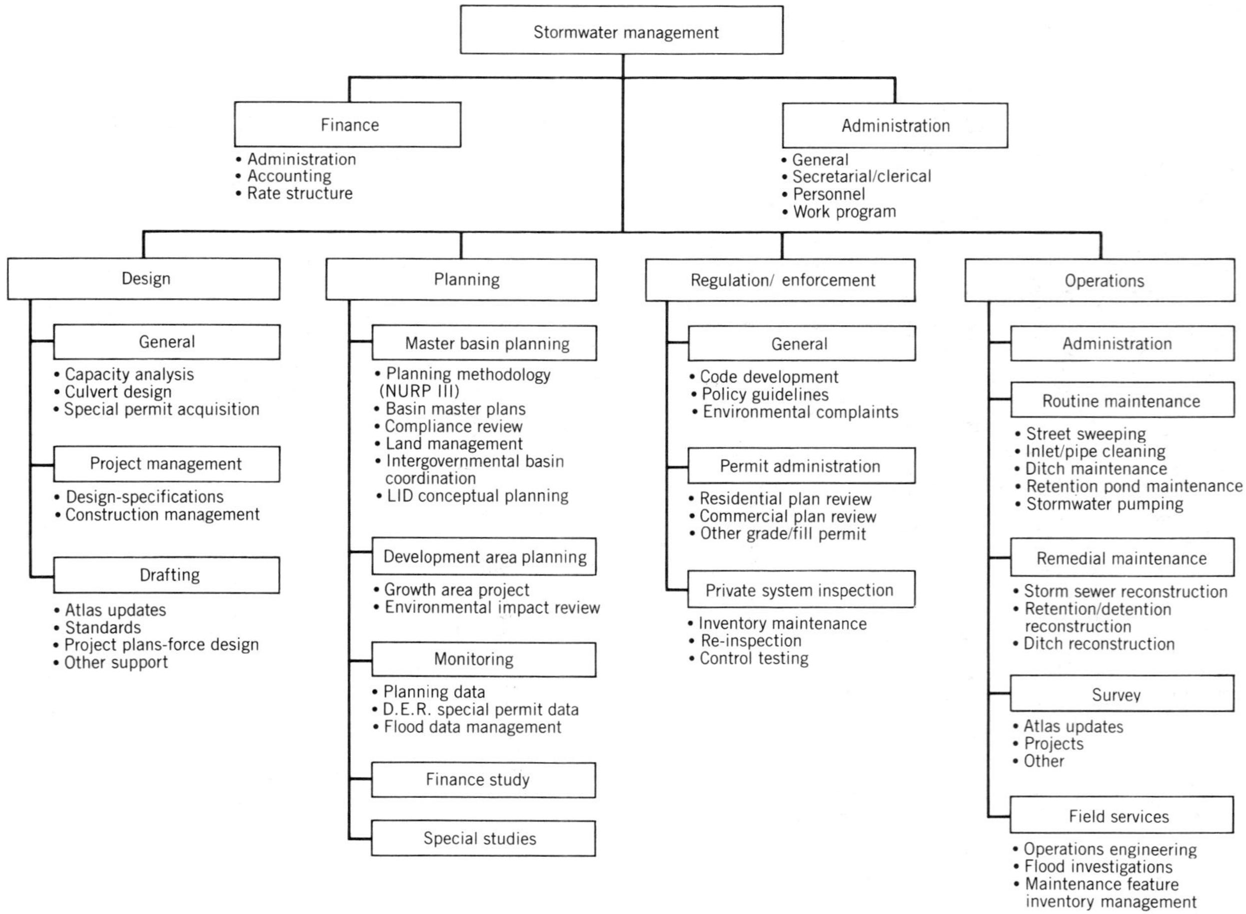

FIGURE 5.9 Organization of a stormwater utility (9).

The most interesting development in stormwater finance of recent years has been the advent of experiments with the stormwater utility concept. Basically, it is an attempt to overcome the out-of-sight, out-of-mind mindset that has prevented the solution of many difficult stormwater problems. Cyre has made an extensive study of stormwater organizational needs and financing options, presenting his report at the 1982 meeting of the American Public Works Association (9). The functions identified by Cyre as those necessary for the management of the utility are shown in Figure 5.9. It is a rather complete treatment and reflects the functions that would otherwise be found in different parts of a city government when the organization was a traditional public works department.

Financing options were also studied by Cyre and are shown in Table 5.4. Any or all of these could be used by a utility for the approach to financing the stormwater service. Of course, the ultimate guide to feasibility for a stormwater utility or management concept is whether the public will accept it. Although there have been a number of successful reports of the initiation of utilities, there have also been some rejections. For example, in late 1983 the voters in Lakewood, Colorado, rejected a carefully prepared plan. It was based on a complete needs study that showed that there were significant stormwater needs in the city and a

TABLE 5.4 Financing Options (9)

A. Revenue for annual operating expenses
   1. General fund
   2. Drainage utility service charges
   3. Interfund loans to drainage utility (transition only)

B. Funding for major capital improvements
   1. General obligation bonding repaid by property taxes
   2. Revenue bonding repaid by utility service charges
   3. Utility tax revenues
   4. Community development block grant funds

C. Fees and charges
   1. Plan review and inspection fees
   2. On-site detention/retention system inspection fees
   3. Impact fees
   4. System development charges
   5. General facilities charges
   6. In-lieu of construction charges
   7. Latecomer fees for developer extensions

D. Funding for special services and projects
   1. Local improvement districts
   2. Utility local improvement districts
   3. Area of special benefit financing
   4. Special-purpose taxing districts

financing scheme that considered equitable and fair charging schemes based on impervious area. The charges were even made lower in the first years to promote voter acceptance, and a voter education program preceded the election (10).

The utility concept itself and the failure of some attempts illustrate the continuing search among cities for approaches that can be implemented in the different areas of the country, and ones that will prove workable and durable over the long term. Each situation is unique and requires careful collaboration between the drainage staff management and the elected council.

Since the stormwater utility concept has elicited substantial interest, and since it offers potential help for cities looking for long-term solutions to the stormwater financing problems, I would like to report on the specific experiences of my community, Fort Collins. The public works department studied potential solutions to the financing problem and has implemented the utility concept. The article by Engemoen and Krempel (11) contains a comprehensive description of the program, and the discussion that follows is adopted from there.

In justifying a consideration of the utility concept, the city found that stormwater management was in a transition period and difficult to manage properly without some changes. There were four basic reasons: stormwater problems are only visible during floods and forgotten afterwards, the utility concept is new and is not familiar to many managers, technical advances in stormwater management are only about one decade old, and the evolution of drainage law, environmental practices, and new designs have caused a need for reevaluation of programs. The city has also found that traditional methods for financing stormwater programs were becoming inadequate. There were three approaches in use: stormwater improvement districts, which were meeting opposition; the use of sales taxes, which were difficult to maintain; and construction by developers, which had many off-site problems in the growing city.

The utility has its basis in a 1976 ordinance and was initiated in 1980. The activities included are the development of a comprehensive capital improvements program and the administration, operation, and maintenance of the city's stormwater facilities. All of this is financed by a stormwater utility fee that is added to the monthly water and sewer bill. With the combined utilities, the city issues one bill with water, sewer, electricity, and stormwater. In general the O&M and the capital parts of the fee are each about one dollar per month per house, and there has not been too much complaint yet. There is also a connection fee for new residential developments, ranging from about \$1200–2000 per acre or \$200–500 per house depending on lot size. In 1985 a related effort to establish a "transportation utility fee" has been challenged in court by some tax-exempt organizations in the city. The outcome is still pending.

The operational procedure for the utility began by dividing the city

TABLE 5.5 Examples of Calculation of Drainage Fees in Fort Collins

	Residential	Commercial
Area, square feet (sf)	7081	22776
Runoff (rate) factor	0.45	0.95
O&M base rate, $/sf/month	0.000336	0.000336
Monthly O&M, $	1.04	7.92
Capital improvement base	0.001013	0.000292
Monthly capital improvement	2.87	6.32
Total monthly fee	3.91	14.42

Calculation of connection or plant investment fee for a commercial development with a rate factor of 0.52 and a reduction factor of 0.85:

Fee = 3.2226 acres × $28777 acre × 0.52 × 0.85 = $4098.01.

into nine basin areas, so that the unique problems of each can be identified and solved. This resulted in a capital program of $40 million. The program is the basis for the capital assessment on the monthly bill. Basically, a 20-year planning period is used with 5-year programming periods. The monthly capital fees are planned to be in the range of $0.50–4.00 within the next 5 years. The capital fee only began in 1982. The operation and maintenance of the facilities is financed by the O&M fee, which averages $0.88 per home per month in 1985.

Calculations for the levels of the fees are demonstrated in (11). The factors considered are the general cost in a particular basin, the causative factor, based on runoff coefficient and lot size, and the reduction factor, representing the extent of mitigation through the use of detention storage. Two brief examples are given in Table 5.5.

LEGAL BASIS AND REGULATION

Generally, the legal basis for governmental involvement in stormwater is the charge to local city and county governments found in state constitutions that authorizes them to provide services to improve the health and welfare of citizens. Special districts, including comprehensive districts, such as Denver's Urban Drainage and Flood Control District, have their own enabling legislation, but the stormwater organizations that do are a distinct minority, as most of the responsibility for managing stormwater falls on multipurpose local government organizations.

In fact, all of the urban water services evolved from their earliest rural roots to their status today, and the reason that there is so much political turmoil about water services is that the country is still looking for the best arrangements. Most of the law regarding water services is still being written (and revised). The result is that we are still working out what the best arrangement for services are, especially services such as stormwater that are not used everyday and widely recognized as essential. The problem of the new subdivision on top of the hill that causes flooding problems in the old development lower down is not appreciated as much by the new development as by those below who suffer the consequences. As a result, many of these problems are not solved by planning so much as by litigation and political action. The problems that are being solved are on a case-by-case basis.

There is not much regulation of stormwater going on as of the mid-1980s. For the most part the regulation that is taking place is by local governments, and it is through land-use regulation and requirements for the quality of construction of new developments. The possibility of regulation of stormwater through the Clean Water Act continues to loom large over the country, however.

Most local governments have some stormwater control programs and a few have very effective ones. Typical program elements include the following, generally arranged in order of priority.

The Flood Control Program. This program enables the protection of life and property. Recent trends have been for this program to be mostly nonstructural, and as a result, not too costly to administer.

The Stormwater Control Program. This program has several elements, all emphasizing control of the impacts of urbanization. A beginning point would be some sort of general drainage ordinance. This might include statements of the responsibilities for drainage, the cost allocations, and the performance standards. Included in this general program area would be the subdivision regulatory provisions or the "subdivision regulations."

The Stormwater Quality Control Program. This program is third in priority even though not many cities even have programs at this time. The importance of this general responsibility of government warrants attention, at least some minimum attempts to control hazardous substances and other problems that arise from wash off of contaminants from the urban area.

The Erosion Control and General Land Quality Programs. In some parts of the country the control of erosion is especially important in urban areas and in construction sites. Construction site runoff, especially in mountain areas, is an important environmental problem, again one that can be controlled as part of the general stormwater strategy.

General Programs for the Control and Beautification of Urban Areas. General programs can be integrated into the stormwater program area. Some examples of these programs might be stream restoration, greenbelt construction, recreation, and environmental education.

There has been considerable attention given to the question of how a local government should regulate stormwater. Professor Tom Debo prepared a Ph.D. dissertation on the topic a few years ago in the Georgia Tech Departments of Civil Engineering and City Planning, and he listed a number of sections that could be included in an ordinance, essentially the following (12):

1. Statutory authorization
2. Definitions
3. Statement of objectives
4. Establishment of flood plain districts
5. Floodway district uses
6. Floodfringe district uses
7. Special use permits
8. Nonconforming uses
9. Hydrologic and hydraulic studies
10. Improvements required
11. Cost of drainage improvements
12. Erosion and sedimentation control
13. Grading and drainage plans required
14. Maintenance
15. Subdivision plats
16. Bonding
17. Repeal of conflicting ordinances
18. Violation and penalty
19. Appeal
20. Effective date
21. Termination date

This obviously is a rather comprehensive approach to ordinance writing and can be considered as sort of a menu from which the desirable elements of a tailor-made program can be selected.

Although there has been a lot of discussion of stormwater quality control, there have been no comprehensive programs enacted with any teeth in them. Some states, such as Pennsylvania, New Jersey, and Florida, have enacted stormwater legislation. However, the mood in the country since the legislation was passed has generally been antiregulatory, and there has not been much news about the results of the legislation. At the most, one would expect that the results would be the checking of development plans to see that they conformed to requirements. Regulation of stormwater quality and the checking of the effectiveness of system operation is not yet contemplated on a serious basis.

The essence of state stormwater legislation, as opposed to regulatory programs for erosion control, such as Maryland's or North Carolina's, is to require local governments and perhaps soil and water conservation districts, to get into the stormwater planning and, to some extent, land-use control business. This is the essence of Pennsylvania's approach, for example, which requires each county to develop a plan for designated watersheds. The plans would include the following elements (13):

1. Survey of existing runoff characteristics.
2. Survey of existing obstructions.
3. Assessment of projected and alternative land development patterns and the potential impact of runoff quantity and quality.
4. Analysis of present and projected development in flood hazard areas and its sensitivity to damages from future flooding.
5. Survey of existing damage problems and proposed solutions.
6. Review of existing and proposed stormwater collection systems and their impacts.
7. Assessment of alternative runoff control techniques.
8. Identification of existing and proposed flood control projects.
9. Designation of areas to be served by stormwater facilities within 10 years, and who will construct and operate the facilities.
10. Identification of the flood plains within the watersheds.
11. Development of criteria and standards.
12. Establishment of priorities for implementation of action within each plan.
13. Setting forth provisions for periodically reviewing and updating the plan.

The New Jersey approach also requires plans, but to be prepared by municipalities rather than the counties; however, the counties are in a review role. The state also participates in the role of providing grants as well as administering the program. The program apparently was enacted as much for the purpose of pollution control as for drainage.

The whole question of the role of government in stormwater programs is another matter. In addition to state governments, there are numerous federal agencies with responsibilities, mostly to provide assistance to state and local governments. Examples include the National Weather Service, the U.S. Geological Survey, the Soil Conservation Service, the Corps of Engineers, and the flood insurance offices within the Federal Emergency Management Agency. In January 1985 a federal–state–local task force issued a final report on recommended nonpoint source policies (14). The report assigns the logical tasks to each of the levels of government, with the lead role going to local government. There are no teeth in any of the requirements, however, and this has been the difficulty in coming to grips with the nonpoint source problems from the beginning of the studies under the Clean Water Act.

The Environmental Protection Agency moved in 1984 and 1985 to clarify its procedures for regulating stormwater discharges. The rules issued September 26, 1984, in the *Federal Register* stated that stormwater discharges that come from urbanized, commercial, or industrial areas or those "designated by the Director" had to apply for permits, and the *Federal Register* of March 7, 1985, stated that the deadline for applying for the permits would be extended to December 31, 1985. Time will tell about the effectiveness of this process compared to the additional government expenses that it will generate.

COST FUNCTIONS FOR STORMWATER MANAGEMENT

Cost functions for stormwater systems are not very well developed at all. There are not many papers in the literature on the subject, probably because each system is so different. We know in general that as the design level increases, the cost will increase, and we know that the more extras a system includes, the higher will be the cost.

In recent years there has been a lot of emphasis on nonstructural approaches to stormwater management, and these approaches carry the promise of effectiveness at a bargain. The concept of the "best management practice" as a water quality management tool for urban stormwater systems is an example. However, the concept of "cost" for these approaches is entirely different than for a simple structural system. The kinds of costs that must be included for nonstructural systems would be represented by program costs for administration, opportunity cost of not using land, and flood warning system acquisition and operation.

A review of the literature on costs of stormwater systems shows that the cost is closely related to design storm frequency (15). The results of

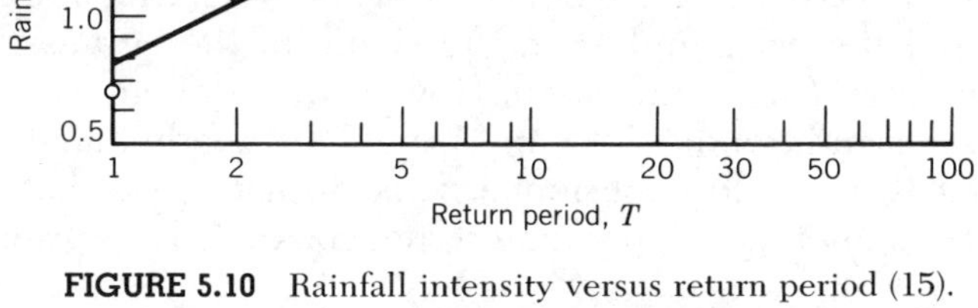

FIGURE 5.10 Rainfall intensity versus return period (15).

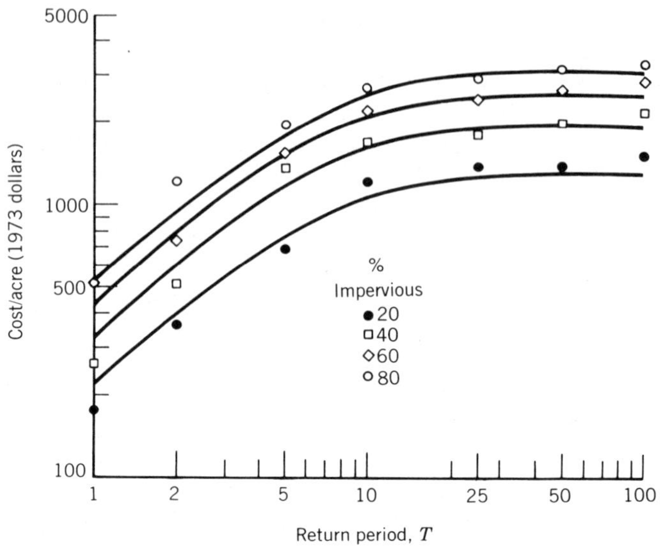

FIGURE 5.11 Stormwater system costs (15).

the study, as shown in Figures 5.10 and 5.11, suggest that costs per acre for minor structural systems rise rapidly up to the 10-year design frequency and then level off. This conclusion, based on stormwater designs in Colorado, seems dependent on the shape of the rainfall curves for the area, and the same kind of study needs to be done in other areas. It suggests, however, that the design decision for storm drainage facilities is very critical in the 2- to 10-year range, and that before standards are adopted the costs should be carefully considered. Since most new storm drains in this range are installed by developers, the result of standards that are too high will be reflected in higher home prices.

In Figure 5.11 t refers to basin time of concentration, taken equal to rainfall duration, and x and T are from the equation

$$I = K\left(\frac{T^x}{t_c^b}\right)$$

where I equals rainfall intensity in inches per hour, T equals return period of design, x equals a rainfall parameter that varies with return period, and b equals a regional constant.

The above analysis was based on some very generalized data and assumptions, and the costs and even the shape of the curves cannot be used elsewhere without modification. However, it is often useful to study shapes of curves and trends to get an idea of how another analysis might compare, that is, to build experience. It is in that sense that some additional cost parameters are presented, not for their use in making engineering estimates but just as experience factors. They might be useful, for example, for engineering students seeking to get some practice in estimating.

The first cost data is simply the variation of total installed cost with size of pipe, as shown in Figure 5.12. This demonstrates how rapidly the cost increases with pipe diameter. These were the initial costs estimated in a storm drainage study for the City of Englewood, Colorado, and since much of the planned system has now been built, there are additional cost data available (15,16).

The second set of cost parameters were developed for the purpose of general master planning of storm drainage systems in Pueblo, Colorado, and reflect cost data of 1975 (17). In this study we tried to anticipate the main cost items and to get very generalized estimates of them. The results are shown in Table 5.6.

Although these costs are not up to date as the individual files of an operating stormwater agency, they do demonstrate the kind of generalized planning costs that can be used for master planning and the establishment of basin cost allocations.

Further estimates were available from the 1982 report (18), and they can be used to compare with other estimates, such as those above, and to demonstrate a way to approach estimating procedures. These are given in Table 5.7, both for the capital costs and for O&M. The maintenance aspects are covered further in Chapter 9, which covers operations and maintenance.

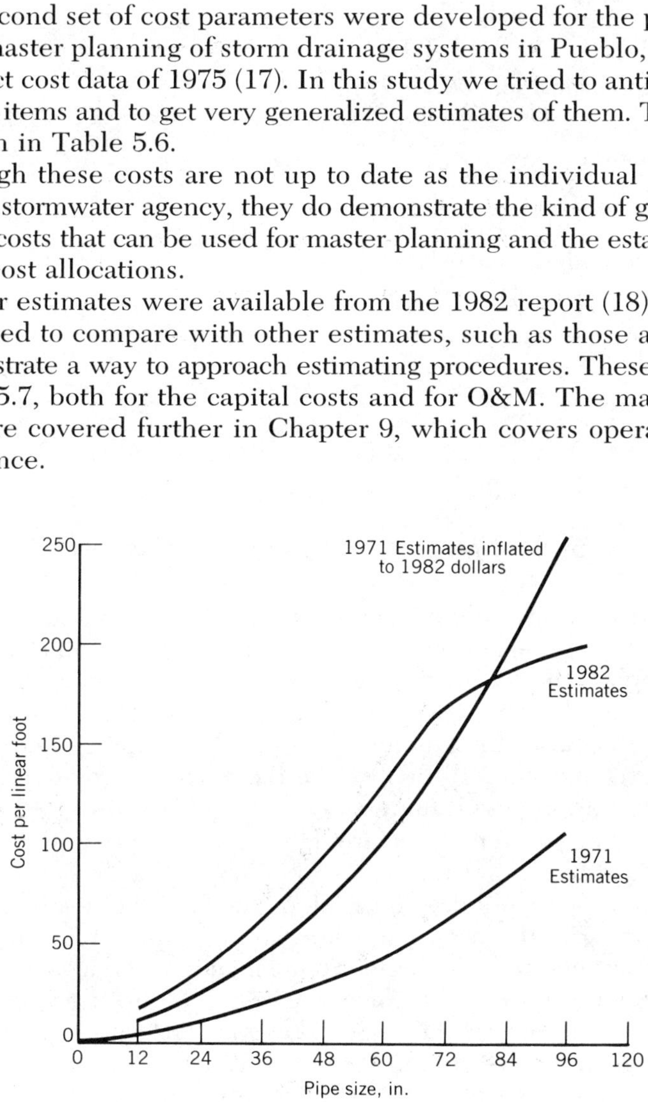

FIGURE 5.12 Cost data for stormwater pipes (13, 14).

TABLE 5.6 Some Drainage Facility Costs (17)

Open-channel construction
Land acquisition: \$4,000/acre (highly site-specific)
Earthwork: \$2.00/cubic yard
Seeding cost: \$0.14/ft^2 (incl. sprinkler system)
Low-flow channel: \$12/linear foot
Check structure cost
Each structure: \$2,000
Closed-conduit construction
Used similar approach to Figure 5.12
Detention pond construction
Each pond was calculated separately (they averaged about \$75,000 each)
Highway crossing estimates
Each crossing: \$7,000
Bridges
Each 2-lane bridge: \$120,000
Each 4-lane bridge: \$180,000
Channel rehabilitation and clearing
\$300/hundred feet
Irrigation crossings
Each crossing: \$9,600

CAUSES AND RATE OF DETERIORATION

Until recently stormwater systems received little care and attention in most cities, and as a result there is little data on their rate of deterioration. Poertner reports that in his survey, most communities reported that the local governing boards were "reluctant" to spend funds for O&M of stormwater systems (19). It is the same problem experienced in wastewater systems, but more difficult since the services provided by stormwater systems are not well understood.

The stormwater system differs from the other two urban water systems in its makeup. It generally consists of concrete, earth, grass, and miscellaneous structures and does not include much in the way of machinery and processes. For the most part there is no regulatory agency making periodic checks nor does the public complain about service except during flooding, when it is too late to make repairs. The public complaints that do occur abate as soon as the flooding is gone. It is a situation that encourages neglect. As a result the attitude toward inventory, maintenance, and general upkeep will be different than for other municipal facilities.

Denver's Urban Drainage and Flood Control District has made a de-

TABLE 5.7 Additional System Drainage Costs (18)

Channel work	
Clearing	\$5000/acre
Excavation	4.50/cubic yard
Embankment	2.60–4.75/cubic yard
Reinforced concrete	275/cubic yard
Riprap	25/ton
Seed, mulch, soil preparation	3200/acre
Highway bridge replacement	60/ft^2
Railroad bridge replacement	100/ft^2
Road replacement	
Asphalt	50/ton
Base course	20/ton
Utility relocation	
Gas and water	5/inch diameter/linear foot
Sanitary sewer, 8-in. PVC	23/linear foot
Electric and telephone crossing	3000/each
Annual maintenance	
Grass-lined channel	
Mowing	650/acre/year
Cleaning	0.5/linear foot
Rock-lined channel	
Rock replacement	1.5/linear foot
Debris removal	0.5/linear foot
Concrete-lined channel	
Minor repair	1.5/linear foot
Debris removal	0.5/linear foot
Storm sewer cleaning	600 lump sum
Detention pond	
Mowing	200/acre
Cleaning	200/lump sum
Weed control	70/acre

finitive study of the kinds of maintenance required to keep up stormwater facilities and reported on the classification of work involved in their contract maintenance program (20). The district's philosophy has been to carry out maintenance by contract rather than to staff up for it. This approach seems very cost effective for an agency that lacks other forces that can be used for stormwater maintenance.

The district divides the work into the following categories: routine, restoration, and rehabilitative. Routine work includes vegetation mowing, trash and debris cleanup, weed control, and revegetation. Restoration includes detention pond mucking, trash rack cleaning, rebuilding steep rundowns, tree thinning and clearing, extending trickle channels, repairing local erosion problems, and doing local grading and shaping. Rehabilitation, the most costly strategy, includes reconstruction items such as replacing drop structures, installing trickle channels, reshaping channels, installing riprap, installing maintenance access, and protecting existing drainage features.

The district found that the different types of work lent themselves to different contracting procedures. The routine work lends itself to bidding much as building maintenance would. That is, the work is divided into packages, and qualified bidders are solicited to enter into the work for periods of time. The restoration work is so individual in character that it does not lend itself to bidding, so individual approaches to problems are taken. The rehabilitation work is extensive enough that it requires engineering drawings and contracting in the same way as new construction does.

It is evident to stormwater managers that there is quite a gap between the theory and the practice of stormwater management, especially in the gap between what is planned and designed and what is actually built. Denver's Urban Drainage and Flood Control District recently sponsored a study of the implementation of detention storage policies in the region, for example. They found that almost none of the communities in the region had systematic maintenance programs and that many of the ponds that were required or even actually built were not functioning for one reason or another (21).

It must be understood that stormwater facilities will degrade rapidly when not properly maintained. Service conditions are severe under practically all circumstances, even in very dry areas. In humid areas the weeds and sediments that can be expected to choke the systems require constant attention. Since these areas are wet, the maintenance requirements are essentially constant. On the other hand, in dryer areas the maintenance requirements are not so frequent, but when severe storms come, on their characteristically infrequent but reliable basis, then the erosion and other damage demands attention.

DATA BASES AND INFORMATION NEEDED FOR MANAGEMENT AND DESIGN

The system characteristics for stormwater are essentially the same as for wastewater collection systems, and the data that is needed is the location, type, and age of each system component. Periodic condition assessments should be conducted and records kept to support the design of the routine, restoration, and rehabilitative work. Operational data for the stormwater system would include stage discharge data, system capacities, problem records, and hydrologic records.

Professional interest in stormwater systems has centered in recent years on design questions since there were so many of them that were somewhat unsettled. The fields of hydrology, hydraulics, and water quality, especially nonpoint sources, have been changing so rapidly in recent years that it has been somewhat difficult to keep up. As a result, it has been the exception for a local government stormwater agency to be well organized in collecting and managing the information needed for any of the phases of stormwater management, planning, design, or operation. There are some exceptions, of course, mostly in large city agencies and with some of the innovative drainage districts.

The driving force for urban drainage is precipitation, and our understanding of stormwater systems begins with it. It was the starting point for the establishment of risk and the performance standards for stormwater systems, and it is also the starting point for design since the system behavior depends on the rainfall inputs. There are three basic approaches to be understood: the statistical, the use of actual rainfall records, and the probable maximum storm approach.

The statistical approach begins with historical rainfall records, analyzes them, and results in a regular presentation of rainfall duration, depth, and frequency. The information can also be presented as intensity of rainfall, and both the depth and the intensity are used in certain runoff calculations. This approach ends up with what was described earlier as a synthetic design storm, but since some parts of the world are still developing the data they need to learn about these, they will remain very valuable design tools for the foreseeable future.

The use of actual rainfall records as an alternative to the synthetic design storm is a rather advanced technique that has only become available with the advent of greater computer power. It seems reasonable to expect now, however, that every stormwater agency should have the local historical rainfall records available for analysis. The testing of the potential performance of existing systems by running the continuous historical rainfall record through hydrologic models is a reasonable assignment for the engineering staff. The historical rainfall record thus is a desirable file of information to be kept in the drainage section. It would seem that keeping 100 years of historical hourly rainfall for one station should only take a few disks for a small computer and just a little space on a tape.

The probable maximum approach is used to recognize that statistical analyses are limited because rains are possible that have never been witnessed before in history. There are some economic consequencies of the PMP approach that are often severe, and unexpected, but when disasterous consequences are possible, hydrologists are often cautious. The basic information needed about the PMP for design and management is what the estimates of PMP are and what the consequences would be if one occurred.

Calculation of runoff is an aspect of stormwater management that has attracted much recent attention, probably because the basic runoff calculation sets the size of facilities that are planned and designed, and thus the economics of the entire system.

While the subject of calculating runoff has certainly received a lot of attention, it is still (and will no doubt continue to be) an inexact science. The reasons are the complexity of the systems involved, the random variation of the rainfall inputs, and the diseconomies involved in trying to calculate runoff any closer than we do. After all there is certainly a point of diminishing returns in making calculations over and over again when we are not sure each time that we really improved our estimates.

It appears that the basic starting point for calculating runoff should be to make sure that upper and lower bounds to the solution are known. After some idea of the solution is known, then models and other calculation techniques can be used for the next stages of the design.

In setting the upper and lower bounds of a problem, a starting point is the amount of rainfall itself, and the rate at which it falls. As strange as it may seem, there are sometimes estimates of runoff that exceed this obvious maximum. The time parameters will have to be estimated to get peak rates as well. One estimate of the lower bound is, of course, zero, but the estimate I suggest is the lowest of the results of several calculation methods utilized by the analyst.

Methods that are easy to use are coming to be called "desktop" methods, and they can be very convenient for first approximations. The most familiar of these is the rational method, which endures in spite of all the criticism. Going beyond these, we get into the different computer-based models, and there are many of them. Some of them are excellent and have had considerable use. The best example seems to be EPA's Stormwater Management Model (SWMM), which had the advantage of being adequately supported with research funds in the first place, with follow-up support over a long period while it was tested. Without all this support, the SWMM would have joined many others on the shelf gathering dust.

A key problem in estimating runoff is the estimation of losses such as infiltration and interception, and the lack of actual data of this nature for design and planning situations is a real limitation in the management of systems and another reason that not too much faith should be placed on any estimates that are not based on actual observations.

In designing and managing systems a knowledge of hydraulics becomes more important all the time because of the increasing complexity of systems and their connectivity. Actually, it is really "hydraulics and routing" that we are concerned with due to the need to calculate the time-varying nature of flows. This requires both the knowledge of hydraulics and the capability to consider the dynamic changes with time. This is the key to answering many of the stormwater management enigmas, such as: is detention storage a good idea; why do we have overloads here in one storm and not another; and why are not stormwater quality measurements consistent?

The first skill necessary is open channel hydraulics, including the capability to calculate flow profiles for steady flow in natural and artificial channels. The skill to calculate flow in pipes is used less often, but comes into play in some applications, such as flow in culverts. Flow in culverts is well understood and reported through the publications of the Federal Highway Administration, formerly the Bureau of Public Roads.

Routing flows is becoming more important all the time, especially with the new requirements for using the more sophisticated models. Understanding the more advanced routing models is facilitated by starting with an understanding of the earlier and simpler techniques, such as simply lagging and combining hydrographs. This is a starting point for understanding routing and constitutes a common-sense check to all the computer jargon that surrounds the solution of the differential equations of flow routing.

In recent years detention storage has become one of the most popular techniques for dealing with peak flows caused by urbanization. It also has benefits for water quality, and while water quality regulations have not really reached into stormwater yet, it seems inevitable that they will.

There are a number of design issues that affect the choice of detention storage. The major ones are size and location of the ponds, arrangements for maintenance, spillway arrangements, safety, multiple-use arrangements, and pollution control benefits.

It is sometimes feared that the presence of detention storage will aggravate rather than improve the very flood peaks they are meant to reduce. This fear must be assuaged by careful design studies.

The skill necessary to plan and design detention ponds is basically one of storage routing, much as in a larger reservoir. A few years ago I listed several "principles" for the use of detention storage, and they seem especially applicable for management understanding of the use of detention as a policy in stormwater management programs, especially given that detention storage is currently widely accepted. These are the principles:

1. Detention ponds may or may not reduce peak flows as they are intended to do. Their effectiveness here depends on several hydrologic factors that may vary from storm to storm.
2. The entire hydrologic network must be evaluated in the planning phase to really know what to expect from detention storage, and this requires, of course, a computer simulation capability.
3. Detention pond design should be multipurpose, as the use of land for detention offers several attractive opportunities for use of open space.

4. Safety and maintenance requirements for detention storage must be fully considered.
5. Ponds must be used only in conjunction with a totally integrated stormwater management program.

The above principles may seem very general and not too helpful at first glance, but their real message is that a stormwater management program is not static and something to be built and forgotten, but it is a dynamic and important part of the urban system and needs to be understood to be made effective. And detention storage is a complex addition that often does not act like it might be expected to, so efforts must be made to improve the understanding of it by the managers.

The subject of water quality as part of stormwater management comes as a surprise to some, since the main concern of stormwater programs, such as they have been, has been drainage and flood control, and, of course, these are still the main objectives of urban drainage and flood control. However, if we are to successfully deal with the emerging urban environmental problems of today, both in developed and developing countries, then the general problem of nonpoint source control must be faced, and stormwater management programs are an important place to initiate control strategies.

There has been a lot of research on the general water quality aspects of urban runoff, and today we know a lot about the subject. However, the subject is so complex that we lack any general theories about it, and the indicated action programs are not as scientific as they are practical.

Actually, the nonpoint source problem goes beyond urban stormwater runoff and extends to all land-disturbing activities. The recognition of this is a healthy development since the general stormwater problem also covers areas outside of cities. In other words stormwater protection also has to be extended to industrial plants, complex agricultural developments, mining activities, and other land-use changes.

The skills necessary to incorporate water quality considerations into stormwater plans include the following: a general knowledge of water quality parameters, understanding of the state of knowledge about prediction technologies, knowledge of the legal status of control programs, knowledge of the technology and economics of treatment processes, and finally, a good understanding of the concepts of "best management practices."

There is a built-in problem, of course, that inhibits public works managers from getting involved in stormwater, and that is, what are the incentives and where is the political support. It seems that the best starting place is with the part of the local government concerned with environmental quality. This group should bring the possibilities for environmental improvement to the attention of the city's top management, who in turn might assign the task of program development to the stormwater group. In newer cities with lots of open space, stormwater quality may not be a problem at all, but in older cities and those with dense concentrations of people and industry, it can be a serious but little-understood problem.

It should be clear that the data bases and information necessary to understand the stormwater system are at the heart of the city's land–water interface and affect the city's quality of life in several ways. Although there is not much known right now about how to design management information systems and there are even problems in mounting any kind of stormwater programs at all in some areas, stormwater is a problem area that will continue to require attention from water managers.

PRINCIPAL STORMWATER INTEREST GROUPS

In contrast to water and wastewater there is no single association that has a clear monopoly on stormwater concerns. There has seemed to be adequate attention given to the research and education requirements of stormwater, and several associations have contributed to the publications and training courses that have been developed. For the most part there has been no federal agency that has directed itself primarily to stormwater, but several have made significant contributions, such as the Federal Highway Administration in street drainage, the Corps of Engineers in computer modeling and scientific procedures, the Soil Conservation Service in runoff modeling, and the Environmental Protection Agency in water quality concerns. No state agencies are very much involved in stormwater since it is normally not much regarded as a state concern.

The association that has given the most attention to stormwater in the last 20 years has been ASCE through its Urban Water Resources Research Program. Over 100 reports were issued covering every conceivable stormwater research topic. The American Public Works Association has also undertaken some very useful investigations concerning stormwater, and its report entitled "Urban Stormwater Management" is a very useful reference document (22).

Because of the interface with water quality, there are a number of environmental groups that have an interest in stormwater in general. All of the agencies and associations that are concerned with water quality have at least a passing interest in the control of stormwater, but their interest is not the same in general as those that provide stormwater service.

The group of flood management agencies with a common interest in policies has banded together to form the National Association of Urban Flood Management Agencies (NAUFMA), and it has been active on a national basis in studying and advising on policies that relate to urban flood management.

LONG-TERM ISSUES

Due to the considerable interest given to stormwater in the 1970s, the state of the practice has settled down quite a bit. The following seem to be some of the main issues that are still not very well settled: stormwater program strategies, techniques for dealing with urbanization and growth, maintenance and operation, and water quality.

The fundamental program issue is how to have a viable urban stormwater service in the face of all the disincentives that have been discussed, especially the lack of interest of elected officials. The stormwater utility concept is one approach, but it needs further review and attention. Long-term answers are needed.

Techniques for dealing with urbanization and growth require attention both for the growth that still will occur and the growth that took place without much attention to stormwater. In some cities there has been so little attention given to land-use control and to building quality developments that the stormwater systems are not in good shape at all. In some communities the attitude is pro-development at the expense of quality long-term living space. In these cities it has been very difficult for stormwater innovations such as detention storage and reserved open space to be considered.

The subject of maintenance may be the most important of all for ensuring the future effectiveness of systems. Reasons for lack of attention to maintenance were given already, and they stand as powerful deterrents to progress toward solving stormwater problems.

Water quality is usually thought of as someone else's problem, and few stormwater managers feel any responsibility to get involved in it. Why should they? It is difficult enough to maintain effective drainage and flood control systems, but the subject of nonpoint sources of pollution still must be dealt with on a national basis, and stormwater managers will eventually be called on to play a key role in solving the problem. The presence of toxic and hazardous materials in catchments means that there are some serious stormwater problems out there waiting to be tackled.

REFERENCES

1. Neil S. Grigg, Evaluation and Implementation of Urban Drainage Projects, *J. Urban Planning Development Division*, ASCE, May 1975.
2. L. A. Roesner, H. M. Nichandros, R. P. Shubinski, A. D. Feldman, J. W. Abbott, and A. O. Friedland, A Model for Evaluating Runoff-Quality in Metropolitan Master Planning. ASCE Urban Water TM 23, New York, April 1974.
3. John A. Lager and Metcalf and Eddy, Inc., Urban Stormwater Management and Technology: Update and User's Guide, Environmental Protection Agency, September 1977.
4. Neil S. Grigg and T. Sriburi, Order Classification of Urban Catchments, *Water Resources Bulletin*, February 1978.

5. R. P. Shubinski and S. N. Nelson, Effects of Urbanization on Water Quality, ASCE Urban Water Resources Research Program TM 26, New York, 1975.
6. Wright-McLaughlin Engineers, Urban Storm Drainage Criteria Manual, Urban Drainage and Flood Control District, Denver, 1969.
7. Murray B. McPherson, Urban Runoff Control Planning, Environmental Protection Agency, October 1978.
8. Office of the Chief of Engineers, Standard Project Flood Determination, *CE Bulletin* 52–8, March 1965.
9. Hector J. Cyre, *Stormwater Management Financing,* APWA Congress, Houston, September 1982.
10. Richard J. Plastino, Death Knell Sounds for a Storm Drainage Utility Fee, *Public Works,* April, 1984.
11. Marc Engemoen and Roger E. Krempel, Utility Approach to Comprehensive Stormwater Management, *Public Works,* April 1985.
12. Thomas N. Debo, N. S. Grigg, and L. S. Tucker, *Flood Plain, Detention and Erosion and Sedimentation Ordinances,* Water Management Science Inc., Atlanta, November 1981.
13. Commonwealth of Pennsylvania, Storm Water Management Act, PL 864, No. 167, SB 744, 1978.
14. Environmental Protection Agency, Office of Water, Final Report on the Federal/State/Local Nonpoint Source Task Force and Recommended Nonpoint Source Policy, January 1985.
15. Neil S. Grigg and J. P. O'Hearn, Storm Drainage Cost Functions, *J. Hydraulics Division,* ASCE, April, 1976.
16. Sellards & Grigg, Inc., Storm Drainage Plan for the City of Englewood, Colorado, 1971.
17. Sellards & Grigg, Inc., Lake Minnequa Storm Drainage and Financing Plan, Pueblo, Colorado, June 1975.
18. Muller Engineering Company, Drainage Report for Tuckers Gulch and Kenney's Run, Urban Drainage and Flood Control District, Denver, November 1982.
19. Herbert G. Poertner, Stormwater Management in the United Stated, Report for the Office of Water Research and Technology, Washington, September 1980.
20. Mark R. Hunter and L. Scott Tucker, Contract Maintenance for Drainage and Flood Control Facilities, *Flood Hazard News,* Denver, December 1982.
21. Bruce Prommersberger, Implementation of Stormwater Detention Policies in the Denver Metropolitan Area, *Flood Hazard News,* December 1984.
22. American Public Works Association, Urban Stormwater Management, Chicago, 1981.

CHAPTER SIX

PLANNING AND MANAGEMENT ESSENTIALS

Studies of the infrastructure problem have shown that it is planning and management rather than the availability of new technologies that are critical to success. This is in contrast to some other problems, such as international competitiveness, where the continued development of new technologies rather than management of existing systems is the key to success. Financial management is probably the most important element since, if there is enough money, the technology and the skill will be made available, but applying other valid management principles to the infrastructure is also critical. The purpose of this chapter is to discuss and explain the most important of the planning and management principles, including finance, as they apply to urban water systems.

The terms *planning* and *management* are not automatically understood by engineers in urban water systems since they are not universally used to describe the work that goes on. These terms are becoming increasingly used, however, to describe the inescapable fact that it is often the management decisions that make the difference in the quality of service rather than the technical decisions. But the management decisions referred to are technical in nature; therefore, it is important that water engineers have an appreciation for planning and management skills. At Colorado State University we have initiated a graduate program in "water resources planning and management," and it is becoming one of the most important of the graduate water programs in the university. The American Society of Civil Engineers has created a division with the same name, and it is becoming one of the most active in the society.

It is an ambitious task to discuss the planning and management of urban water systems in one chapter. Actually, this chapter is meant to cover the essentials of the topic in a general way, since the entire book is about planning and management. Still, there are so many essentials that this has grown to be the longest chapter in the book, and there is much more to say than can be covered in the single chapter.

In searching for the right coverage in this chapter, I was led to two volumes that were produced by the Council of State Planning Agencies in 1983 and 1984 (1,2). These are entitled, "Rebuilding America: Planning, Managing and Financing Public Works in the 1980's." In effect, these volumes show the topics that the authors considered important in solving the problem of "rebuilding America" as falling mostly in the realm of planning, managing, and financing. In effect, most of the topics covered in these two volumes fall into the general category of financial management since topics covered under capital budgeting and investments are so close to finance. This chapter takes a broader view of planning and management and includes topics such as productivity and organization. For a more detailed view of finance the reader can refer to Vaughan and Pollard (1) and Vaughan (2) and a number of others listed at the end of this chapter.

In beginning to write this chapter I was impressed that there are really two main problems that have to be solved in planning and management: getting enough money for operation and maintenance, and doing an adequate job in capital budgeting and financing. If these two problems are solved, then the talented management that is available can solve the rest of the problems. We launch now into a discussion of what these problems are, and we conclude with our section on the most important elements of financing.

PROCESSES OF MANAGEMENT

The success of a venture depends on the external conditions surrounding it and on the skill of the group responsible for it. This is certainly true of water systems, and the importance of management has been emphasized in recent years as we have seen more and more attention to the use of enterprise management for public systems such as water.

The audience for this chapter will be the engineers and managers who would like to study management topics for suggestions about how to improve the performance of urban water systems. Planners and managers who have a need to understand urban water problems will also have an interest in the chapter, as well as engineering students who would like to understand urban water problems from the management viewpoint. Since there is no shortage of articles and books on management in general, the contribution of this chapter is to interpret management practice for the urban water system. Accordingly, it does not cover the principles of management from the beginning, but how they apply to the urban water system.

The use of the term *management* here is rather broad, and it will be taken to include engineering, executive, and administrative management. The focus is on two main areas: the skills needed by managers in water organizations and the processes of planning and problem solving. These are the root determinants of success in urban water organizations, if the external conditions will allow success.

If external conditions are not right for success, management can still succeed, but the definition of success may be different. For example, let's say a water utility director wants to implement enterprise management and self-finance to improve the financial management of a utility, but all of the top management, including the governing board, is against it. These would be very difficult conditions indeed for success, but the manager might be able to settle for an improved system of internal accounting that would lead later to better total financial management. Two reasons that wastewater treatment plants do not work better are inadequately trained personnel and inadequate operating budgets. Both of these problems go back to management. Management has the responsibility to present the case for the budgets and for training personnel. It is true that some external circumstances make it very difficult for management to obtain adequate budgets, but it is still a management responsibility.

When we speak of management in the context of an urban water system, we may lose some of the audience since the word is so general and abstract. We need to be specific when we use the term so that it is clear what we mean. Many studies of management identify planning, organizing, directing, and controlling as the most important processes of management. These have special meanings when applied to public water systems. Actually, there are two lines of these processes: the line of building physical systems and the line of operations of the systems after they are built. Sometimes they are confused. When we speak of planning, for example, we may either be speaking of planning systems or planning operations. It is important to make the distinction.

A way to make the distinction is shown on Figure 6.1, which shows the steps of planning, organizing, and controlling the urban water system. The process of planning the physical facilities of the urban water system is seen as the preparation of the plans for the system itself or the planning of the capital facilities. The process of organizing then becomes the design and construction activities. Controlling the system, however, is the O&M process, and it requires the processes of planning, organizing, and controlling for success in making the original goals of the system, as it was planned, become a reality. This is an important distinction and merits further elaboration.

Our operations researchers tell us that when we plan a system we must make sure it will operate properly before the planning is completed. In the language of planning, they say that embedded in every planning

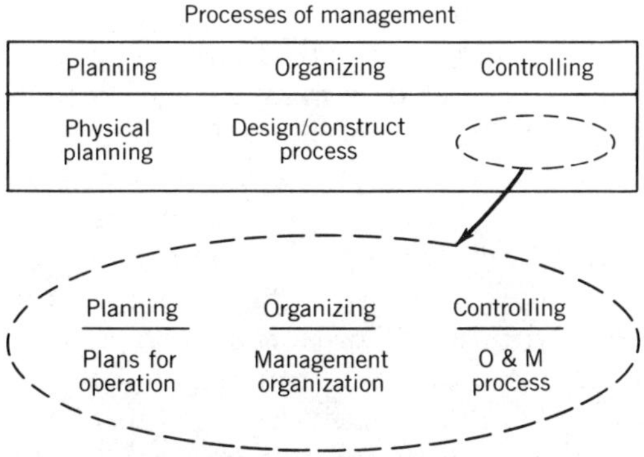

FIGURE 6.1 Processes of managing urban water systems showing the operational problems embedded in the planning problem.

problem is an operational problem. This means, for example, that in designing a new industrial product, such as an automobile, we require the construction of a full-scale operational model that can be tested. The way the prototype is operated will depend on the outcome of the tests, which may result in further planning before the final version is fully planned and constructed. In the water field we are not able to normally construct these full-scale operational models so we rely on computer simulations and on engineering judgment. Figure 6.1 illustrates a little of this process, whereby the plans for operating the system, including the organization and the O&M process, are imbedded in the management process, and the planning for the physical system includes planning for the success of the O&M phase.

In general, the tasks of management are taken to be planning, organizing, directing, and controlling. There are many ways to present these tasks, but they include most of what managers do. One person, when asked what management was, is reported to have replied, "management is what managers do." This refers to the difficulty in explaining management. In order to fulfill their responsibilities managers must exercise skill in these tasks, and that is what the next few paragraphs are about. In the next section we will discuss the unique requirements of planning in urban water systems.

The manager in an urban water organization must have skills in three generic roles: leader, supervisor, and technical manager. In addition, the manager must have certain personal skills, some by cause of habit and training and some by virtue of the knowledge gained through education, training, and experience.

Personal skills needed include time management, interpersonal relations, and good work habits. Knowledge-based skills include financial management, public relations techniques, and numerous technical skills as, for example, those involved in working with computers. This book is not about these skills, but the manager reading it should make a personal check as to his or her preparedness in these areas. There are also other traits that are essential in a manager, but which are beyond the scope of this discussion. One of the most important is integrity. It seems that the single greatest cause in the breakdown of some public systems is the absence of personal integrity in management. In some places this appears as corruption, a characteristic of public management that is absolutely deadly. It makes no sense to talk of improving management or productivity when corruption is present in management systems.

Leadership is an important determinant of success in water organizations, as well as all other organizations. One of the most important skills for the manager in a water organization is that leadership be knowledge based. Technical workers and the public respect knowledge, and the manager should make sure that he is always prepared. One of the secrets of General George Marshall as an effective leader in peace and war was his awesome knowledge about the systems he was managing. Many water managers have never thought about leadership as one of their main responsibilities, considering that they were managers instead, but leadership in water organizations is a necessity that deserves more attention.

Leadership is not really a topic that can be taught or even analyzed very well. However, a recitation of leadership qualities can help the manager inventory his or her own status as a leader. There are many lists of such qualities. One includes such topics as the following: fluency of speech, dominance, knowledge, emotional balance, creativity, self-confidence, achievement drive, drive for responsibility, interest in work, and sociability (3). Again, the value of citing the list here may help an individual manager take a personal inventory.

One of the requirements of leadership is to be able to make effective and timely decisions. It falls under the category of the skill of decision making, covered in detail in Chapter 8, which covers decision support systems, but it is also applicable under the topic of leadership. Making decisions in an organization is one of the most important jobs of the leader. It is an executive function, and when it is done right there is a "zing" in the organization since the staff knows they can depend on quick and effective action. When it is done in a hazy way, the morale may suffer, and rumors feed on the uncertainty. Decision making is really an art and science. It can only partially be taught; it is one of those skills that is best learned by experience. The topic gets a full chapter in a recent book on operations management (4). The author presents the skill as being necessary for effective operations at all levels in an organization. The process he presents is similar to the problem-solving process. Another writer, Moody, has prepared an entire book on decision making (5). In his book the techniques of quantitative decision making such as statistics, utility theory, and some other analysis techniques are presented. The approach to decision making that is most appealing, however, is one that takes into account all the intangible factors that have to do with human nature and the fact that everything cannot be predicted and quantified. One text that gets at that is by Behn and Vaupel at Duke University (6). In their approach they recognize the "gut-level" weighing of alternatives that takes place in the real world of management and present credible ways to analyze the alternatives quickly. There is, of course, much more to say about decision making; what is emphasized here is its importance in the overall management of the enterprise.

The management of water organizations also requires exceptional supervisory skills in day-to-day operations. The fact that the quality of a city's drinking water supply depends on the success of the operational managers places an awesome responsibility on them. Also, lives and property may be at stake in the operation of dams and reservoirs, and the swimmability and fishability of streams depends on success in wastewater management. Supervision is a control responsibility of management. After the work is planned and organized, it must be controlled. This requires reports, information systems, direct supervision, and numerous other processes. Later in the chapter we discuss organizational effectiveness briefly, and one of the determinants of it is the success of supervisory management.

Supervisory management involves work planning, direction, coaching, and performance review. A good organization will work symbiotically, with the different levels and functional units working smoothly together. Techniques such as "management by objectives" are designed to help achieve this, but as today's management literature will attest, the organization and management of work is a topic that needs constant attention.

The area of technical management is the one where urban water management requires unique skills. The manager may not be highly skilled in the areas of administrative management required, but he should be able to command respect by a solid base of knowledge, the ability to learn, and the ability to listen to experts and learn from them. This latter ability carries with it the caveat that the manager should be able to distinguish between good advice and bad advice.

There are three general categories of technical management that need special attention: scientific, financial, and technological.

Scientific management capabilities would include the general areas of health and environmental concerns. In recent years both of these have stressed the capabilities of water managers to operate effectively. For example, if confronted with a regulatory demand to add special treatment processes to clean up wastewater to remove a certain constituent about which little is known, should the manager "cave-in" or investigate what is known about the health effects of the constituent and take a careful management approach? The manager should take an independent stance, including a personal and/or staff investigation, before making a decision.

An example of this situation would be the current debate over trihalomethanes (THM). In the environmental area some rulings of environmental boards have made little sense in recent years while rules and regulations were being sorted out, and large sums have been saved by wise managers before complying with rulings, and large sums have been wasted by those that acted hastily. The coin has another side as well: large amounts have been wasted by fighting rulings that should have been complied with rapidly.

The second technical area, financial, is one that seems to be missed by engineers who migrate to positions of management without any training in finance. However, today is certainly a time to be aware of financial management needs in the public sector. Debt financing, tax rollback initiatives, capital budgeting, utility fees, financial impact statements, and infrastructure banks are just a few of the terms that have been in use lately that illustrate the newly confusing field of water financing. Water managers can learn about financing by paying attention to development in their own utility and by following the business news. It helps to read a financial periodical such as the *Wall Street Journal* because almost all of the different approaches to finance pass through the pages over a period of time. The topics of financial management that seem important are covered later in the chapter.

The third area, technological, is intended to represent the engineering category. The term *technology* is used instead of engineering to recognize that managers may have a valid grasp of the technical issues involved in certain decisions even if they are not engineers. In other words technology has become too complex to leave to engineers! As an engineer I make the previous statement with tongue in cheek, but I have found that some managers defer too easily to engineers, just as engineers defer too easily to lawyers and accountants. None of our disciplines has any monopoly on understanding.

Necessary skills in the technological area include being current in topics such as computers, treatment processes, methods for system maintenance, and operational schemes. The manager's responsibility is to have a good idea of the appropriateness of the technological alternatives, especially to enable a valid evaluation of the recommendations that are presented by experts or consultants.

To underline the reasons for stressing the importance of management, let us consider a few examples of situations where talented management can make the difference. These are situations such as that described by a friend of mine when he said, "You know, I sure waste a lot of time, but there are some situations where I earn my entire year's pay in 15 minutes." He was referring to closing a sale or solving a tough problem or similar situations. Many examples could be cited, but the discussion will be limited to those that might confront the urban water manager.

One example might be relations with a regulatory agency. Water and wastewater organizations must deal with several regulatory agencies, at all three levels of government. The view of the water organization held by the regulatory agency depends to a certain extent on their assessment of the credibility of the management. For example, when I was responsible for regulating some wastewater agencies, I would view the reports about violations in the light of our expectations about management's intention to solve the problem. If the utility management would say, "we recognize this problem and we are working to solve it," that would elicit a different response than if there would be no communication, or if management would treat our officers as a nuisance rather than as reasonable people to work with. This is an important distinction. One utility our inspectors dealt with had a very high reputation. Our people would say, "when XYZ finds a problem, they can be expected to fix it." When a related organization, one of about the same size, found a problem, the reaction was the opposite. Our inspectors expected a hard time, little cooperation, and then to deal with lawyers rather than technical people who were interested in solving problems not litigating them.

Another common situation is dealing with an irate public. As any public works manager knows, this is a tough problem since the world is filled with unreasonable people. And some will be unreasonable no matter how well they are treated. People respect credibility and honesty, however, and if management gives straight answers and if the people perceive that management has their interests in mind, the path will be clear for success. Skill in dealing with the public is one of the most important areas for water managers.

Another area where it is important to be skilled in management, but with emphasis on leadership, is in the solving of regional problems that require cooperation. For example, situations exist where the managers have not talked with their counterparts in nearby communities, and the result might be proposals for expensive systems where some cooperation could have improved the situation for everyone.

Another area where it is important for management to be competent is in dealing with technical problems, of all kinds. Examples include acquisition of new management technologies, such as a computer system. The chaos created by incompetence in dealing with this challenge is legion! Managing the design and construction of a complex treatment plant is another area requiring real technical skill and experience. Another example is rate setting, with all the attendant considerations involved. Another is union–management negotiations, with the many contract and human relations considerations involved. Another is dealing with the whole question of environmental rules and regulations.

PRODUCTIVITY IN URBAN WATER SYSTEMS

Of course, an urban water manager should constantly be seeking higher productivity in the delivery of urban water services. We know what the term *productivity* normally means in the industrial context (i.e., the output per work hour), but what does it mean in urban water systems? This is an important question since without a definition of productivity, it will not be possible to apply productivity improvement techniques.

Generally, productivity means the quality of being productive, and that means to produce something of value. This is the connection between productivity and management: the something of value has to be defined or prescribed in the context of the goals of the organization. This is the reason for spelling out the services offered by the urban water subsystems in the chapters on each system.

In studying productivity the goals of the service must be viewed in the context of the organization and the activities that are necessary to achieve the goal, and each part of the organization must be analyzed separately, then the organization must be viewed as a whole. By breaking down the tasks of the different parts of the organization into pieces that can be studied separately, the working of the systems becomes amenable to analysis.

It is fairly clear that the administrative parts of a water organization can be studied for productivity in much the same way that offices everywhere can be, so the emphasis here is on the productivity of the operational parts of the urban water systems. This includes the productivity of labor, capital, and energy; these seem to be the major components of the cost and effectiveness of urban water systems of all kinds.

More specifically, productivity can be defined as some measure of output compared to input, where either of these variables can be measured in different ways. A common approach might be dollars per million gallons of water produced, a number that appears in the chapter on water supply. Another, more representative number would be work hours per million gallons produced, as that would be independent of debt and special historical factors of the water system. Mali, in writing about productivity, states that it is a combination of effectiveness and efficiency, and in that sense it subsumes much of the concerns of management science: how to measure and raise effectiveness (7). He uses a definition that essentially uses productivity to measure how well resources are utilized in achieving an objective. According to Mali, there is a hierarchy of measures: the overall goal, the performance objective, the performance standard, and the productivity objective. Used in this way the concept of productivity is revealed for what it is: a way to break the objectives of work into successively smaller pieces so that its effectiveness can be measured.

In my opinion the important thing about productivity measurement for urban water managers is that it highlights the objectives of the "business" of water and increases cost consciousness. By studying productivity, a manager becomes better at the responsibility function, at the control of the operation. Breaking down the business of water management into the work tasks will also improve the organization and sharpen management altogether.

In applying productivity concepts to urban water systems, both service and cost have to be quantified. In quantifying service, the water produced or sold, the wastewater handled, the constituents removed by treatment, or the damages prevented must be calculated. Not much has been done to develop these productivity parameters yet. On the cost side, the capital and O&M costs must be broken down by subsystem, by function, and by category of input, that is, by construction cost, debt service, labor, energy, and materials. After all of these categories of information are available, the appropriate ratios can be calculated.

The concept for using productivity measures as a way to measure the progress toward goals is illustrated in Figure 6.2. There is a hierarchy of goals and objectives for an organization, and if every unit achieves the goals set for it, the organizational purpose will be achieved and the mission accomplished. The units of the organization include the individual worker, and the achievement of the individual is normally measured by some sort of productivity measure, as are the achievements of sections and work units. This is a normative approach to the use of productivity targets to work toward achieving organizational purpose. The process must be humanized, of course. I have been a fan of scientific management, using principles espoused by the gurus of management, such as Frederick Taylor and Frank and Lillian Gilbreath. I have been shocked to find these principles criticized as being ones that lead to exploitation of workers, however, and the right balance point must be reached.

Actually, what many writers mean by productivity is better manage-

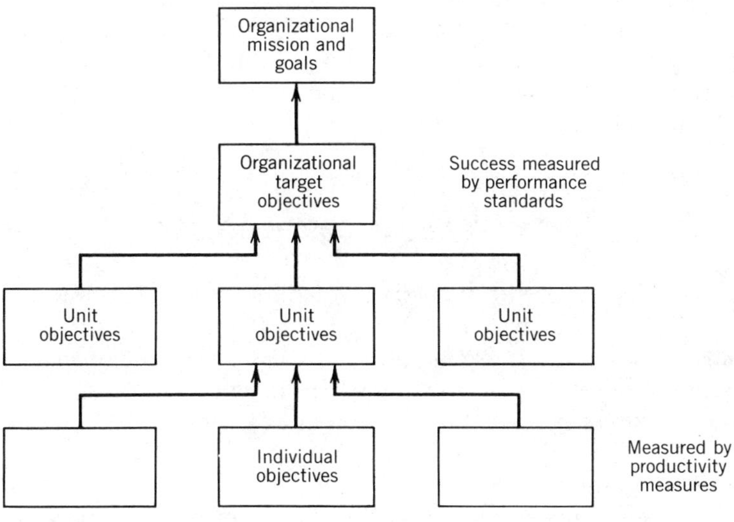

FIGURE 6.2 Flow of objectives, goals, and productivity measures.

ment. For example, a recent article about cost control in water utilities emphasized management principles that would make the utility more effective (8). The authors, from the Dallas Water Utility, listed five principles necessary to make utilities self-sustaining, and therefore more productive: to generate sufficient income through user charges, to follow sound financial practices, to cover the full cost of operations and not be falling behind, to generate capital improvement funds, and to innovate to improve productivity and lower costs. There are 12 components of their program to achieve higher productivity, most of which can be classified as management essentials:

1. Performance pay for managers
2. Work planning and goal setting
3. Use of management consultants
4. Reorganization when required
5. Productivity task forces
6. Use of industrial engineers
7. Recognition of results
8. Employee suggestion program
9. Quality circles
10. Performance audits
11. Financial guidelines and policies
12. Water main replacement program

In another article about productivity, Paul Hersch lists a number of tips to make utility managers more productive (9). His suggestions cover several management topics: how to deal with internal and external communications problems, how to be more personally effective, and how to manage better in the sense of industrial psychology, that is, to get the organization to be more productive.

PLANNING AND PROBLEM SOLVING

This section discusses the principles of planning and problem solving as they apply to urban water organizations. Planning is of extraordinary importance in managing organizations of all kinds, but in many public organizations there is little of it done. Problem solving is the reason for planning (and management), and the manager needs to understand the process of problem solving in order to have a direct approach to handling some difficult issues.

Someone asked me once, "Why are you interested in planning; isn't planning just a dull subject?" I replied that when managers plan it is to solve problems, and there is nothing dull about it. When "planners" plan, it can be dull if the managers are not directly involved and interested in the planning. That kind of planning leads nowhere and will not last, of course. The results of such planning are plans that "stay on the shelf." A basic premise of successful planning is that the managers, the movers and shakers, must be involved in the planning.

There are different phases and different objectives of planning, and planning applications in an urban water organization fit on a two-dimensional matrix. The columns are the stages of management: planning, organizing, design, and operation. The rows are the subdivisions of the organization: water supply acquisition, treatment, wastewater collection, and so on. A third dimension could be the levels within each of the subdivisions of the organization. Everything has to be planned, and that includes the physical capital facilities as well as the operating systems. Actually, this view of planning, while a complete picture, is somewhat difficult to grasp conceptually unless we view planning as a responsibility of the entire organization.

Definition of Planning

Planning is, in general, the act of making a plan, and a plan is a guide for action. This view of planning extends to activity plans and construction plans as well. In other words the plan tells you what to do. If management is what managers do, the plan then tells managers what to do, and planning takes its proper place as the first task of management.

Planning is a requirement for problem solving, and problem solving requires decision making; thus, planning, problem solving, and decision making have the same general "process." This process has taken on many names, but I choose to call it the problem-solving process since I believe that both planning and decision making are subactivities of problem solving, rather than vice versa.

This process is shown in Figure 6.3, which involves planning, decision making, and implementing. Understanding the processes related to problem solving is of fundamental importance, especially when we consider that problem solving is a basic activity of humans, one that we engage in continuously. Planning is studying what to do; it enables us to carefully consider our options. Decision making is deciding what to do, a discrete act, one that cannot be avoided by managers, a reality that was summed up by Harry Truman by his sign "The buck stops here." Implementing is making the decision work, and it often is the most time-consuming and difficult part of solving problems. Implementing a bat-

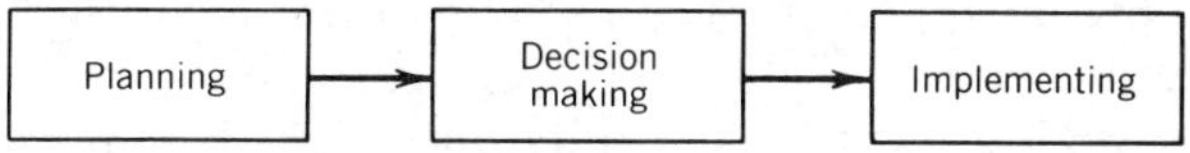

FIGURE 6.3 Problem-solving process.

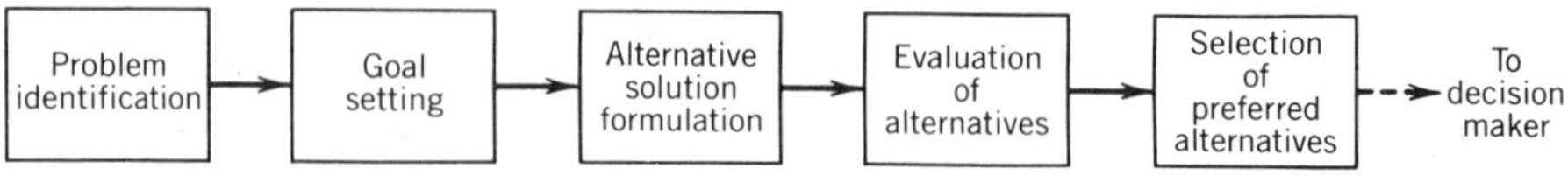

FIGURE 6.4 The planning process.

tlefield commander's decision, for example, can well cost the lives of many soldiers, but not involve the commander very much. The responsibility of decision making, never light, is especially felt in that situation.

The planning process includes several very important steps, as shown in Figure 6.4. Identifying the problem is the first step, and it cannot be skipped. Really, the difference between being "effective" and being "efficient" is summed up here. I like the story about the auto passenger who wakes up to hear the driver say, "I have good news and bad news! The good news is, we're making great time. The bad news is, we're lost." In this case all the plans and actions are lost since the right problem, arriving at the destination, is not being solved.

Goal setting, while a key feature of the planning process, is very complex due to the different value sets of the many actors involved in managing public organizations. This can lead to extreme frustration for managers who are not ready for the challenge of managing the public organization.

Formulating the alternative solutions is the creative step of planning. This is the one that most people like to participate in. The familiar refrain, "Why don't we do it this way" is an expression of the desire to make an input to the formulation of alternatives.

The evaluation of alternatives is a scientific process involving systems analysis, economics, impact analysis, and political awareness. This is the step for the accountants and modelers. Everything should be examined with a fine-toothed comb.

Selection of preferred alternatives means to develop a way to rank and present the alternatives to the decision maker, but it is not the decision making itself. That is a discrete step reserved for the decision maker and is not really part of the planning. For a more complete discussion of the process of planning, see *Water Resources Planning*, which is entirely devoted to the principles and techniques of planning for water resources systems (10).

Planning and Problem Solving in Urban Water Systems

To even discuss such a broad subject as planning for urban water systems we need some classification schemes. The first one, shown in Table 6.1, presents a view of the clusters of responsibility for different aspects of the organizations. Each of the centers of responsibility must plan. For example, organizing the finance function requires planning. Operating

TABLE 6.1 Responsibility Centers for Urban Water Management

	Management Task			
Management Area	Planning	Organizing	Design	Operation and Maintenance
Administration				
Finance	•	•	•	•
Organization/personnel	•	•	•	•
Support systems	•	•	•	•
System management				
Capital plant	•	•	•	•
Operating systems	•	•	•	•

the capital plant requires an operational plan, and this element is the same as planning the operational systems.

Table 6.2 presents the functional breakdown for the three major urban water systems: water supply, wastewater management, and stormwater. In effect, this classification represents the third management dimension of urban water management, the other two being management area and management task, as were shown in Table 6.1. The items in Table 6.2 are generally self-explanatory. They show the way to view the three urban water services from a functional viewpoint. What we have, by combining Tables 6.1 and 6.2, is a three-dimensional view of the management problem categories. Each of the functions in Table 6.2 must be dealt with for each of the categories and each of the stages of management shown on Table 6.1.

There are also several ways to look at the planning required in terms of time horizon or stage of planning, by jurisdiction, by function, and by purpose. Table 6.3 gives some examples of each. There are other ways to describe these classifications, and Table 6.3 should be considered just as examples.

The subject of how to classify planning requirements illustrates the art of planning, rather than the science. There is no way to rigorously specify the planning strategies of organizations in general since they are

TABLE 6.2 Functions of Urban Water Systems

Water Supply	Wastewater	Stormwater
Supply acquisition	Collection	Collection
Transmission	Transmission	Damage reduction
Treatment	Treatment	Transmission
Distribution	Wastewater disposal	Disposal
Use	Residuals disposal	

TABLE 6.3 Examples of Types of Planning

Stage of Planning	Jurisdiction of Planning
Policy planning	Areawide planning
Framework planning	Urban planning
Strategic planning	Organizational planning
Implementation planning	Division or unit planning
Function of Planning	**Planning Purposes**
Water supply planning	Master planning
Wastewater planning	Preliminary design
Stormwater planning	Action planning
	Program planning

all so different. The manager needs to develop an approach that will be appropriate for the situation at hand and the problems that must be solved.

ORGANIZING URBAN WATER SYSTEMS

Since organizing is one of the basic tasks of management, its importance should be readily recognized. Organization is particularly important in facilitating decision support. There is a direct correspondence between the form of the organization, the development of information systems, and the tasks to be performed. The successful achievement of the tasks to be performed is the key to making an organization productive, and, as a result, some managers believe that task assignment is the most important mechanical function of management. In other words once you have figured out how to delegate the tasks to achieve the purpose of the overall organization, the rest is easier.

The matter of task assignment fits with the organizational form. You wouldn't send the electrical department or the computer analyst to repair a broken distribution line. As obvious as this simple example seems, any student of organizational theory knows that there is no perfect organizational form. Developing an organization is an art, and like other art forms, the considerations and skills involved are so complex that they cannot be routinized.

This section begins by describing the principles of organizational theory, with special attention to urban water organizations. Then we will turn to an analysis of how the organizational design can be improved to aid in operational productivity.

The starting point for describing organizations is usually to define the structures as line organizations, line and staff organizations, or matrix organizations. Other variations of these such as "functional organizations" or "programmatic organizations" also appear.

In recent years the behavioral aspects of organizations have received increased attention. This is an important trend since the human aspects of organizations are normally the determinants of success or failure, rather than the pure organizational form.

In spite of the lack of uniqueness of organizations, there is a unifying principle behind every effective organization with an operating mission, and that is that the principle that the path of communications, control, command, and intelligence, sometimes called C^3I by the military, should be straight and unencumbered by too many levels of supervision or staff. Figure 6.5 illustrates the concept. The concept of C^3I normally refers to military applications, but it helps clarify management procedures for water systems (11).

Consulting the literature on public works organizations reveals that most are variations of the basic structure shown in Figure 6.5, with differences that can be explained by personal preferences of management and by the historical development of the organizations. Organizations grow incrementally, and the present form of one can usually be explained by the collection of past incremental changes. They are like the proverbial picture painted by a committee of artists, reflecting different styles and lacking the brilliance of a single creator. A driving force behind the incrementalism is the aversion to reorganization that experienced managers have, an aversion that is learned from the frequent unexpected negative results of reorganization.

The staff aspect of an organization is illustrated by Figure 6.6 where the simple form of Figure 6.5 is expanded to two operational departments and staff functions are added. Figure 6.5 could be a simple combined water and wastewater utility. Figure 6.6 shows staff at three locations. It is always desirable to minimize staff to hold down overhead, but some staff functions are absolutely necessary. An example would be the lab-

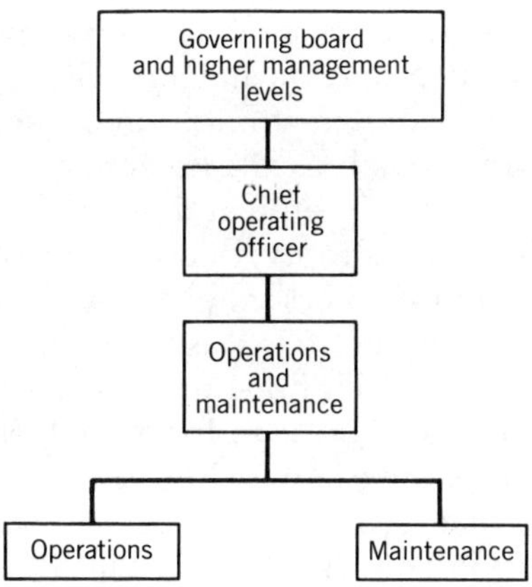

FIGURE 6.5 Basic operating organization.

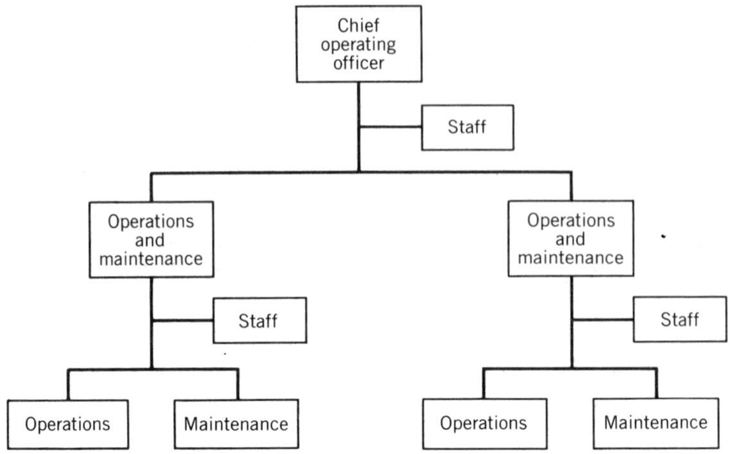

FIGURE 6.6 A two-division line and staff organization.

oratory support for a water and wastewater utility. This could be semantics, of course, since the laboratory could be put on the line as a line function, but it really is a support service.

As an actual example of the kind of organization shown in Figure 6.6, the organizational chart from the Fort Collins, Colorado, water and wastewater utility is shown in Figure 6.7 (12). In this organization three line units are designated: water treatment and supply, wastewater treatment, and water and wastewater systems. Each has line and staff activities. The laboratory functions are shown as a staff unit reporting to the general manager.

The need for coordination within organizations has created a search for better forms that facilitate communication and break down barriers. The matrix organization is an example.

Figure 6.8 shows one example of a matrix organization. In it the organization has three departments, each having a specified mission such as water supply, wastewater, and stormwater. Since the requirements of the departments are often similar, it might make sense to create special units for similar functions, such as those shown on Figure 6.8. A good example of the need would be a project design team for a multipurpose reservoir that furnished water supply, flood protection, and water quality enhancement by flow control. Personnel from each of the departments could be assigned to the project team as a second assignment, and the project team could be assigned a full-time manager.

The matrix concept can also be seen from the fact that each employee may have two or more reporting points. Figure 6.9 illustrates this concept, and the split reporting that it reveals illustrates one of the disadvantages of the matrix organization concept, that is, that it violates the principle of unity of command.

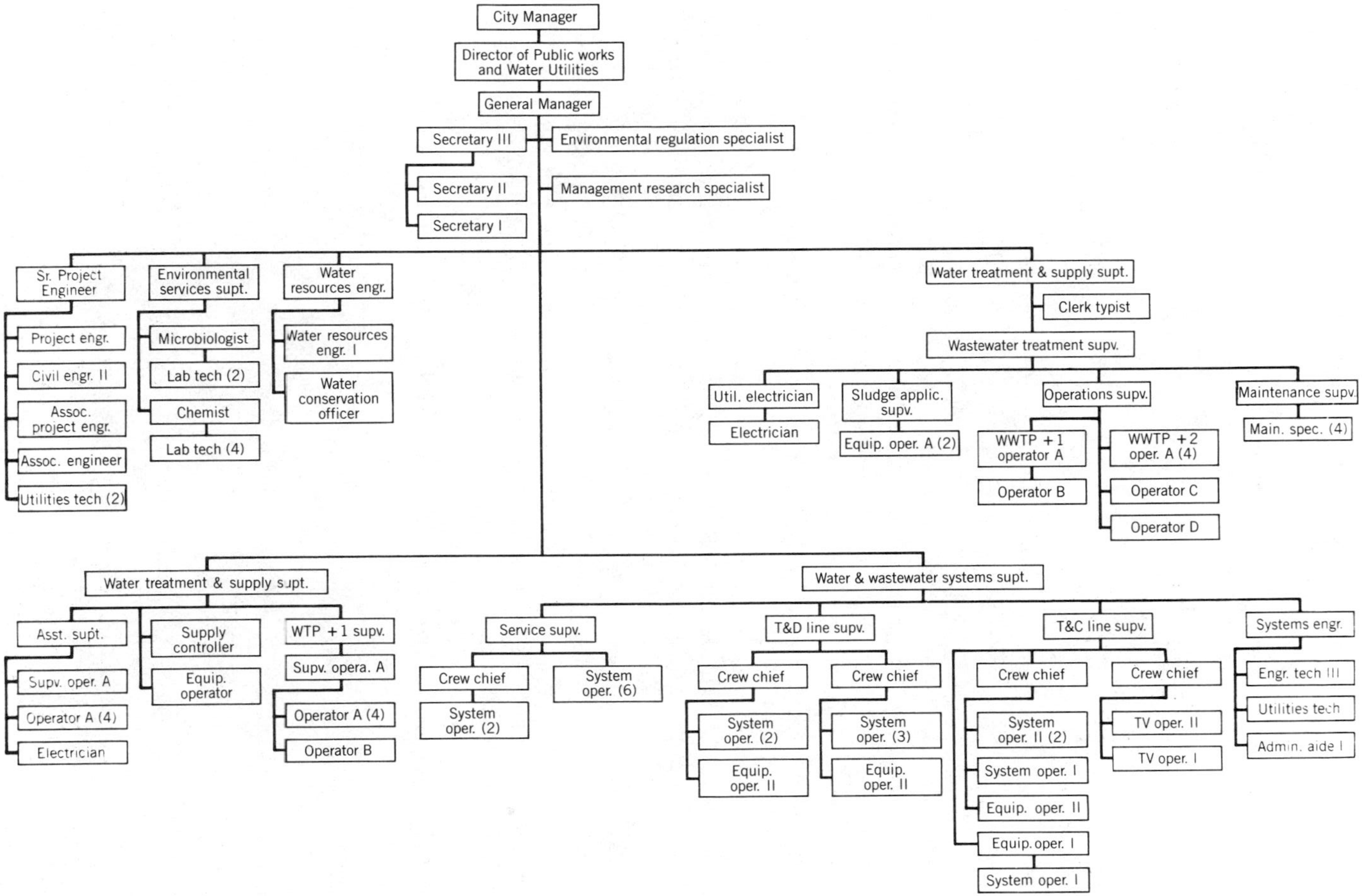

FIGURE 6.7 Water utilities organizational structure (12).

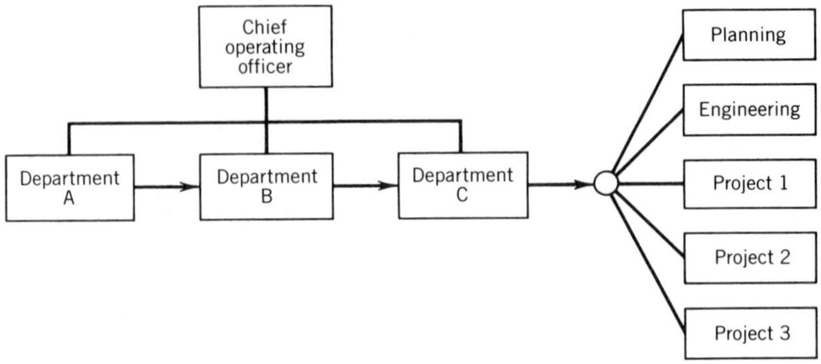

FIGURE 6.8 A matrix organization.

References on public works or utility management refer to three general types of water organizations: city departments, authorities, and private companies. Practically any water service organization can be classified as one of these three. These forms really refer to the governance of the organization and do not really affect the internal organizational form in any major way.

The internal form of water and wastewater organizations generally converges toward the form shown in Figure 6.10. The various study and design functions can be absorbed by the engineering department. Other technical staff can be dispersed around the organization, depending on the size and complexity of the agency.

Each of the operating departments would have both operating and maintenance functions, and maintenance staff would likely be dispersed throughout the subdivisions of the operating units. The third water service, stormwater, requires a simpler organization since normally there is no treatment required. Figure 6.11 is a suggested form. It shows maintenance as the main operating function. The first staff function—planning, engineering, and construction—is the main technical support area. The

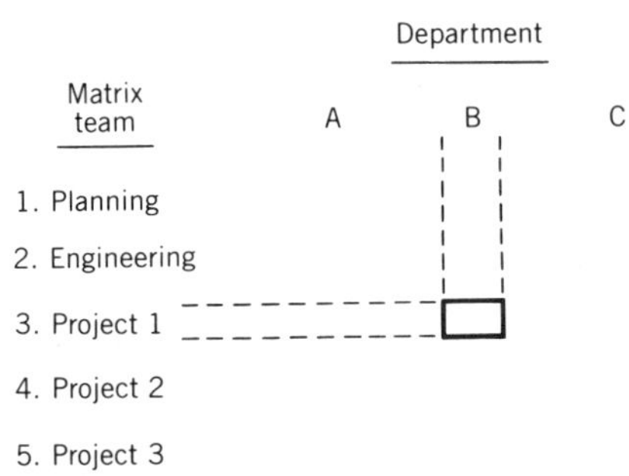

FIGURE 6.9 The matrix concept showing workers in department B assigned also to project 1.

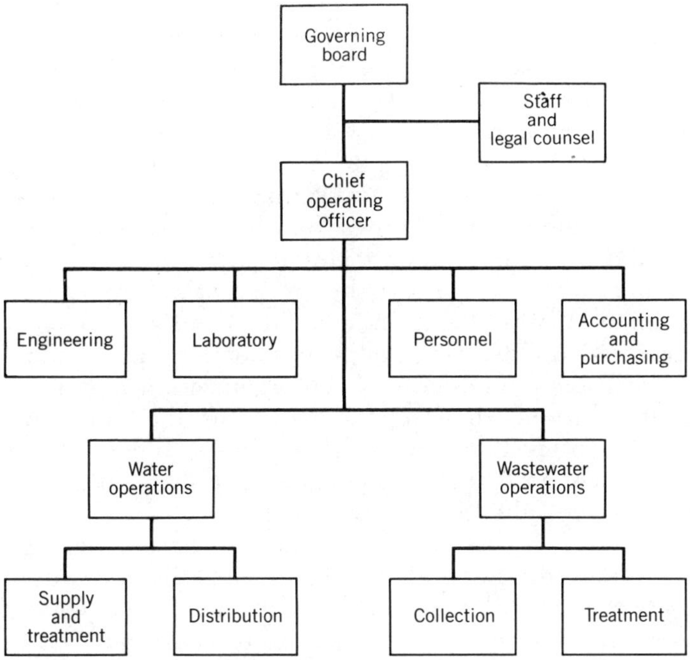

FIGURE 6.10 Generalized internal organization of water and wastewater utility.

second staff function—finance—reflects the increasing need to emphasize the development of financial support to develop and maintain stormwater systems.

The effectiveness of organizations depends on complex behavioral factors more than the structure of the organization itself. Modern industrial psychology is rich with studies of factors such as motivation, the design of work, influence and power, communications, decision processes, performance evaluation, and numerous other management concerns (3).

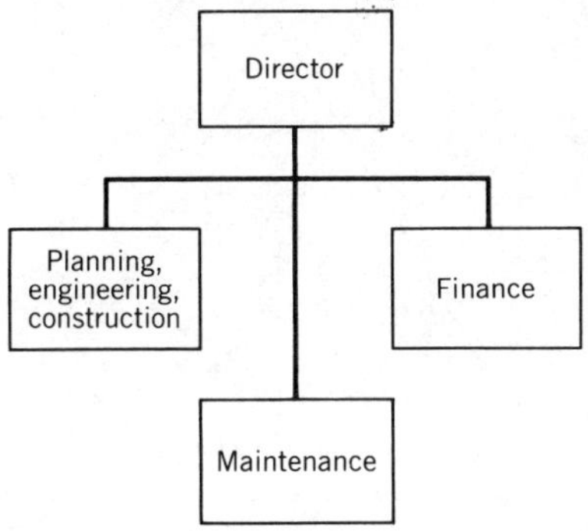

FIGURE 6.11 A stormwater organizational model.

Managers in government organizations are familiar with systems such as job classifications, position descriptions, work plans, management-by-objectives, and salary schedules, but these are only the beginning of the set of techniques necessary to make an organization effective. The design and the development of the organization are of fundamental importance in setting it up for effectiveness in the age of computers. For this reason I would like to outline the essential organizational factors normally considered by management and then discuss how they must adapt to the use of computers and information technology.

The first organizational task is the design of a properly structured working group to carry out the purpose of the organization. Organizational theory begins with the definition of purpose of the organization, and the important structural components of the organization begin with those that enable it to survive, then to prosper. That is, the bones and muscle of the organization are those units that enable it to carry out its direct mission. This is why in Figure 6.1 operations and maintenance are shown as the direct, not peripheral, organizational units.

After the structure is worked out, it is necessary to staff the organization. This means to identify the kinds and levels of skills necessary to make it work smoothly. Job titles, position descriptions, and work plans for each position are required. The setting of objectives and the design of the work for each job should be set in a hierarchial fashion that enables each worker to meet his or her needs through achievement of the job objectives, and for the combined effect of meeting the individual objectives to lead to achievement of organizational purpose and objectives. This is illustrated in Figure 6.12.

Communication and coordination are key requirements of organizational effectiveness. They are part of the C^3I model. In C^3I we have the four critical functions of the organization: command to ensure control; communications to ensure coordination; and intelligence to gather information. In modern public works organizations the emphasis will in-

FIGURE 6.12 Hierarchy of objectives.

creasingly be on the communications and intelligence, both being functions of the success of implementing computers in management.

Imagine an organization of workers in a factory such as is shown in Figure 6.13*a* being replaced by computers and robots as in Figure 6.13*b*. Such a replacement is now being partially experienced in manufacturing plants that are being automated. In this change the workers are being replaced by robots and the supervisors by computers. A new class of worker, the "computer and robot maintenance worker" will be needed. From the standpoint of efficiency the shift shown in Figure 6.13 promises a higher standard of living to humans. From the standpoint of human cost, it threatens to create a generation of lost workers, a sad social phenomenon.

The interesting thing about Figure 6.13 is that it clearly demonstrates the similarity of human organizations to computer organizations, with the common thread being the flow of information and control.

At the lowest level of Figure 6.13*b* the action element is task completion. Above that it is task assignment and supervision. Above that is more complex planning, task assignment, and supervision. These are both

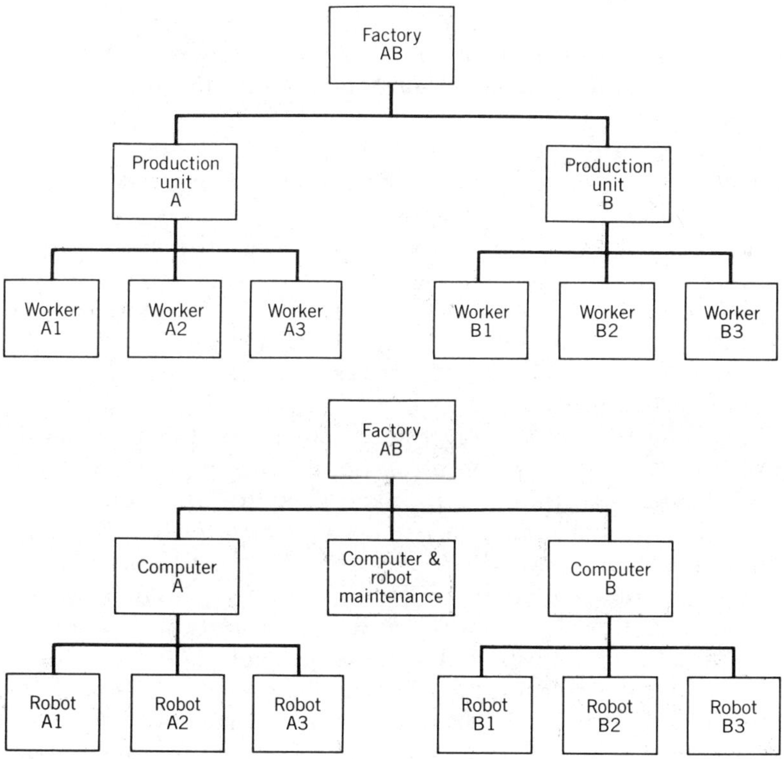

FIGURE 6.13 Production organizations.

human and computer functions. What is happening in the computer revolution is that most routine tasks and task assignments are being turned over to robots and computers. The more complex ones are being retained by humans with their infinitely more intelligent capabilities. There is research going on today about the impact of computers on organizations, and this promises to be an important issue for the future (17).

The automated production organization shown in Figure 6.13*b* can be compared to the human body. At the robot level are the body's functions: eating, working, moving, and so on. The supervisory computers are like the parts of the brain and nervous system that control individual functions such as the sensation of touch at a fingertip, and the factory supervisory level is like the human brain itself (18). The important aspect of Figure 6.13 is that it shows the need to have information flows organized properly, regardless of the stage of automation of the organization.

PROJECT MANAGEMENT

In planning and management of urban water systems, one of the most important management activities is the management of projects. The project manager role is one of solving all of the undefined and defined problems that serve as obstacles to successful completion of a project. The activity of project management encompasses the full span of time fr the conceptual brainstorming about what a project will be to the final inspection and testing. Since project management is so important, some urban water system managers have given special attention to the preparation of engineers for the role.

Fort Collins, Colorado, has developed a special manual for project managers to instruct them about the procedures necessary to successfully complete a project (19). Since the experience as a project manager is the most important ingredient to success, the manual cannot be substituted for experience on the job, but it can really help in the initial training and to standardize procedures.

In explaining the process of project management as they see it, the Fort Collins city engineer begins by describing the phases of projects as follows: the conceptual, preliminary, final design, preconstruction, construction, and close-out phases. These are described in additional detail in the next chapter, which discusses the design and construction process. In each of these phases there are certain important points where go—no go decisions are possible, and it is important that the project manager have these identified. There are also certain established levels of authority to make decisions that have to be respected. The whole process is controlled by the required reports and inspections.

The city engineer stresses the role of the project manager in making things happen, and the importance of knowing when to ask questions and when to just go ahead and proceed. In the role of making things happen, a lot of people are depending on the project manager, so the position is one with great deal of responsibility, and it is an excellent way to learn how a system works. Not surprisingly, many higher-level managers begin with experience as project managers.

Project management is such an important skill that a special association has been formed to exchange information. Called the Project Management Institute, it attracts managers and engineers with responsibilities such as capital project construction management.

FINANCIAL MANAGEMENT IN URBAN WATER SYSTEMS

Inadequate budgets are the cause of many operational problems in urban water systems, and the cause for inadequate budgets is inadequate management at executive levels. In periods of tight money, and most water managers probably think that is all the time, financial managers ascend to higher levels of management. This is because their skills are so much in demand in that period. In times of plenty, financial managers are needed to guard against waste and corruption. Financial management is one of the most useful skills for water managers in general, especially those at the upper levels of management.

The purpose of this section is to describe the most important principles of financial management as it is needed by executives. The section is not intended to provide the kind of detail that an accountant would find useful, but to provide a perspective for the engineer or manager who needs an overview of the principles of financial management for urban water systems.

Financial management can be viewed much as water management. There is a supply, a demand, and a need to keep an inventory. Planning and control is done through budgeting and financial reporting. We will describe these techniques as well as sources of revenue and some special topics including the inflationary environment for financial management today. The emergence of "privatization" as a new possibility to aid in financial management and to provide new possibilities for urban water management is also covered in this section.

During the recent period of increasing attention to water management, we passed through a period of experiment in financial management. Some utilities practiced tight management in which services paid for themselves, but many of them practiced a looser style of financial management in which various subsidies were relied upon. The active role of the federal government, particularly in construction grants for wastewater treatment, was considered by some to be a principal cause of the lack of attention to basic strength in financial management. Now the trend seems to be reversing, and many utilities practice self-finance and the enterprise approach to management or the complete reliance on charges to finance operations. Going even further, there is more attention today on privatization of services, following the theory that services can be offered more efficiently and at lower cost by the private sector.

Self-finance is the more widespread practice. It simply means that charges are adjusted to cover costs, and the customer is expected to bear the full costs. Debates go on about "equity," and these debates require considerable judgment about values, judgments which must be made through the political process. Privatization has been very common among water companies, although more have recently gone public than private. The trend in earlier years was to offer water supply through private companies, much the same way that telephone, electricity, and gas are offered privately. The water companies are regulated monopolies.

Subsidies have been available for some water programs. The most abundant source in recent years was the wastewater construction grants program initiated through the Clean Water Act of 1972. Expenditures from this source were about $4.5 billion per year in the heyday of the program. In the early 1980s the annual outlays were reduced to about $2.5 billion, and inflation had considerably eroded the buying power of the grants. Most municipalities were getting grants for 75% of construction costs. This subsidy was reduced to 55% in 1984, reflecting the trend to move toward self-reliance in wastewater treatment. Subsidies have come from other sources as well. Some state governments were assisting water and wastewater utilities, particularly when there were public health or economic development objectives involved.

In the early 1980s there was considerable discussion of privatization, but there may be more talk than action. Some of the incentives depend on the vagaries of tax law, and this period has been characterized by much political debate about reform in tax law. This has the effect of creating considerable uncertainty in privatization plans.

The budgets of an agency have the effect of displaying its programs. The operating budget shows the plans to operate and maintain the agency. A manager should be able to examine the operating budget, the organization chart, and the job descriptions and tell what will happen during the year. By the same token the performance of the organization should be reviewable by comparing the output of the organization with its potential as revealed by these documents.

The Enterprise Principle

A recent speech by the mayor of Tampa, Florida, describes their approach to managing their infrastructure. He stated that the city was becoming a "holding company" where all of the services were becoming self-sufficient in revenues, and the reliance on the general tax base was diminishing. This is another way to describe the increasing enchantment in the United States with the "enterprise" approach to offering public services. This is described in detail in a publication offered by the Government Finance Research Center (20).

The general concept is that services should be self-supporting and that they should charge users according to the benefits they receive from the services. The use of pricing through user charges is the basis for the control of the allocation of the services and for raising revenue. The equity issue is central to the philosophy, that the charging schemes should be fair.

The enterprise principle seems to be behind many of the new suggestions for managing utilities. They should be run like businesses, and they should pay their own ways. Financial management principles follow the trend toward the enterprise approach, since if the decision is made that a service will be self-supporting, then revenue generation approaches and the financial control function are brought under the control of the manager rather than the political process. This is the major advantage of the enterprise approach for managers—that they gain the control they need to make decisions and to implement innovations in financial management. Still, there is a built-in political component in water management, as well as urban management, and the desires of the local population must be factored into decision making. Thus the financial management process becomes a political-administrative approach. That is what public administration is in general, so the change here is toward the practice of public administration and away from management by politics alone.

Financial Analysis

The urban water manager must become an astute financial analyst to get the best deal among all of the approaches that are necessary and available today. This means that the subject of financial management should not be foreign to the manager, and that topics such as the bond market, interest rates, inflation, and rate structures must be familiar. Tax law is increasingly necessary also, since the options available to the manager may include access to private capital or services, and the tax implications can well be an important factor.

Financial analysis is really a subset of the general topic of planning. We might call it financial planning. The Government Finance Research Center prepared a special brochure about it to aid those involved in wastewater management planning (21), and it presents the elements shown in Figure 6.14. These elements are discussed in some detail later in this chapter since they are all critical to the success of the venture.

In the case of institutional analysis, the example given in the brochure shows some of the institutional characteristics that must be present to have success. These are shown in Figure 6.15.

The cost analysis would consider the construction costs, the operating and maintenance costs, and other costs such as the cost of regulatory

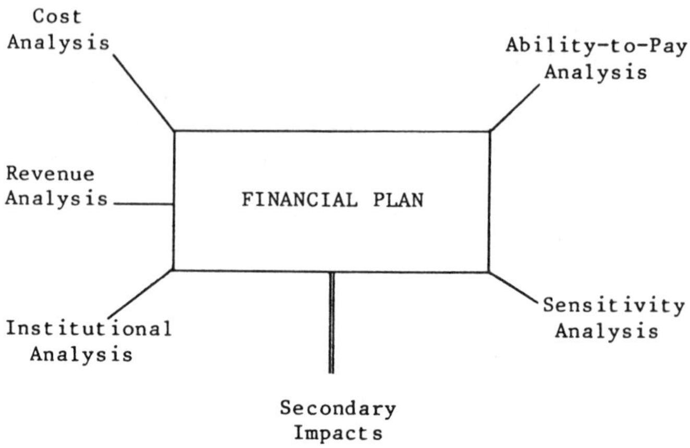

FIGURE 6.14 Example of the composition and derivation of a financial plan (21).

programs and the continuation of planning. The revenue analysis considers the sources that will be available, including the federal grants, which are the subject of the brochure. The ability-to-pay analysis is one that some planners are not accustomed to and must be done with some guidelines in mind. In the case of the brochure, two examples are presented.

In the first case, where a 75% federal grant is available, the average annual cost of the system comes out to be 0.6% of the median household income of the community. This is deemed to be acceptable since it falls

FIGURE 6.15 Example of analysis of institutional capabilities (21).

Town	County	Special district	State agency	Private entity	Partial list of institutional characteristics to be considered
X	X	X			1. Politically accountable at the local level.
X	X	X		X	2. Ability to assign responsibility for day-to-day operations.
X			X		3. Commitment to program goals.
X	X	X	X	X	4. Perpetual (continuous) in nature.
X	X	X	X	X	5. Administrative accountability.
X	X	X			6. Authority to incur short-term and long-term indebtedness.
X	X	X			7. Ability to retire existing debt.
X	X	X	X		8. Authority to accept and utilize grants.
		X	X		9. Capacity to ensure proportional cost sharing.
X	X				10. Capacity to control local taxes and user charges.
X	X	X			11. Enforcement power.

within some EPA guidelines that require that the cost be under 1% of median household income (MHI) when the MHI is less than $10,000, less than 1.5% when MHI is between $10,000 and $17,000, and 2% when MHI is greater than $17,000. In the case of the sensitivity analysis, the example showed that the cost would be 1.3% if there was no federal grant available, and this violated the guidelines. The secondary impacts would look at three categories: economic, social, and environmental.

Financial Control and Reporting

Financial management has two very important purposes for the water manager. First, it enables the success of the venture by providing the funds that are necessary to plan, build, operate, and maintain the system. Second, it provides the information and the "handles" necessary to control the organization and the system itself. To accomplish these purposes, two features of financial management deserve initial attention: the control and the reporting features. After we discuss these, we will discuss the tools that make financing and control possible.

Control is accomplished through accounting and auditing, with its reports, checks, and balances. The reports furnish the management information needed to make adjustments and to report to boards of directors, the customers, and to regulatory agencies. Utilities and enterprise funds should use the "accrual" method of accounting, according to the National Council for Governmental Accounting (20). This means that expenses and revenues are credited when they occur, and that the financial reports will reflect a picture of the actual financial health at all times, allowing for appropriate adjustments when unexpected events occur. The reports that are usually prescribed are the balance sheet, the statement of revenues, expenses and changes in retained earnings, and the statement of change in financial position. There are other reports that must be dealt with by the manager, of course, especially those that have to do with management of the budget.

Financial control infers more than accounting. The built-in checks of the organization for control of purchasing, the fixed assets records, inventory, and the control of hiring through staff activities all support the management of the financial health of the organization.

One presentation of these approaches used a three-way classification of the financial management responsibilities: liability management, assets management, and fund accounting and management. Both liability and asset management require long- and short-term viewpoints (22). Funds management means that we must have a budget and it must be managed. This is the next subject we turn to.

Budgeting

Everyone has some idea about what budgeting is, but most have no idea of how important it is in the management of a public enterprise, nor how much there is to making it a success. The control of the budget can be said to be the most important single element in managing an organization; at least it is the point where most power and influence are exercised.

The two approaches to budgeting are the annual operating budget and the longer-term capital budget. Both are critical to the success of a water enterprise. The items that are included in each cannot be summarized easily, but in general, the operating budget takes care of regularly occurring expenses such as salaries, expendable materials, and operational expenses, and the capital budget is linked with long-term plant and equipment and the physical water system itself. In terms of everyday concepts the operating budget would include the groceries and phone bill, and the family home and automobile purchases would be financed from the capital budget.

The use of budgets as management tools has two distinctly important lines. First, the planning of the budget sets the planning for what the organization does and when things can happen. Second, the use of the budget as a control device for the parts of the organization reveals how closely the organization is adhering to the plan.

From the standpoint of urban water systems, the greatest criticism of operating budgets has been that they have been inadequate for operations and maintenance. As a result, the systems have become broken down and need considerable upgrading and rehabilitation. This is an area where management needs to pay attention: to prepare budgets that are adequate to do the kind of job that needs doing, whatever it takes. In urban water management it is expedient for political leaders to defer maintenance since it has no political visibility. Responsible management, however, will be thinking about the long-term good of the organization and the public welfare, and the result should be adequate maintenance.

The operating budget must be planned on a definite cycle. In the year that the budget is spent, the funds that are available are those that were approved during the previous fiscal year. The approval in the previous fiscal year may mean that the planning for those funds took place even earlier. Figure 6.16 shows a budget exercise that requires 3 years to plan, prepare, and spend the budget. In the budget planning year the organization might be required to submit some estimates of funds needed, say for a multiyear period. In the budget preparation year there would be detailed planning, leading up to the approval by the board of directors or governing board of the next year's budget. It is easy to see that in any given year there would be at least three budget years in the manager's life: one for planning, another for approval, and the third for operating.

The federal budget is an example of how complex budgeting can get to be. Referring to Figure 6.16, the year of budget planning would be an intense one where agencies battled out the right to even submit budget requests. This process culminates with review by the Office of Management and Budget of the requests, and often results in fierce battles as to

Budget	Budget	Budget
planning	preparation	year

FIGURE 6.16 A three-year budget phase.

whether the budget office can refuse to allow the budget requests to go ahead to the president. All of this culminates with the president's budget message to Congress in early February, which would be in the "budget preparation" year, according to Figure 6.16. This is only the beginning of the struggles with Congress, which consume the rest of the time until the beginning of the federal fiscal year on October 1.

The budget preparation should follow some variation of a planning–programming–budgeting format. This means that the planning comes first, then the budgets have to be programmed, and then budgeted. Budgeting, then, should conform to the overall plans for the unit's activities, considering the goals, objectives, and priorities of the organization.

Budgets follow the organizational chart. Each "budget center" will have a budget, and some of the conflict over organizational change is related to the struggles over money and power that come from being in charge of a budget center. The budgets from the smallest budget centers are compiled into larger and larger ones until the compilation reaches that of the entire organization. This is the budget that goes before the board of directors for approval.

A good reference on the procedures and other aspects of operating budgeting would be any of several excellent texts on local government finances, especially the one by Lennox and Hillhouse (23). This text has a full chapter devoted to the operating budget process.

The politics of budgeting is another subject that is of concern to managers. After being in an organization for a few years, the manager becomes aware of the political nature of budgeting, especially in terms of achieving status and power within a public organization. This refers to the competitive aspects of budgeting. On a larger scale, say for the U.S. budget, this problem becomes political on a larger scale. Budget process scholars may be interested in studying this subject further, and a good reference is by Wildavsky (24).

Whereas the main criticism of operating budgets has been that they have allowed for deferring maintenance, the criticism of capital budgets is that they have been nonexistent. This means that urban water systems have been overloaded and not upgraded for today's demands, and that the future generations of users have been burdened with the debt that was incurred by the present and the past. This problem can be overcome by effective approaches to capital budgeting.

The capital budget should also be tied to a plan, but rather than a program plan it should be a facilities plan. There should be master development plans for all aspects of the physical facilities of an urban water system, and the capital requirements should be programmed accordingly. This means that capital budgets should be planned for several years in advance, say for a 5-year period.

For the most part urban water systems are forbidden to charge present customers for the costs of future growth, so no reserves for capital expansion are accumulated. Growth is normally finance through debt, and this created a lively market for municipal bonds. We will discuss debt financing more later in the chapter.

The important thing to remember about capital budgeting is that it should be linked to valid expansion plans. If water systems are run like businesses, it will be fairly obvious which capital facilities are needed. If they are subjected to changing government regulations and uncertainties about future demands, then the process of capital budgeting is made more difficult. Even something that might seem like a blessing, such as a federal grant, may distort this process.

Revenue Sources

The operating and capital budgets have to be financed from logical sources, and providing enough financial capability to operate and maintain the system is one of the most important jobs of the top management. Sources that are normally used in urban water systems are rates and user charges, property and other taxes, system development charges, bond proceeds, loans, and grants.

The operating budget has to be renewed each year, and it is logical that it should be financed from recurring revenues. In past years the property tax was used to finance operating budgets in some utilities, but with the "new" emphasis on enterprise budgets, the trend is definitely away from tax revenues to finance urban water services and toward user charges. Emphasis is placed on "new" above since some well-managed utilities have never become reliant on tax revenues, but have been self-supporting for a long time.

The property tax is called an ad valorem tax since it is calculated according to the value of the property. Different states have different formulas. The Colorado formula, for example, provided at one time that assessed valuations would be one-third of market values and that taxes would be levied on the assessed values. A water district that had a 10-mill levy for service on some 10,000 properties with an average value of $75,000 each would then collect

$$\text{Revenue} = 10/1000 \times (75{,}000/3) \times 10{,}000 = \$2.5 \text{ million}$$

These funds could then be used to help operate the district. This would be a revenue of $250 per property per year, which would be a reasonable service charge for water service if that route for financing was chosen instead of taxes.

Sometimes other forms of taxes have been used for water system operation or construction. Income taxes are fairly common as well as sales taxes. Normally these would be used on a one-time basis for capital improvements rather than to finance continued operations. There are always exceptions to every generalization, of course. An example of the use of these taxes might be to use the revenues for the construction of capital improvements that were caused by growth that benefited an entire community.

Rates and user charges deserve a separate discussion since they are so important, and the next full section will be devoted to them.

System development charges have become an increasingly important part of overall financing strategies, since they provide a way to isolate the cost to serve a particular segment of a system and to levy charges for it. In effect, they allow new users to "buy into" an existing system by paying their fair share of it. As a simple example, consider the community that has a water supply system already built, complete with adequate capacity for new developments. When the new development comes in, the fair cost of serving it is calculated and it will be charged a system development fee to pay for its share of the system. This fee would, of course, be passed on to the purchasers of developed property in the form of higher costs for land or they would be required to pay the fee themselves. Ultimately, it passes the cost of the infrastructure on to the property owner.

Grants have been an important part of the overall strategy of paying for urban water systems. The best known grant program is the wastewater construction grants program, which has financed some $40 billion in

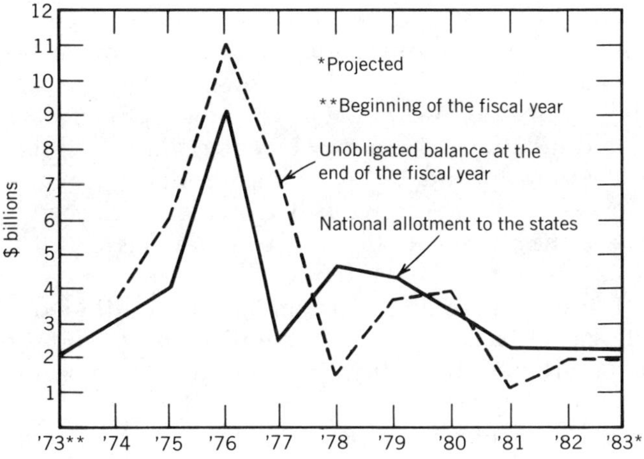

FIGURE 6.17 Fiscal management in the municipal construction grants program (25).

treatment facilities over the past 10–15 years. This program, along with most others, is in decline now, and it appears that the emphasis will be on the enterprise concept and self-finance for a long time. Some of the other water-related grants have been with the Farmer's Home Administration, the Economic Development Administration, and the Department of Housing and Urban Development. There have also been a number of state programs.

The importance of taking a realistic view of the availability of government grants can be noted from an inpection of Figure 6.17, which shows the history of funding of the construction grants program (25). The "wave" of funding looks like a flood hydrograph as it moves through the system. This kind of distortion of ways of doing business seems to be a permanent part of our political system, but it makes financial planning a real challenge.

Rates and User Charges

The setting of rates and user charges has become an important problem for all types of urban water systems, and as a result, a great deal has been written about it. The model for water supply utilities has evolved through the experience of the water industry itself and is generally reflected in publications of the AWWA (26–28). The procedures for setting charges for wastewater are not as well established and have been driven by the requirements of the EPA for user charge systems that have to be in place prior to the awarding of a federal grant. Stormwater systems are the newest water service to use user charges as a financing vehicle, and they are just getting started in a few locations through the stormwater utility concept and user charges based on parameters such as the amount of paved surface for a property.

According to AWWA procedures, the rate-setting process consists of three parts. First is the determination of revenue requirements, or the cost to operate the system. This will include cost estimates for a few years into the future, say 5–10 years, with appropriate allowances for depreciation and inflation. There may be some funds built in for capital improvements and major repairs, as well as for debt retirement that has been made necessary by the previous borrowing. The second part is the determination of the cost of service by customer classes. The third part is the actual design of the rate structure itself (27).

The specification of revenue requirements is conditioned on whether a "utility" or "cash" basis of determinations is used. The difference is basically that of whether the utility's rates are regulated or not, such that a rate "base" and rate of return are specified. If they are, the utility basis is used; if not, the cash basis is adequate. The rate base is the value of the property on which the utility is entitled to earn a return. The revenue requirements on the utility basis is made up of the operation and maintenance expense, the depreciation, taxes, and a return on the rate base. Since publically owned utilities do not make a profit as such, they can generally take a more flexible approach to the setting of revenue requirements, but with the problems with restoring infrastructure, it is important that realistic practices be followed and that the full cost of doing business be recovered as revenue.

In the cash basis of doing business, the revenue requirements are easier to explain since they are much like those of other city departments. According to *AWWA Manual M1,* they consist of basic items—such as operation and maintenance, debt service, plant replacements, and normal improvements—and optional items such as payments in lieu of taxes, appropriations or reserves for major improvements, and contributions to other city departments such as free water (27). The basic items will be found to be the same in many utilities, and there will be some wide variation in the optional items.

In determining the cost of service, several points of view can be introduced. The AWWA rates and management manuals specify two basic approaches: the commodity-demand method and the base-extra capacity method (27, 28). The difference between these two methods is essentially the way to classify the costs. In the commodity-demand method costs are subdivided into the following three categories: demand costs, commodity costs, and customer costs. Customer costs are the costs incurred as a result of serving the customers. They include items such as meter reading and billing. The commodity costs are those that are incurred in the function of producing and selling the water. Items such as power and chemicals would be included here. Demand costs are incurred in the course of meeting the peak rates of demand and can include capital items such as extra capacity and operating costs that are incurred in the course of meeting peak demands.

In the base-extra capacity method the allocation of cost is again into three categories, but in this case the three are a little different. The customer cost category is the same, but there is a little difference in the base and extra capacity categories. The category of base costs includes the costs associated with producing the average demands, and the extra capacity category includes the costs associated with meeting the peaks.

AWWA's rates manual states that the difference between the two methods is in the costs that go into the "base" category (27). They would include the commodity costs plus some of the demand category that might be necessary to meet the average annual demands. This explanation is not clear; the real differences between the two methods would only become clear after some practice in using them, and there would be some discretion for the analyst in choosing the categories.

The rate-making procedures have been receiving a test in my own community recently. Fort Collins has been trying to decide how to make the charge system more equitable and, at the same time, how to be more conserving and promote green lawns in the city on a reasonable basis. The study continues as this is written, and it began with an analysis of the rate structure. The engineer followed existing procedures and used a general approach such as that shown on Figure 6.18 (29). It will be recognized as a variation of the basic AWWA procedures already discussed.

In the case of wastewater systems, procedures for collecting user charges are not so well worked out as in water supply systems. The major recent influence has been federal regulations, but in prior years procedures were being worked out on a city-by-city basis. A report prepared in 1973 by a joint committee of three professional associations described the state of the practice at that time (30). In effect, the practice envisioned in this report is a split of costs among property taxes and user fees. The procedures for allocation between property and users are well developed in the report, and they reflect the ad hoc nature of wastewater charges that was in effect at the time the report was written. For example, the report refers to any collection of storm drainage that enters the system

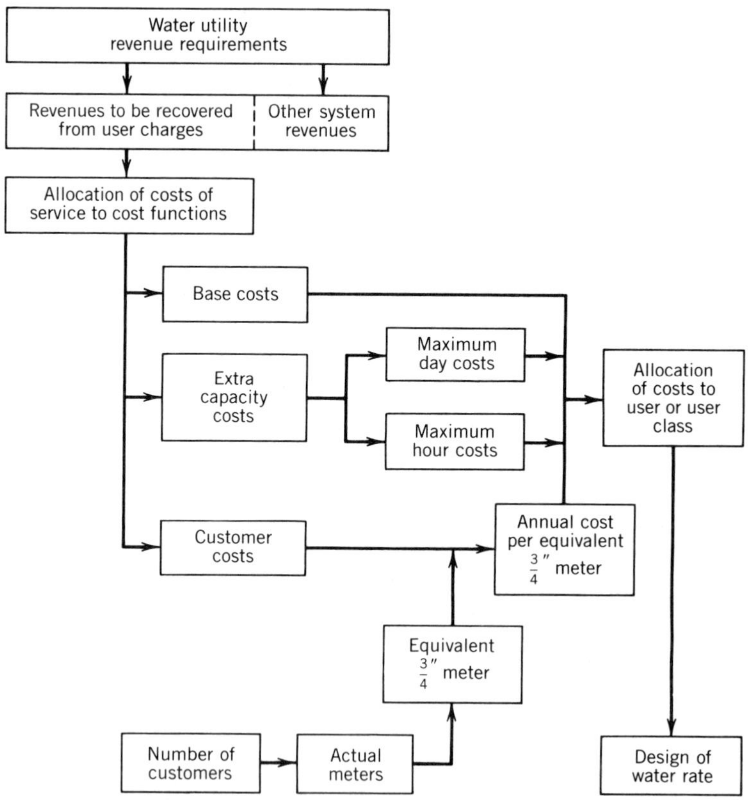

FIGURE 6.18 Schematic development of water rates (29).

to be chargeable against property, and therefore to the property tax, whereas the thinking today in some quarters is that it is possible to charge users for storm drainage. The influence of EPA in the charging schemes for wastewater seems to have resulted in there being three approaches that will be acceptable. One, based on volume of wastewater, would be appropriate for communities where the strength of the wastewater would not vary much among the users. Where there was considerable variation, a second approach would be used, with a surcharge for the strength of wastewater. A third model is described that would consider both quantity and quality of wastewater, and it is considered to be the most complex of the three (20).

In the case of the industrial component of wastewater, there was considerable controversy about the original provisions of the Clean Water Act that provided for industrial "cost recovery," meaning that industries would pay back the parts of the federal grants that went to pay for their wastewater above certain minimums. There were successive moratoriums and changes in the approach, and in 1980 the industrial cost recovery requirement was repealed, with a new provision, the "industrial cost exclusion" implemented. In effect this new provision disallows the federal grant for that part of the public wastewater plants with more than 50,000 gallons of sanitary waste and presents a method for calculating the federal share. There are still many controversial elements to this industrial component of treatment, and each community must approach its own situation with a view to the industries it has and their impact on the system. EPA made a major report to Congress on the impact of the industrial cost exclusion program (31). Another reference that may be helpful in establishing wastewater rate structures is the Texas Water Utilities report, which describes the principles involved and presents a sample wastewater charge ordinance (32).

In the case of the stormwater service, rate making is in its infancy and is mostly based on parameters such as lot size and runoff coefficient. In that sense we might consider the charges somewhat related to the property value, so it is still somewhat like a property tax. The future of stormwater user charges is still to be worked out. A good reference is by Cyre (33).

In general, the rate-setting process is an integral approach in financial management, and much is known about how to do it; the variations come in how it is applied in specific situations. The general approach given for water supply can be followed for wastewater and stormwater also; that is, to know the revenue requirements, the cost of service allocation, and the general design of the rate structure.

Cost Allocation

In the rate-setting process there is included a generalized need for allocating costs, but cost allocation goes beyond the concept of the rate, and extends to political questions between levels of government and between the government and private parties, about how to allocate and share costs. In recent years this has become a very important topic as the level of government subsidies has continued to fall. This section presents some of the generalized fundamentals of cost allocation.

Basically, cost allocation means to find ways to assess costs in a fair way in proportion to how different parties benefit from a project or a service. Most projects and service have costs that are necessary to run the service in general, sometimes called "joint costs," and costs that are clearly identifiable with beneficiaries, sometimes called "separable costs."

In multipurpose water resources projects, the use of the above terms has led to the frequent use of a method called the "separable costs–remaining benefits methods" of cost allocation for projects. It is described in water resources texts such as that by Goodman (34). Basically, the method presents a way to separate joint and separable costs and allocate the separable costs according to benefits received. Usually, the parties are distinctly identifiable levels of government or industries.

The case with cost allocation to rate payers is different in the sense that so many different parties are involved. To overcome this difficulty, they are divided into customer classes for the assessment of costs, both on a one-time basis and for recurring expenses.

There are many examples we could give of cost allocation, three of which are often encountered in urban water management. The first is the case of allocation of costs among levels of government to pay for a multipurpose water project. Take, for example, a Corps of Engineers multipurpose reservoir located near some urban areas in the eastern part of the United States. The project might have, for example, three purposes: water supply, flood control, and hydropower. The federal government has been debating its policies for cost sharing of these purposes for several years, and it is not settled, so the example will be hypothetical in terms of policies. We might say that the hydropower will be produced by the government and sold to utilities on a wholesale basis, so the separable costs of this purpose are entirely financed by user fees. The flood control might be jointly financed between the federal and state governments, with appropriate allocation by negotiation. The water supplies might be financed through sale of the water to local governments through long-term contracts. In this example, then, the allocation of costs is done mostly through the political and negotiation process.

In the next example the project might be a drainage and flood control project that is necessary to develop a particular part of an urban area. There will be some land developers who will benefit through the improvements to their property, but some of the benefit will accrue to the public at large. The city will decide on the allocation of costs through negotiation with the developers and with reference to the policies and goals of the city administration. This situation can benefit from more analysis of costs and benefits than it usually receives. The state of the art of analyzing problems of this kind has been improving through the technique of financial impact analysis.

In a final example the allocation will be according to customer classes, say for a water and wastewater utility. The water cost allocation would be according to principles such as those found in the AWWA manual discussed earlier, and the wastewater allocation would have to carefully consider the impact of industries. A variation of this that has not been used very much is that of zonal allocation of costs. In other words, if it is more expensive to serve some zones of a city than others, the appropriate rates would be assessed. There are, however, variations in charges between central cities and suburban areas, and this is often the source of distress in water management. When the local considerations of rate needs are added to the local political situation, the result is often complex total rate structures and rules.

It would be nice if cost allocation could be organized so that no negotiations were ever necessary; that would greatly simplify water management. However, it will never happen since there are too many actors and profit margins involved.

Debt Financing

The use of debt financing is becoming more and more popular with the increase in attention to the concept of the enterprise management approach. The traditional view is one of the use of general obligation bonds and revenue bonds, but the trends are toward different methods that will adapt to the debt markets of today.

In fact, the debt market is not nearly as simple as it once was. With the increase in interest rates of the late 1970's, there was a strong move toward what is known as "creative financing." Much of this is not new, but different ways to package the same concepts. It can be confusing, however, since so many new terms are in use. Terms such as *commercial paper, leasing, bonds with warrants, floating-rate instruments,* and *put option bonds* are mentioned in a recent "primer" on creative financing for public entities (35).

Regardless of the trends toward creative financing, the emphasis seems still to be toward an increasing tendency to use revenue-secured debt as the vehicle to obtain capital financing. Vaughan reports that revenue debt is now nearly three times general obligation debt (2). This trend seems perfectly consistent with the increasing emphasis on enterprise management. The overall picture of debt is one that requires attention to our capacity to repay without difficulties, and that subject ought to be of concern to the boards and managers of public enterprises.

In effect, the approach to debt financing for managers is to determine how much money is needed and when, and to find the best deal for it. This will require professional advice, of course. There are many firms today vying for the right and business of providing this advice. The repayment of the debt will require funds to be allocated from revenues, usually from the operating budget, to retire the bonds.

The subject of debt financing is directly tied to capital budgeting and decision making. This is one of the most important subjects of financial management in all organizations, even the U.S. government. In discussing the topic, it is necessary to consider additional topics such as investments, inflation, interest rates, and currency exchange rates, especially when projects are financed in other than the home countries' currency. Since inflation is so important, a special appendix on cost indexes has been included (Appendix 1).

Privatization

In any consideration of financial management today, the subject of privatization must be considered. There are many apparent advantages, and with the climate for financing assistance from the federal government diminishing, privatization looks better all the time.

Privatization simply means for the public entity to turn over the operation of all or some part of an enterprise to a private company. The easiest way to visualize it is to think about the nationalization of industries in some countries. In effect, they take private businesses and begin to operate them by the government. Privatization is just the opposite: the public functions are taken over by the private firm.

In the case of water supply services, many are familiar with private companies already since there are so many private water companies, particularly in the Northeast. They have worked fine, and there have been no apparent problems with this provision for private enterprise for urban water service. In fact, during the Great Depression, water companies parently were regarded as among the safest of investments and held their values quite well (36).

The subject of discussion in the past few years has been how to privatize wastewater treatment plants as a way to solve the financing dilemma since the federal government is getting out of the construction grants business. There have been many articles, reports, and meetings on the concept in the past several years, and some experiments are beginning; thus, it looks like privatization is a concept that is more than a flash in the pan.

Privatization can be implemented in several ways, and they can be classified according to how facilities are financed and how they are operated. The following classification is my own, in the simplest terms possible, to avoid all the financing jargon which is difficult to understand:

1. Complete private ownership and management. In this arrangement the private company or consortium would build or buy the facility and operate it under a service contract with the city.
2. Private financing of construction and lease back to the city. In this arrangement the capitalization is by the private investor, and the operation is by the city.
3. Contract operations by the private firm. This situation would have the complete responsibility for operations assumed by the private firm.

In effect, these arrangements make up a spectrum of public–private involvement in projects and systems. This spectrum was illustrated in a recent publication of the Association of State and Interstate Water Pollution Control Administrators (37). In Figure 6.19 this spectrum is illustrated by a graph of the level of concern or involvement in management versus the degree of private involvement.

There are many variations of the above scenarios, and many different legal arrangements. The first and third of the above would probably have the common aspect that the city would continue to handle billing and pay the private operator a service fee for the management of the facilities. The private firm would assume the responsibility for the quality of the service.

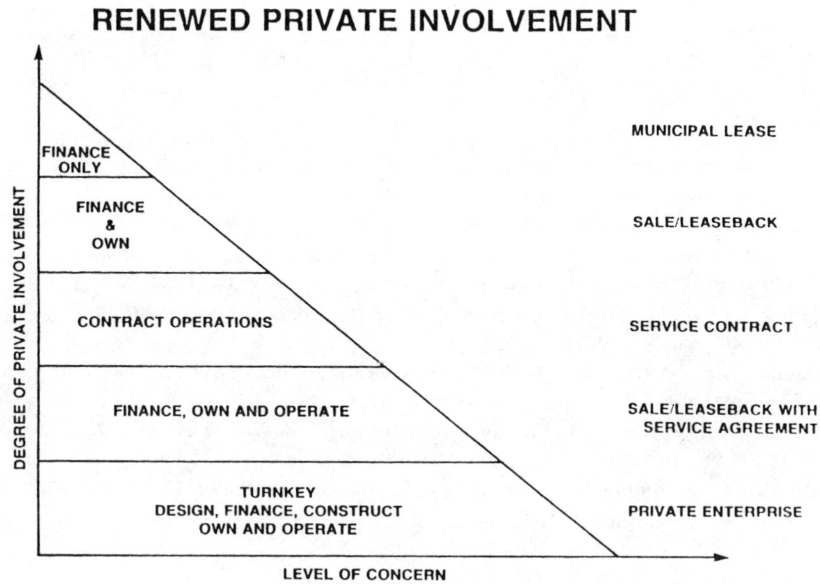

FIGURE 6.19 Renewed private involvement (37).

There are many obvious advantages to privatization. In the case of financing, it takes the load off the municipality and puts it with the private capital markets. Private firms have inherent advantages over public enterprises in being able to operate efficiently; however, some public enterprises can operate very efficiently when they are well managed. The private firms can qualify for tax benefits and depreciation credits that are not available to the public entity. The private entity might be able to qualify for industrial development bonds and get lower than market interest rates for its capital financing. If the private firm operates many facilities, it might be able to gain some economies of scale in training and purchasing that would not be available to the public enterprise. These potential advantages should translate into lower rates for customers, but the experience base to verify that is not in place for wastewater yet, and I am not aware of lower rates among private water companies. What we can be sure of, however, is that there will be considerable interest among the private sector in privatization since there may be business opportunities. This has been borne out by the number of bidders in some early calls for bids and by the type of advertising that is now appearing offering services in contract operations.

We have discussed how to privatize wastewater plants, but there is no reason that the subject of privatization cannot be extended to all components of urban water systems. The future will tell how successful the various approaches will be.

REFERENCES

1. Roger J. Vaughan and P. Pollard, *Rebuilding America*, Vol 1, *Planning and Managing Public Works in the 1980's*, Council of State Planning Agencies, Washington, 1984.
2. Roger J. Vaughan, *Rebuilding America*, Vol 2, *Financing Public Works in the 1980s*, Council of State Planning Agencies, Washingt, 1983.
3. David R. Hampton, C. E. Summer, and R. A. Weber, *Organizational Behavior and the Practice of Management*, Scott, Foresman, Glenview, IL, 1984.
4. James H. Greene, *Operations Management: Productivity and Profit*, Reston Publishing, Reston, VA, 1984.
5. Paul E. Moody, *Decision Making: Proven Methods for Better Decisions*, McGraw-Hill, New York, 1983.
6. Robert D. Behn and J. W. Vaupel, *Quick Analysis for Busy Decision Makers*, Basic Books, New York, 1982.
7. Paul Mali, *Improving Total Productivity*, Wiley-Interscience, New York, 1978.
8. Thomas E. Taylor and J. Renfroe, A Management System for Cost Control, *Water Engineering Management*, May 1984.
9. Paul Hersch, Productivity Tips for Utility Manager, Management in Action, *Water Engineering Management*, March 1985.
10. Neil S. Grigg, *Water Resources Planning*, McGraw-Hill, New York, 1985.

11. Charles A. Zraket, Strategic Command, Control, Communications and Intelligence, *Science,* June 1984.
12. City of Fort Collins, Colorado, Operating Report, Water Utilities, 1983.
13. William E. Korbitz, Ed., Urban Public Works Administration, International City Management Association, Washington D.C., 1976.
14. William E. Korbitz, *Modern Management of Water and Wastewater Utilities,* Garland Press, New York, 1981.
15. American Water Works Association, Water Utility Management, *AWWA Manual M5,* Denver 1980.
16. Government Finance Research Center and Peat, Marwick and Mitchell & Co., Wastewater Utility Management Manual (draft), U.S. Environmental Protection Agency, Washington, July 1981.
17. Thomas W. Malone and S. A. Smith, Tradeoffs in Designing Organizations: Implications for New Forms of Human Organizations and Computer Systems, Sloan School of Management, Working Paper 1541-84, MIT, Cambridge, MA, March 1984.
18. James W. Kalat, *Biological Psychology,* Wadsworth Publishing, Belmont, CA, 1984.
19. City of Fort Collins, Capital Project Management Control System Manual, January 1981.
20. Arthur Gitajn, Creating and Financing Public Enterprises, Government Finance Research Center, Washington, D.C., May 1984.
21. Government Finance Research Center, *Financial Management Assistance Program, Planning for Clean Water Programs, the Role of Financial Analysis,* U.S. Government Printing Office, Washington, D.C., 1981.
22. Lyndell Graven, Reporting and Control Procedures for Financial Management, in AWWA Seminar Proceedings on Financial Planning and the Use of Financial Information for General Management Personnel, May 16, 1982.
23. Lennox Moak and A. M. Hillhouse, *Local Government Finance,* Municipal Finance Officer's Association, Chicago, 1975.
24. Aaron Wildavsky, *The Politics of the Budgetary Process,* Little, Brown and Company, Boston, 1984.
25. Association of State and Interstate Water Pollution Control Administrators, Innovative Financing for the Clean Water Program, the State Perspective, Washington, August, 1983.
26. American Water Works Association and Municipal Finance Officers Association, *Water Utility Accounting,* Edwards Brothers, Ann Arbor, 1970.
27. American Water Works Association, Water Rates, *AWWA Manual M1,* Denver, 1972.
28. AWWA, Water Utility Management, Manual M5, Denver, 1980.
29. James H. Stewart, Consulting Engineers, City of Fort Collins, Water Rate Study, December 1984.
30. APWA, ASCE, and WPCF, Financing and Charges for Wastewater Systems, 1973.
31. Environmental Protection Agency, Report to Congress on Industrial Cost Exclusion, WH-547, March 1981.
32. Clayton H. Billings, Ed, *Manual of Wastewater Collection,* Texas Water Utilities Association, Austin, 1981.
33. Hector J. Cyre, Stormwater Management Financing, APWA Congress, September 1982, Houston.
34. Alvin S. Goodman, *Principles of Water Resources Planning,* Prentice-Hall, Englewood Cliffs, NJ, 1984.

35. Lawrence Shubnell and W. W. Cobbs, Creative Capital Financing: A Primer for State and Local Governments, Resources in Review, Government Finance Research Center, Washington, May 1982.

36. Harold Shipman, A Word from Your President, *Newsletter*, Inter-American Association of Sanitary Engineering, March 1983.

37. Association of State and Interstate Water Pollution Control Administrators, Proceedings Innovative Financing Seminar, Washington D.C., April 7–8, 1983.

CHAPTER SEVEN

ENGINEERING ESSENTIALS

The effective management of urban water systems requires engineering skills that span the full range of disciplines. Some of these skills are acquired by managers that are not engineers but who have quantitative abilities and an interest in engineering matters. The purpose of this chapter is to describe the main skills required so that the technical requirements will be clearly understood by engineers themselves and by those managing them.

The range of skills required includes the following kinds of engineering: civil, mechanical, environmental and chemical, electrical, and industrial. There is a close relationship between the way some of these skills are applied, such as in project management, and management itself, which is also very important, so this chapter should be considered together with the previous one.

The civil engineer is thought to have the primary role in managing water systems since the original development of urban water systems was by civil engineers. The knowledge areas required are hydrology, hydraulics, water quality control, and design engineering, including structural and geotechnical engineering; in short, the full range of civil engineering is involved.

The mechanical engineering skills are important because of the machinery and controls that are used in urban water systems. Systems include valves, pumps, tanks, various controls, and a very important maintenance requirement (covered in Chapter 9) that is familiar to engineers with skill in manufacturing and process control. A very important part of urban water system management is the civil/mechanical engineering subject of plumbing engineering, at the small and large scales. For example, protecting pipelines from corrosion would fit in this category, or protecting against surge and water hammer. The role of the electrical engineer is similar, as we will see later, due to the increasing importance of computers and electronics in the instrumentation and control of systems (covered in Chapter 10). The chemical engineering role is usually taken by the environmental engineer, and the kinds of problems faced, unit operations, are very similar to those faced by chemical engineers in the manufacturing and process industries.

The industrial engineer is listed since the attention given to cost control and productivity, although common to all the engineering disciplines, made most of its progress through its identity with industrial engineering. These concepts are covered in Chapter 6.

The contruction process is an absolutely essential part of engineering since the ultimate performance of the systems that are planned and managed is so dependent on the quality of construction.

In the following sections we will discuss some of the important engineering skills related to hydrology, hydraulics, water quality, and the construction process. The discussions are not meant to be exhaustive but to point out the main areas needing attention by the urban water manager.

HYDROLOGY AND WATER BALANCES

The beginning of understanding of hydrology is the water cycle and, in particular, the impact of urban areas on it. Figure 7.1 shows a simple portrayal of the water cycle, with the basic hydrologic processes of evaporation, precipitation, runoff, and infiltration shown. There are several things to notice about the impacts of the urban area on the water cycle: the increased percentage of runoff that occurs due to the impervious surface, the reduced infiltration due to the same effect, the change in quality

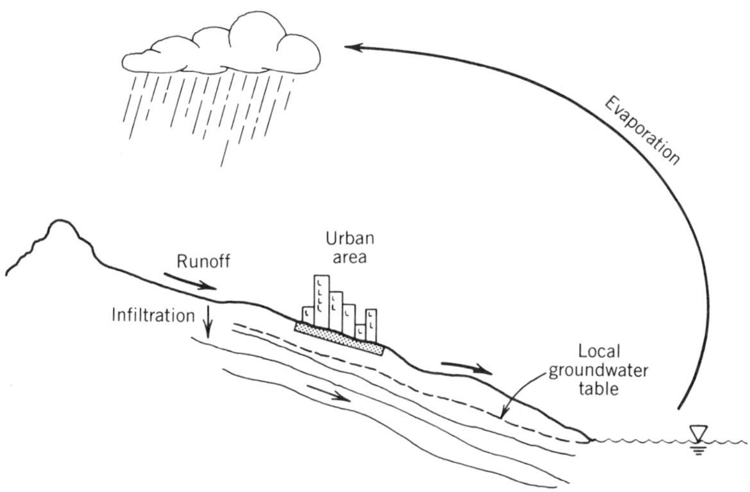

FIGURE 7.1 Simple water cycle.

of runoff water due to the urban washoff, and the localized effects on the water table due to both quality and quantity effects. For a full discussion of urban hydrology the reader should refer to the appropriate texts, of course, but it can be said here that the field is still very much in the research stage. There are some specialty texts on the subject, one being a recent text by Lazaro (1), but most of the really current information is in journals, research reports, and design guides for stormwater management. A source that is often quoted is the report by Luna Leopold, a well-known hydrologist with the U.S. Geological Survey (2). In this report Leopold describes the main concerns of urban hydrology: increases in storm runoff, need for detention storage sediment production, and water quality deterioration. The reports of the ASCE Urban Water Resources Research Program listed in Appendix 2 provide an even more complete overview of the literature available.

McPherson called for more studies of the hydrology of urban area, in particular, for work on developing a metropolitan water balance inventory (3). What McPherson was calling for was the application of systems analysis and modeling so that managers knew what was going on in the entire urban water system. They still do not know today, but progress is being made, and the way to make progress is one of the main subjects of this book. By implementing the information systems described later in the book, we can begin to approach the complete metropolitan water balance inventory or water budget, considering both the quality and quantity of the water.

While precise figures for the impact of urban areas on the water cycle cannot be given, the following paragraphs will present some of the generalized numerical values.

The first consideration might be the water needs of an urban area. Per capita demands are discussed in the chapter on water supply, and a hypothetical water balance for an urban area is given. Following the example given there, a hypothetical urban area of one million inhabitants could be expected to demand a continuous water supply of 150 mgd at the common value of 150 mgd average demand. This would translate into a need of the entire flow of a stream of 230 cfs. While 230 cfs sounds like a modest stream, we have to remember that the demand must be met at the dryest times of the year, and that the stream must retain an in-stream flow and meet a number of demands other than just the water supply of the city next to it.

The demands on the stream will be distributed across several uses, and there will be return flows that depend on the distribution of the uses. A figure that has been recently used quite a bit for planning is 70 gpcd for domestic uses, but the total flow in a wastewater system is apt to be greater than that due to the other return flows, infiltration, and leakage, so the return to the stream may be nearly 100% of the original diversion. There will be quite a change in quality, of course, but the effects should be acceptable if the requirements of the Clean Water Act (discussed later) are met. In the summer there will be quite a consumptive use of the water if there is considerable lawn watering, as there is in the dryer parts of the country.

The general impacts of the urban area on storm peaks and volumes has been studied quite a bit, but no one can report the exact magnitudes of effects since they vary with so many condition parameters. We know in general that the effects are higher peak flows, less infiltration, and reduced quality of runoff. These are discussed in more detail in the next section.

HYDROLOGIC CALCULATIONS

Some of the hydrologic calculations that are common in urban water management are water supply yield analysis, safe groundwater yield, stormwater and flood calculations, and drought analysis. These are not all, of course, but a knowledge of these is certainly helpful in urban water management.

Yield of Water Supplies

The basic calculation of the expected yield of a surface water supply source is an exercise in probability and statistics. It involves knowing what the historical experience of yield is and applying statistical analysis to the data. Although statistical analysis can get to be quite complex, enough knowledge to apply it usefully is simple. The usual place to start is with the normal probability curve.

Let's say we have a series of some 50 years of annual records at a water supply reservoir. The yields for the 50 years in acre-feet for a small city water supply might be as shown below:

13,228	13,582	9,393	14,807	15,645
16,027	19,121	23,402	12,856	23,275
14,248	16,028	20,520	14,473	21,222
14,027	19,201	24,037	15,364	12,462
21,951	23,015	18,305	18,269	14,201
13,033	15,853	18,616	17,279	21,428
14,885	7,769	17,243	11,356	5,640
11,549	13,800	13,740	17,545	11,075
8,601	18,148	19,125	13,103	15,154
13,410	10,426	15,473	15,613	16,376

These data can be arrayed in several ways to enable the study of their statistical behavior. Using simple statistical analysis, the easiest two approaches are to make a histogram and a cumulative distribution. First, the histogram is a sorting of the data into categories, as shown below:

Range	Number of Data
4,000–8,000	2
8,000–12,000	6
12,000–16,000	21
16,000–20,000	13
20,000–24,000	7
24,000–28,000	1

These data make a histogram as shown in Figure 7.2. This histogram has the general shape of the bell curve associated with a normal distribution and suggests that we can fit the normal curve to it well. This is accomplished by calculating the mean and standard deviation of the data, which can be easily done on any scientific calculator. This yields a mean of 15,698 acre-feet and a standard deviation of 4150 acre-feet. Using these values we can consult any table of normal distribution ordinates and plot the fitted bell curve also shown in Figure 7.2. There are, of course, more

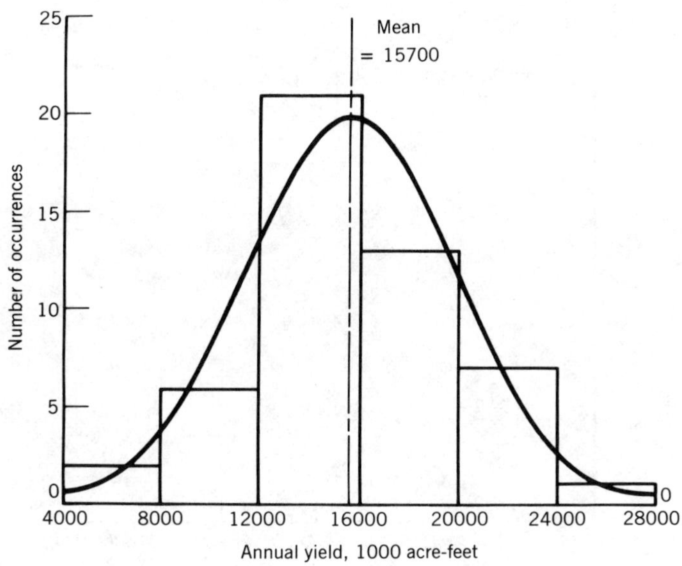

FIGURE 7.2 Distribution of annual water supply.

advanced statistical techniques to test how well the data fit the normal curve, and these can be applied to the data if the circumstances warrant. For our purposes we observe that the normal curve fits fairly well.

To actually use the information from the distribution, we must plot it in its cumulative form. This is facilitated by arraying the data in order from the greatest to the smallest value, as can easily be carried out by the reader. The highest value is observed to be 24,037 and the lowest to be 5640. These are plotted as shown in Figure 7.3, which represents the cumulative numbers of observations. The fitted normal curve is shown along with the plotted data from the sample, and it is seen that the fit looks guite good. The way to portray the data if any extrapolation is desired is to plot it on probability paper. This is shown in Figure 7.4, which also demonstrates a good normal fit. On this kind of probability paper the normal distribution plots as a straight line as shown. There is some data manipulation required to plot the values shown in the figures. Most normal distribution tables will have the values in terms of a variable z or another symbol, which represents a normalized value calculated as

$$z = \frac{x - \mu}{\sigma},$$

where x is the value of the annual yield, μ is the mean value of the yields over the period of record and σ is the standard deviation of the values. In Figure 7.2 to convert the normal distribution ordinates to the number of occurrences the ordinates would be multiplied by the number of ob-

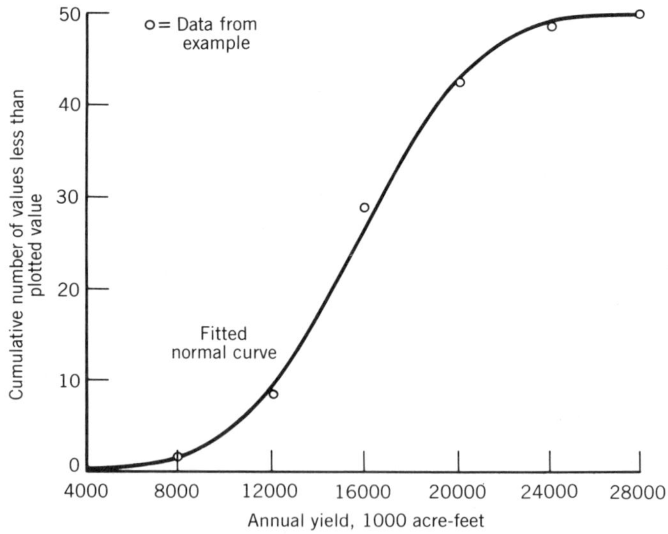

FIGURE 7.3 Cumulative distribution of water supply.

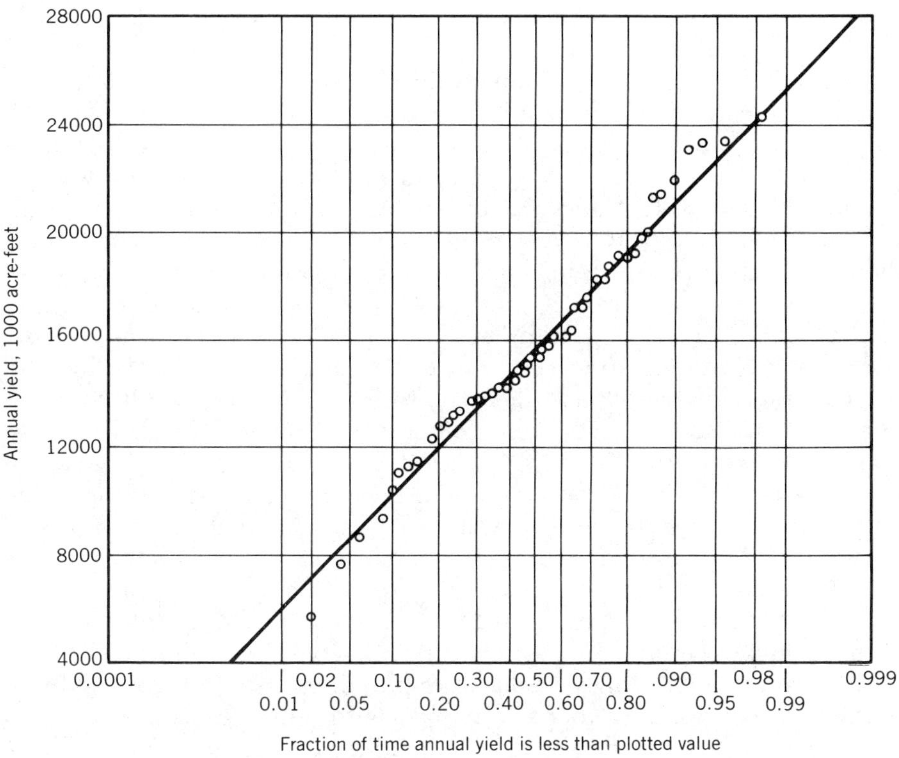

FIGURE 7.4 Water supply yield diagram.

servations, that is, 50. In Figure 7.3 to fit the theoretical curve to the data the mean value of the yields is plotted at 50% of the cumulative number of values, or 25, and the rest of the curve is plotted relative to that value in terms of standard deviations from the mean. For example, the value at one standard deviation greater than the mean, 15698 + 4150 = 19848 is plotted at 50% + 34.13% (from a normal variate table) of the values or (0.8413) (50) = 42 cumulative values. In Figure 7.4, where the probability plot is made, the yields are ranked from 1 to 50 with 1 being the smallest, or vice versa. This is similar to the procedure for flood frequency analysis using rankings of the annual floods. This example uses a "plotting position" of $n/(n + 1)$ for each yield, also similar to the flood frequency analysis. This means, for example, that the largest yield, 24,037, would have a rank of 50 and a plotting position of 50/51 = 0.9804. The smallest, 5650, has a rank of 1 and a plotting position of 0.0196.

The concepts above can be used to make a good analysis of the yield of streams where the annual values are of interest. The planner has to make sure the statistics are viable, and there are, of course, many tests and other ways to do that. Once the statistics are available, the planning report will be much improved since information on safe yield and risk can be compiled.

In general, the concept of safe yield means the minimum yield in a period of record. In the example above that could mean the observed value of 5640 acre-feet in 50 years of record. Another way to look at it might be to use the probability plot and say that the 100-year low flow is about 6000 acre-feet and that the observed value was not a 50-year value but more like a 100-year event. This assumes that the data follow the normal distribution. This analysis begins to look at drought probabilities, which are normally analyzed in terms of persistence of the drought. The values plotted in Figures 7.3 and 7.4 can be used to look at the 1-year droughts, meaning those that last 1 year only. To look at the more serious events, the data has to be arrayed into multiple-year periods. Most texts on hydrology demonstrate how to do this.

The concept of the drought flow is also used in water quality calculations when the discharge from a wastewater treatment plant is to be limited to the 7–Q–10 conditions. This means the lowest flow to be expected in a 7-day period once in 10 years. Using criteria such as this is a continuous source of difficulty due to the statistical variations among streams. Whereas one stream may have a 7–Q–10 of, say 500 cfs, another with an equal average annual flow but in a dryer area could have a 7–Q–10 of zero. This is an extreme example, of course, but the resulting requirements for wastewater treatment would be vastly different.

Using statistics, it is possible to demonstrate one of the important benefits of having multiple sources of water rather than relying on a single source. To demonstrate it, I present the results of an analysis using the statistical correlation between three sources that might be available to a city. Space does not permit the full presentation of all the analysis, but is very straightforward.

The situation is that three reservoirs are available to a city, as shown in Figure 7.5. The reservoirs have mean yields and standard deviations as shown below:

Reservoir	Mean Yield (acre-feet)	Standard Deviation (acre-feet)
A	20,000	3,000
B	60,000	6,000
C	40,000	6,000

Three hypothetical situations have been analyzed: the first is that the reservoirs are almost independent of each other statistically, meaning that they do not depend on the same storms, watersheds, and so forth. This is expressed by a cross-correlation coefficient of 0.1 between reservoirs. The second case is one where the reservoirs are slightly more

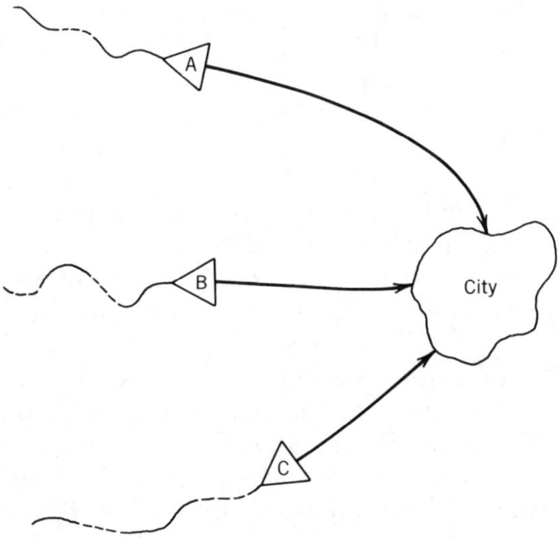

FIGURE 7.5 Hypothetical three-reservoir supply.

dependent, with cross-correlation coefficients of 0.3. The third case is one of dependence, with the cross-correlation coefficients being 1.0.

Using statistical techniques, this situation can be analyzed to determine the mean and standard deviation of the three reservoirs acting together. In effect this is a problem of determining the probability distribution of the sum of three random variables. The expected value of the mean of the three together is the sum of the means, or 120,000 acre-feet. The variance of the sum, however, is the sum of the variances plus twice the summation of the covariances (see text on statistics for explanation), and when the standard deviation of the sum is found by taking the square root of the variance, the difference is significant. In this case, the results are as follows:

Case	Mean	Sum of Standard Deviations	Standard Deviation of Sum
I	120,000	15,000	9,770
II	120,000	15,000	11,140
III	120,000	15,000	15,000

The table shows that the relatively independent sources (case I, r = 0.1) has a standard deviation much less than the other two cases, which have the increasing dependen. It is a quantitative presentation of the old admonition not to put all the eggs in one basket. In a practical sense the reliance on independent sources may allow more reliability at lower cost. If, for example, we have the sources above under the condition that they are dependent (case III), and we have a shortfall of one standard deviation, the yield will be 105,000 acre-feet. If we have the same average yields in the less dependent case I, the yield will be over 110,000 for the same shortfall of one standard deviation. The shortfall of one standard deviation would be expected to occur about once every 6–7 years, following the general rule that the frequency corresponds to the inverse of the cumulative probability. The insurance that is provided by the additional water has a financial value. Just looking at orders of magnitude, the 5000 acre-feet would be worth on the order of $5 million if water rights sold for $1000 per acre-foot, a common value in some areas of the West.

Since much of the nation's water supply is from underground sources, it is necessary to be able to make judgments about groundwater yields. The safe yield of a groundwater source is a matter involving hydrologic and hydraulic calculations since the aquifer is a storage reservoir, and it is a matter of balancing the inflows, outflows, and flow characteristics of the aquifer. Since aquifers are so geologically complex, it is often impossible to calculate safe yield with any certainty, and decisions regarding new wells or pumping allowances may have to be made in court.

Urban Storm Runoff

Another common hydrologic calculation in urban areas is urban runoff. The requirements for calculations are too detailed to be discussed in full here, but it should be pointed out that there is a difference between the approaches necessary for small, medium, and large urban basins. The small basin requires an approach such as the rational method or a variation of it. For the design of systems for catchments up to about 200 acres, this method, although widely criticized for its inadequacies, gives usable results. Between 200 acres and about 10 square miles, it is unusual to have much data to base designs on, and methods that rely on models are mostly used. Examples would be the stormwater management model, the SCS model, and others. Above about 10 square miles the models begin to lose their appeal due to the large amount of data needed to run them with any advantage over the use of estimates based on local data. Approaches based on regression studies of local watersheds or even of the basin under study seem best.

The use of the rational method for estimating runoff from small basins is so widespread that the only thing to say about it in recent years seems to be to discuss its weak points, which are many. The method is valid when it is used properly, and proper use is a condition to the validity of any method, especially more "sophisticated" methods, many of which are not sophisticated at all.

Basically the rational method involves the following equation:

$$Q = CIA,$$

where Q is the rate of flow in cfs, C is the runoff coefficient, I is the rainfall intensity in inches per hour, and A is the watershed area in acres. Although the equation does not appear to be dimensionally homogeneous, it is for all practical purposes, since one inch-acre per hour equals 1.008 cfs. The important considerations in using the equation properly involve proper selection of the runoff coefficient and the rainfall intensity. For a discussion of these values see an appropriate text on hydrology.

Table 7.1, in use by the Urban Drainage and Flood Control District of Denver, illustrates a modern approach to the selection of the C values. They are referenced to percent impervious and rainfall frequency, effectively taken care of some of the objections about the method (4).

TABLE 7.1 Recommended Runoff Coefficients and Percent Impervious (4)

Land Use or Surface Characteristics	Percent Impervious	Frequency 2	5	10	100
Business					
Commercial areas	95	0.87	0.87	0.88	0.89
Neighborhood areas	70	0.60	0.65	0.70	0.80
Residential					
Single-family		0.40	0.45	0.50	0.60
Multiunit (detached)	50	0.45	0.50	0.60	0.70
Multiunit (attached)	70	0.60	0.65	0.70	0.80
½ acre lot or larger		0.30	0.35	0.40	0.60
Apartments	70	0.65	0.70	0.70	0.80
Industrial					
Light areas	80	0.71	0.72	0.76	0.82
Heavy acres	90	0.80	0.80	0.85	0.90
Parks, cemetaries	7	0.10	0.10	0.35	0.60
Playgrounds	13	0.15	0.25	0.35	0.65
Schools	50	0.45	0.50	0.60	0.70
Railroad yard areas	40	0.40	0.45	0.50	0.60
Undeveloped areas					
Historic flow analysis greenbelts, agricultural	2				
Off-site flow analysis (when land use not defined)	45	0.43	0.47	0.55	0.65
Streets					
Paved	100	0.87	0.88	0.90	0.93
Gravel	13	0.15	0.25	0.35	0.65
Drive and walks	96	0.87	0.87	0.88	0.89
Roofs	90	0.80	0.85	0.90	0.90
Lawns, sandy soil	0	0.00	0.01	0.05	0.20
Lawns, clayey soil	0	0.05	0.10	0.20	0.40

The selection of the rainfall intensity involves decisions about both the acceptable risk, through the return period, and about watershed time parameters, through the time of concentration. Proper selection of these values involves considerable engineering judgment.

For the larger basins the selection of a method is arbitrary, and my own preference is to use more than one to get an idea of the sensitivity of the calculation to the method used. In an analysis of two different methods it was shown that the results from a hypothetical watershed with ten subcatchments can vary as much as 40–50% just by changing the amount of detail in the input data. Specifically, I worked with Dr. Bruce Bradford at Georgia Tech to develop an analysis of a hypothetical watershed of 1070 acres with nine subcatchments, each with 200 acres or less (see Figure 7.6). Using a version of the SCS method as published in their technical bulletin, we showed that the calculated runoff varied significantly if it was calculated using the nine subcatchments separately versus being lumped all together. Using another synthetic unit hydrograph method called the Espey method, we showed that the calculated runoff varied still more. The results from the analysis are shown below (5):

Method	Number of Subcatchments	Calculated Q
SCS	1	1501
Espey	1	1930
SCS	3	2458
SCS	9	2579
Espey	9	2546

In the example the main difference is in the results using different numbers of subcatchments, or in the degree of "lumping." By using different methods this way more confidence can be gained in the overall results.

HYDRAULIC CALCULATIONS

While hydrologic calculations involve the earth science, there are also many important engineering hydraulic questions involved in managing urban water systems. They include management concerns such as how to operate facilities, engineering questions of planning and design, legal questions such as where to draw a flood plain line, and financial questions such as how much risk to take.

In the following paragraphs we will discuss the hydraulics problems considered to be the most important for managers to deal with. They will not be discussed in the detail necessary for design but at the conceptual

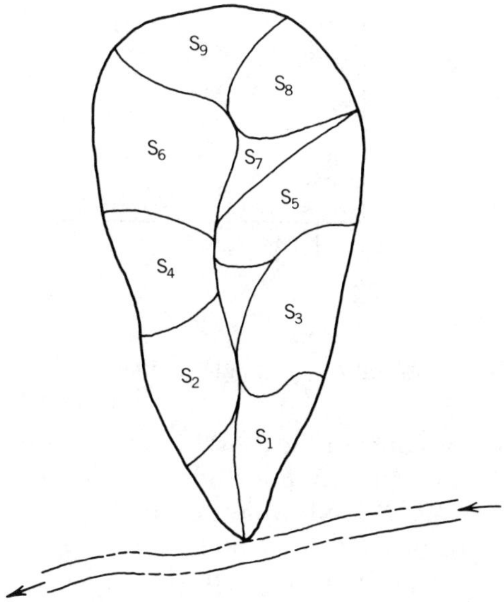

FIGURE 7.6 A typical urban drainage basin.

level for management understanding. In general, they are discussed in the order of how they apply to water supply, wastewater management, anstormwater and flood control.

In water supply systems hydraulic problems occur in the acquisition reservoirs, in the treatment plant, and in the distribution system, and they are all important.

In the acquisition system it is common for there to be systems to reduce head or pressure as the water supply flows from the reservoir into a treatment plant where the treatment processes are conducted at atmospheric pressure. This necessitates some form of pressure reducer or head breaker. Some systems are taking advantage of the need to reduce pressure by generating power with the energy available, but there remain some serious hydraulic problems for some, particularly with cavitation and valving. A common situation is that shown in Figure 7.7 where a needle-

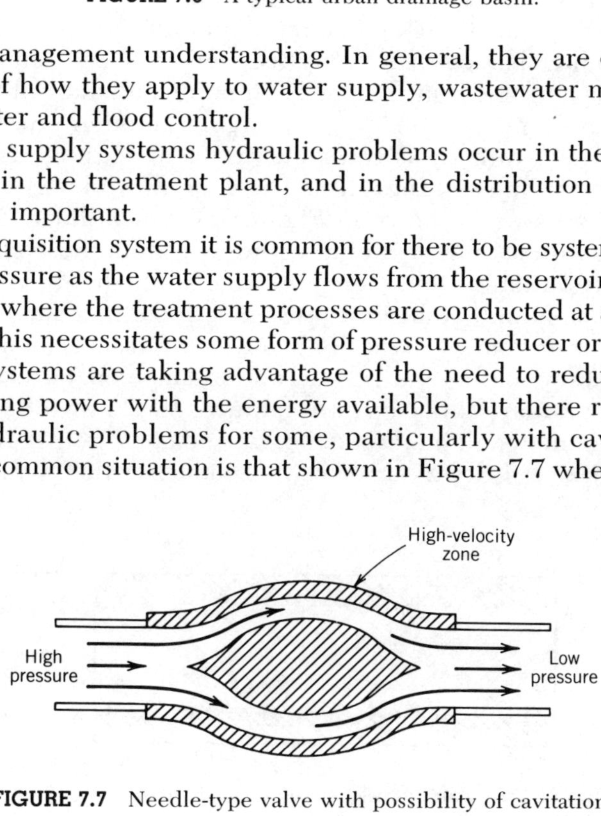

FIGURE 7.7 Needle-type valve with possibility of cavitation.

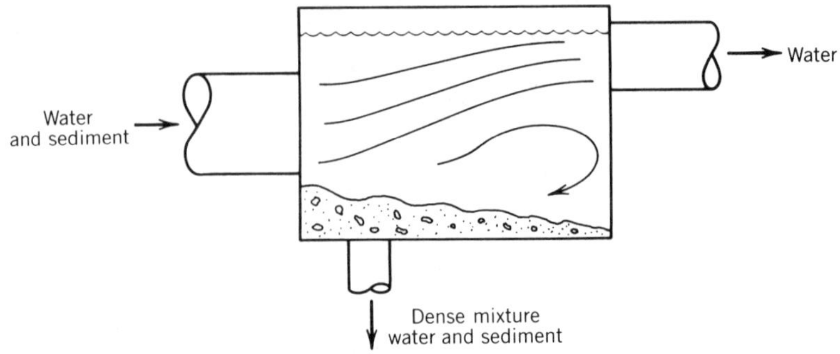

FIGURE 7.8 A simple settling tank.

type valve is shown controlling flow from a high-pressure to a lower-pressure zone. There are a number of different kinds of valve used in this application, and this is only an example. Figure 7.7 shows the zones where the high velocities and cavitation may occur. Along with cavitation, which is a hydraulic problem characteristic of high-pressure zones, is water hammer and surge, which accompany valving under conditions of high velocity that accompany high-pressure zones.

In treatment plants there are a number of unit processes that depend on the hydraulic characteristics for effectiveness. One example is filtration, which is a problem of flow through porous media, with the media characteristics being variable due to clogging. Another problem that involves hydraulics directly is sedimentation and settling, used in sedimentation basins and settling tanks. The general working of a sedimentation unit process is as shown in Figure 7.8, which illustrates a simple settling tank with sediments being withdrawn from the bottom.

In the distribution system there are two main hydraulic concerns: energy management and control of pressures, including surges. The first case, energy management, basically involves the management of storage, pumping, energy losses, and related performance characteristics in the distribution system. This naturally requires the knowledge of system characteristics in detail, including the age and condition of pipes, so that roughness characteristics are known with enough certainty that problems

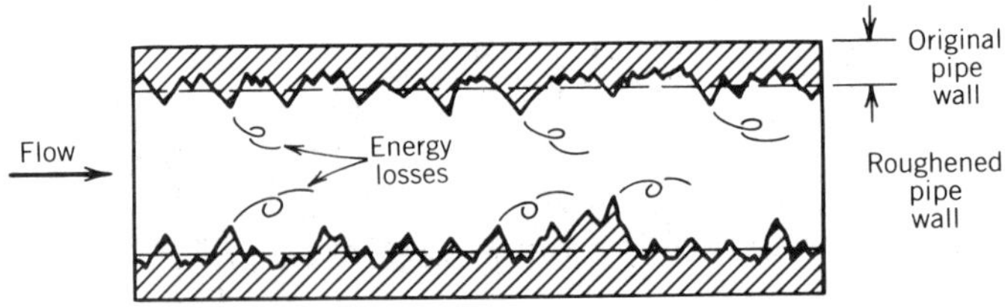

FIGURE 7.9 Energy losses in pipe roughened by corrosion.

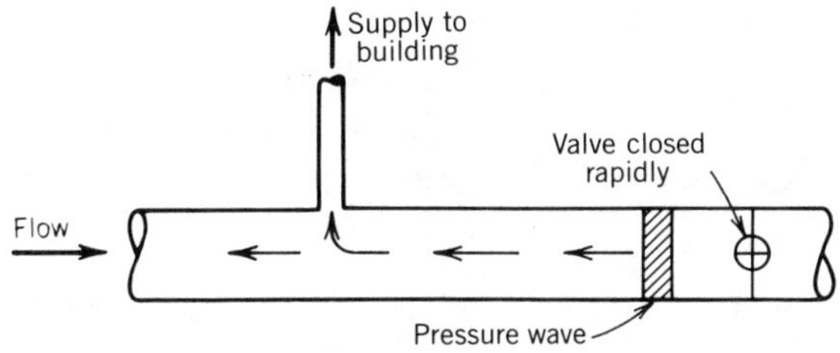

FIGURE 7.10 Pressure wave caused by sudden valve closure.

can be corrected before they become serious. The loss of energy through reduced flow capacity caused by, say corrosion and the formation of tubercles, can be debilitating to the energy budget, as shown in Figure 7.9. The other main problem, surge control, means to ensure the smooth and continuous supply of the right system pressures and flows, avoiding the kind of problem shown in Figure 7.10 where a pressure wave is shown resulting from rapid valve closure and about to travel toward a building, possibly causing damage.

In the wastewater systems the basic flow is in open sewers, resulting in free surface flow, a different variety than in water distribution systems. Since flows are rather continuous and not subject to many surges, there are not many hydraulic problems in ordinary sewer flow. However, it is important to maintain healthy flows with velocities high enough to maintain a fresh sewage, avoiding septic conditions and undue sedimentation. Figure 7.11 illustrates the concepts involved.

In wastewater systems reliance is sometimes placed on pressure sewers or force mains. This means that special provisions must be made for trouble-free pumping of sewage, including all of the sediments, floatables, and other ingredients that may be encountered. Sludge pumping in wastewater treatment plants also requires special pumping equipment.

Stormwater and flood control systems require a more detailed treatment of hydraulics due to the time-varying flows and the many different situations faced. The first situation is the use of streets to convey storm flows. Many drivers do not realize that one of the services provided by streets is to convey both minor and major storm and flood flows. It is a problem of open channel hydraulics where the street is actually a channel with the gutter provided for the minor flows and the street itself being

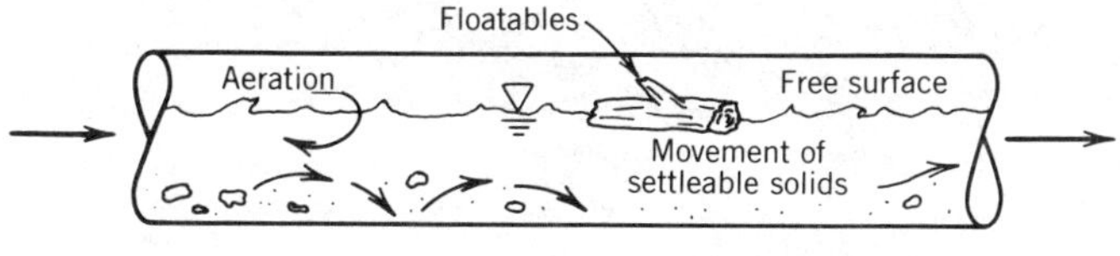

FIGURE 7.11 Flow conditions in wastewater sewer.

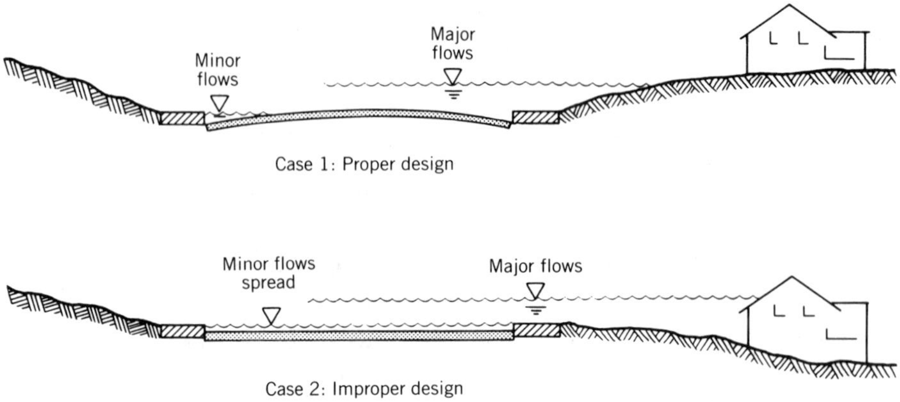

FIGURE 7.12 Street drainage.

available for major flows, without causing damage to the adjacent buildings, if the system works right. Figure 7.12 shows the proper and improper operation. In case 1 the minor flows are contained within the gutter areas without flooding the street, and the major flows are handled, within reason, without flooding the buildings. In case 2, however, the minor flows spread across the roadway creating a driving hazard, and the major flows enter the buildings.

Channel flows of all kinds are frequently encountered in stormwater and flood design, and the designer has to be familiar with the hydraulics of both steady and unsteady flow. Flow in ditches and conveyance channels is a common design problem, particularly to size the channels in a way to avoid adverse problems such as overtopping and erosion. Channels are often protected against erosion by the use of riprap or other materials, and designers must be skilled in specifying the right protection and in making sure that it lasts. It is common for protection to be designed that makes a good first impression but does not last. This situation is only

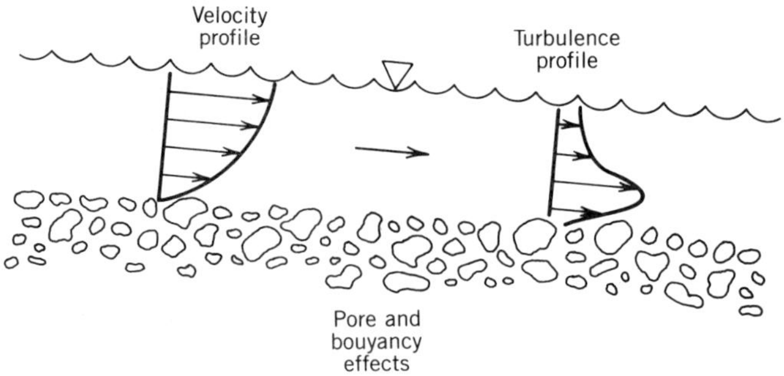

FIGURE 7.13 Flow over a natural channel bed.

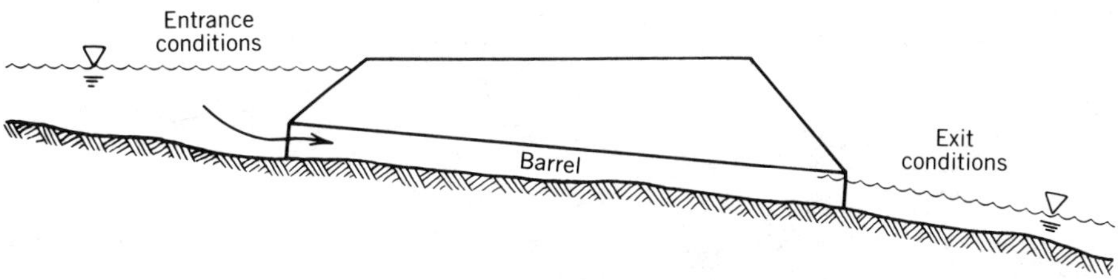

FIGURE 7.14 Flow in culvert.

avoided by experience in design and management. The mechanics of erosion are actually quite complex, involving turbulence, buoyancy, and the intricacies of soil pore mechanics, as shown in Figure 7.13, which illustrates a rather stable channel with pore and buoyancy effects in the bed material, and with the velocity and turbulence profiles illustrated. The turbulence is greatest near the bed as shown and tends to aggravate the transport of the bed material particles.

Another complex problem in stormwater is the proper hydraulic design of culverts. As shown in Figure 7.14 the flow in a culvert has to be checked for entrance conditions, flow in the barrel, and exit conditions; the hydraulic operation can be greatly impeded if the flow paths are not kept clear of obstructions. The flow in the barrel of a culvert is very similar to that in a storm sewer: it can either be free-surface flow or closed-conduit flow, and the mechanics of the flow depend greatly on which it is. The shape of the capacity curve, for example, sometimes indicates a decrease in capacity as the depth increases due to the increasing wetted perimeter.

In stormwater systems it is increasingly common to incorporate detention storage into designs. The proper design of detention storage requires quite sophisticated approaches to hydraulic analysis, including the use of unsteady flow models. On a small scale the problem is the same as the larger reservoir routing problem, which considers the inflow and outflow hydrographs and the change in storage. It is common to require that new subdivisions incorporate enough detention storage to reduce postdevelopment peaks to predevelopment levels, so the system-wide effects must be analyzed. Figure 7.15 illustrates the general behavior of a detention pond and the way if affects a hydrograph.

A hydraulics problem with substantial land-use and financial implications is that of flood plain definition. Since the passage of the Flood Insurance Act of 1968, this has become a very common activity for local government agencies and their consultants. The hydraulics involved are very complex since there are several different kinds of flow involved, since the flow channels are constantly subject to change, and since natural, movable bed channels are usually involved. Figure 7.16 shows the nature of the problem in the typical situation of flow through a bridge opening. At the top of the figure the profile shows changing regimes of

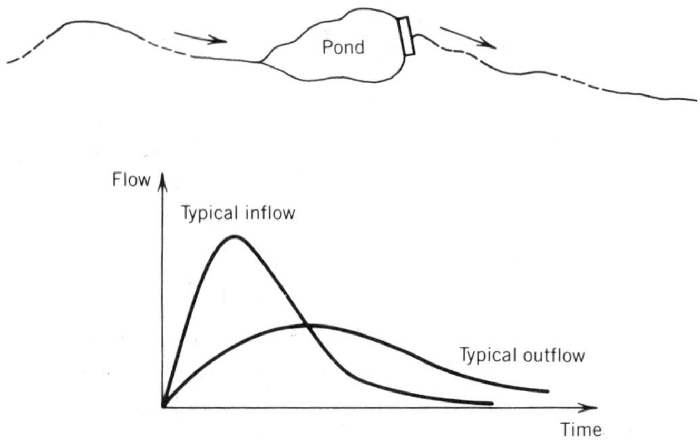

FIGURE 7.15 Behavior of detention pond.

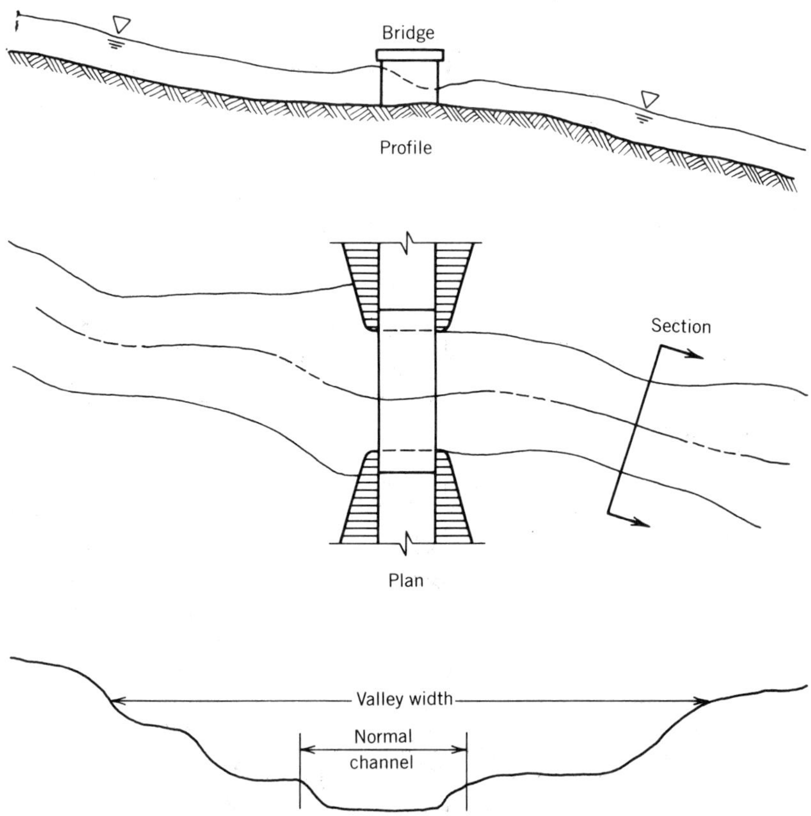

FIGURE 7.16 Hydraulic features of flood channel.

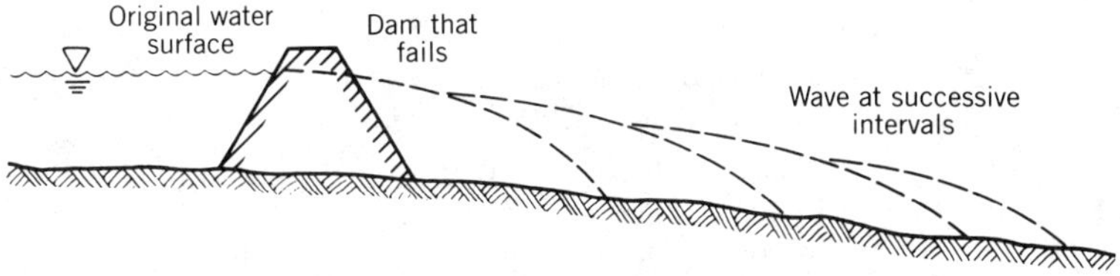

FIGURE 7.17 Dam break hydraulic profiles.

flow from tranquil to rapid, perhaps corresponding to subcritical and supercritical. The plan view shows the flow through a bridge with the backwater that is created, also highlighting the hazard that the bridge may collt debris and become clogged, thereby creating a worse hazard. The lower part of the figure shows a channel section that illustrates the normal division of the channel into a low-flow section and a valley width that allows for some flood flow. This kind of section is typical and accommodates the flood plain regulatory approach described in the chapter on stormwater and flood programs. For the most part, the hydraulic analysis used in these studies is based on a computerized step backwater curve, with the most popular version being HEC-2 by the U.S. Army Corps of Engineers.

With the increasing popularity of computers and more sophisticated numerical techniques, the use of unsteady flow routing is becoming more commonplace. This kind of analysis allows the consideration of the phenomenon that the water spreads out as it proceeds downstream and that the usual assumption of steady flow, meaning constant discharge, is not necessary for all cases. A typical situation here is the dam break analysis shown in Figure 7.17. After the dam breaks (hopefully a hypothetical situation) the water begins its trip down the valley rapidly, slowing down as it fills the valley storage spaces. Heavy rain in the upper end of a basin can cause the same kind of action, but the effects will normally not be as traumatic. This is not always the case, however, as illustrated in Figure 7.18, which shows two flood scenarios that occurred near Estes Park, Colorado. In the summer of 1976 a very heavy rainfall occurred between

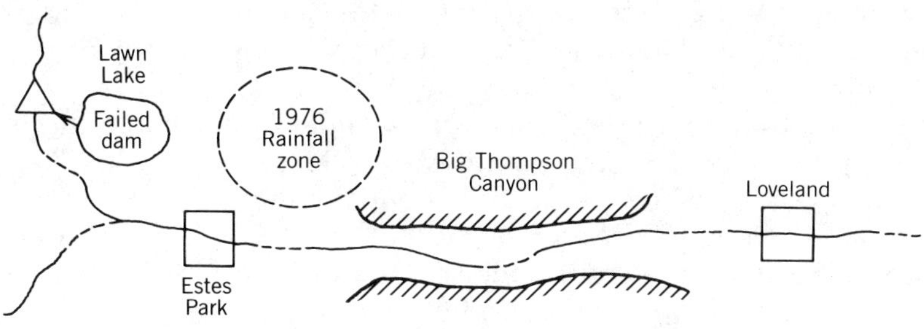

FIGURE 7.18 Dynamics of flooding near Estes Park, Colorado.

the town of Estes Park and the Big Thompson Canyon. It caused a tremendous flood wave that killed about 150 people, mostly those caught in the canyon. In the summer of 1982, an irrigation dam high in the mountains above the town failed, and the resulting flood wave caused several millions of dollars in damage, but killed only about four people. The hydraulic scars left by both these events are noteworthy to observe even today, and they illustrate the serious nature with which flooding must be regarded.

WATER QUALITY ANALYSIS

Water quality has been an important subject in the 1970s and first part of the 1980s, and it certainly promises to continue to attract the attention of water system managers. There are several aspects to the problem: control of drinking water quality, control of in-stream and groundwater quality, and control of effluents from wastewater plants and nonpoint source discharge points. These problems have common parameters, and comprehensive control of water quality requires an integrated approach to all of them.

The overall scheme of things is shown in Figure 7.19, which illustrates the inevitable successive use of water as it moves through its cycle. The figure shows the normal withdrawal of surface water by all three cities shown with city B having a supplemental well supply that is threatened by seepage from a landfill, which also threatens the main stream. City C has a supplemental source from a protected upland watershed. The industry uses self-supplied water and furnishes its own treatment. City A has a problem with combined sewer overflow (CSO) and city C has significant stormwater discharges, which are not controlled for water quality. The withdrawals for drinking water are regulated by the Safe Drinking Water Act, which has one kind of water quality standards. The effluents from the wastewater treatment plants are regulated by the Clean Water Act, which also specifies the guidelines for setting stream water quality standards. The nonpoint sources are not yet effectively regulated. Regulation of groundwater quality is just beginning, and there are not many comprehensive programs to study yet, although there are programs to regulate well construction and to control the injection of wastes underground.

The Safe Drinking Water Act and the Clean Water Act are discussed in the chapters on water supply and wastewater management. The purpose of this section is to discuss the essential beginning points of water quality management that the urban water system manager must know.

It would be nice if we could specify water quality with a single index, but attempts to develop any such index have generally failed since there is not general agreement about what the objectives should be or which

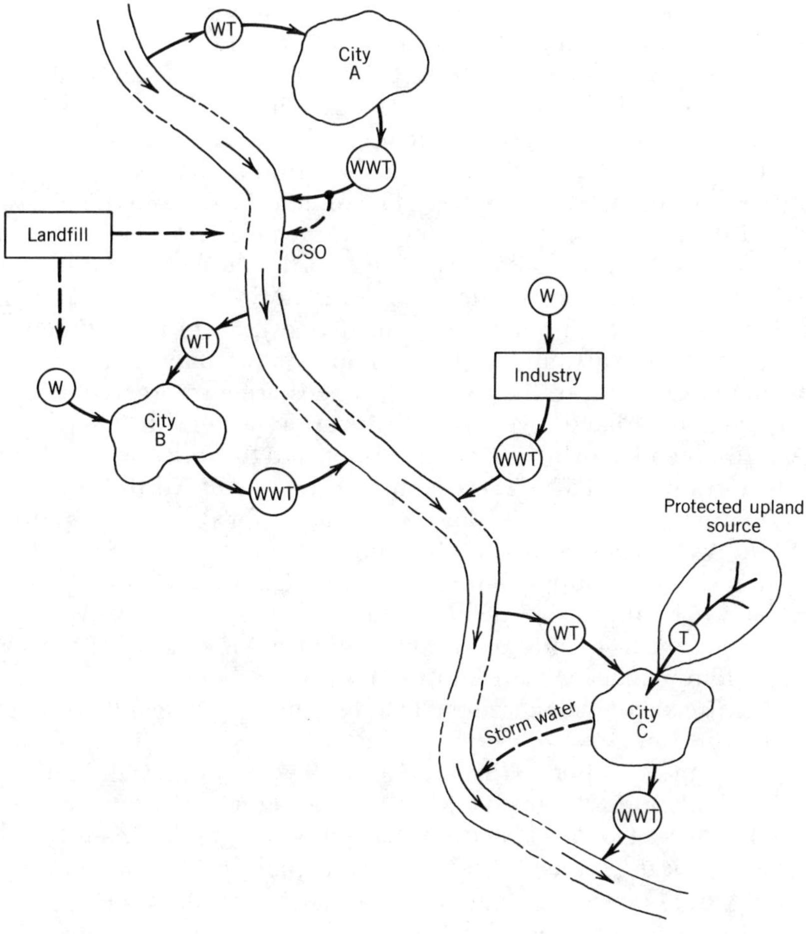

FIGURE 7.19 Some uses of water and land that affect water quality.

parameters should be included in any index. The most general water quality parameters are oxygen levels, bacterial levels, and the concentrations of chemicals and nutrients. Oxygen is measured in two ways: the amount that is in the stream (dissolved oxygen, DO) and the demand for oxygen by wastewater (biochemical oxygen demand, BOD) or other measures. Bacterial levels are generally monitored by specifying the level of the indicator organism, coliforms. Chemicals, especially toxic ones, are the newest addition to the list of contaminants of concern. Nutrients, mostly phosphorus and nitrogen, are generally of concern because they enrich and kill off lakes and estuaries.

As discussed earlier in the chapter on drinking water, there are four phases of the development of water quality standards envisioned in the U.S. drinking water regulations: volatile synthetic organic chemicals

(VOCs); synthetic organic chemicals (SOCs), inorganic chemicals, and microbiological contaminants; radionuclides; and disinfectant by-products. This represents some current thinking about the importance and urgency of these different kinds of contaminants.

In the stream water standards the approach is for each state to set the needed standards, which depend on the unique characteristics of the waters and water uses of the state. The contaminants might be classified a little differently than for drinking water, but by and large the same substances are involved. For example, North Carolina's water quality management plan recommended standards in the following categories: organic wastes (oxygen-demanding substances), bacteria, sediment, nutrients, and toxic substances. These names of contaminants are often confusing since some of the toxic substances are also organic wastes.

It is typical to have a stream classification system that begins with waters suitable for a drinking water supply, and to work down from there to waters of lower quality. For example, the State of North Carolina has the following categories: A-I and A-II waters for drinking supplies; B and SB for water-contact support (S is for salt waters); C and SC for fishing and industrial water supply; and SA for shellfish propogation. There will be unique regulations for each category, and they will be revised from time to time with much debate, often with great stakes for the urban water system manager. For example, I witnessed a case where a group of local citizens mounted a tremendous campaign to maintain a B standard for a small stream in North Carolina rather than degrading it to a C, which was close to the existing reality at the time. The coliform standards were: B, 200/100 mL and C, 1000/100 mL. Wondering why the opposition was so vocal, it soon became apparent that the real agenda was to head off an effort to locate a wastewater treatment plant on the stream. The B standard would make it too expensive to locate the plant there, whereas the C standard made it feasible.

Effluent standards that are set under the general authority of the Clean Water Act prescribe the levels of contaminants that are permissable to discharge by municipalities or industries. They are set after an analysis of the ability of the stream to assimilate the oxygen-demanding wastes and with a view to prohibiting other harmful discharges.

The analysis of the effect of the different wastes on the stream has become greatly complicated in the past few years, especially with the emergence of the tremendous variety of synthetic organic chemicals. Past analysis relied to a large extent on the oxygen sag curve whereby the effect of a waste load on the stream dissolved oxygen could be estimated, as shown on Figure 7.20. This analysis is still in use and has resulted in many wastewater treatment decisions and requirements, but the complexity of the quality control function has gone well beyond this analysis, which is now considered routine. There have been attempts to develop water quality models that will predict the other contaminants as functions of time and distance, but they have not been very useful in management.

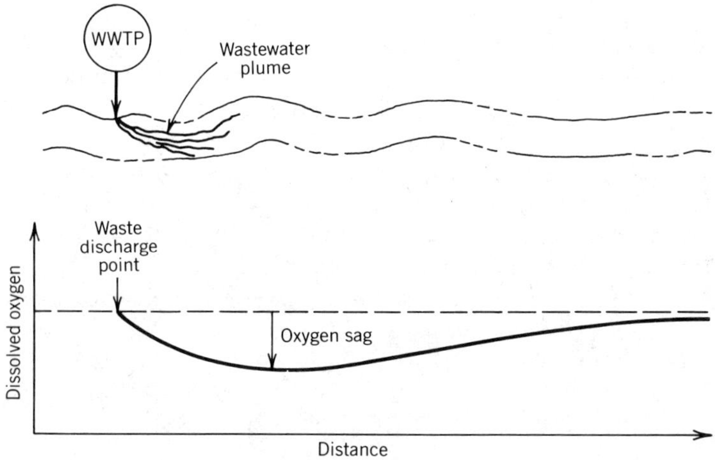

FIGURE 7.20 Nature of oxygen sag curve.

Even a discussion of the various models available gets complex quickly since there are so many modeling scenarios and so many different constituents. An EPA report was prepared to inform managers and planners about them, and it looked at six categories of the models, which will be listed here for the purpose of brief discussion (6).

Group I	Steady-state stream models
	DOSAG 1
	SNOSCI
	Simplified stream models
Group II	Steady-state estuary models
	ESOO1
	Simplified estuary models
Group III	Quasi-dynamic stream models
	QUAL I
	QUAL II
Group IV	Dynamic estuary and stream models
	Dynamic estuary model
	Tidal temperature model
	RECEIV
	SRMSCI
Group V	Dynamic lake models
	Deep reservoir model
	LAKSCI
Group VI	Near field models
	Outfall plume

These models apply to different hydraulic situations and can be used with different water quality constituents. Most can handle oxygen. Many of them can handle conservatives of nondegradable substance. Some can handle nonconservatives with first-order decay. The more complex the reactions involved, the further away the models are from being used in actual regulatory decisions.

All in all, water quality is a very complex field, but its very complexity lends itself to management controls since the experts often cannot agree on the exact impacts and controls that should be used in decisions. This leaves the field to the managers who can use judgment to make the best decisions under the circumstances.

THE CONSTRUCTION PROCESS

Although it is as much management as engineering, the construction process receives most of its attention from the engineers, and the ultimate quality of construction is mostly dependent on engineering inputs. The aspects of construction that are critical to manage properly begin with the design process, and continue through construction itself, including inspection and construction management. The design process includes not only the construction plans but all of the legal and contract documents that are necessary to ensure a quality outcome.

The overall process is somewhat as shown in Figure 7.21. Every situation is unique, of course, but there are some generally applicable procedures that are always followed, as shown in the figure. There are two distinct phases: the design phase and the construction phase. The design phase may either involve the staff of the agency itself or a consulting engineer/architect. The decision about which alternative to use is a management decision. For minor projects where the capability clearly exists in-house, it seems natural to use the agency's engineering staff. Where the projects are large and of a one-time nature, it seems natural to go outside to a design firm.

Figure 7.21 shows four inputs to the design process: design concepts, financing, the selection of an engineer, and a decision to proceed. All of

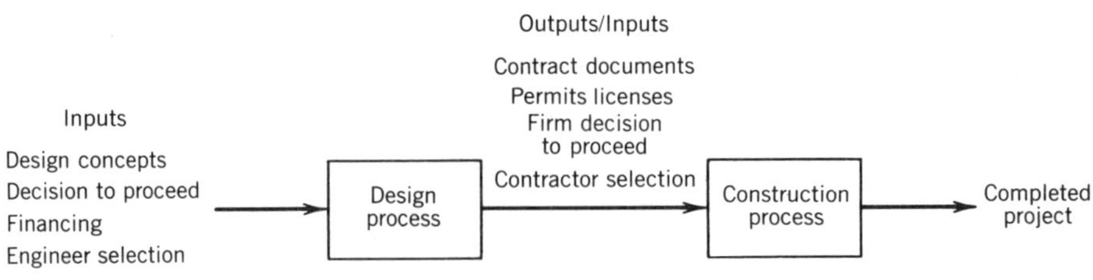

FIGURE 7.21 The design/construct process.

these are critical since considerable money is on the line, even in the design phase. The design concepts may have been worked out by the staff of the agency or by an engineering firm in a preliminary design study, depending on the complexity of the project. It is a healthy sign for the staff of the agency to have enough expertise to know what they want and need. This is particularly important when automation is foreseen, due to the need for the operators to accept the automation and make it succeed. This is discussed more in Chapter 10.

The outputs of the design process are the contract documents, the permits and licenses, the selection of the contractor, and a firm decision to proceed. These are substantial outputs, of course, and represent quite an investment before a spade of earth is turned. The investment at this point of a project can run from 10% of the total project cost upward for complex projects. If difficulties in licenses develop, or problems with something like an environmental impact statement, the cost can be even higher.

Design Process

The design process is the creative stage of determining how to solve the many problems encountered in developing the facility. Good design engineers will have experience in anticipating and overcoming the difficulties that arise. The outcome of the process is a representation of the final project on drawings and plans, with the purpose of communicating to the contractor, the owner, the financiers, the regulators, and the suppliers what is contemplated in the project. The process requires considerable interaction with the staff of the urban water agency to make sure that all of the design decisions are consistent with the overall operation of the system. Good working relationships are essential to the success of the interaction, and the success of the overall venture depends greatly on the outcome. For this reason the owner and the engineer will have project managers, and it is essential that they develop this good working relationship. Later they will work with the contractor's representative, and the need for working together will be even greater.

In the design process it is necessary to use the latest technologies such as computer-aided design. This can facilitate the success and the speed of the design process and improve the results of the investment of public money in the system under design. Still, since design involves creativity, we will never be able to automate everything.

The profession of consulting engineering offers complete design services for urban water systems, and some of the best knowledge about innovations and advances comes from the experiences of these firms. One distinguished professor in the environmental engineering field always recommended to his students that they start their careers in a top consulting engineering firm, since he believed that the influence they would find there would be better than in a public agency. However, the experience available in public agencies can also be outstanding, and the opportunity to work under a really dedicated public water manager can be invaluable.

The net result for urban water managers is that they must know something about the art of design, either for in-house work or to manage the work of consultants. In either case the outcome should be the best set of contruction documents available for the design concept that thas been determined.

There are several ways to test the quality of the design work. The evaluation of the design concepts is the first step, to determine if the plan is the best. This is a problem that has one foot in the planning process and the other in the design process. If the concept is valid, then the actual design must be evaluated. Are the processes and equipment the best? Is the layout good? There are many such questions. To test designs, some have taken to employing an independent firm to evaluate the work of the basic design firm to see if it can be improved. One approach to this is to use the "value engineering" procedure.

Value engineering is an approach to finding the best design of a system to achieve the operational objective of a system. Whereas the concept is applicable in general to construction, it has been applied in urban water systems for specific purposes, the best known of which is to meet the federal requirement that wastewater treatment plants be cost effective. This was a requirement to receive funds from the construction grants program. According to EPA, value engineering is:

> A systematic and creative approach to identify unnecessarily high costs in a project in order to arrive at a cost savings without sacrificing the reliability or efficiency of the project or increasing operating and maintenance costs. (7)

Value engineering is a concept, and there would not be a hard and fast way to implement it, but the EPA report cited lists some steps that would normally be included in a study. They consist of five phases. The first would be the information collection step, where different methods are considered and basic information on the system and the objectives are established. The second phase is the speculative one, where some creative thinking is done about different ways to achieve the objectives of the system. This might be thought of as second guessing the creativity of the original design team. The third phase is the analytical phase, where the suggestions that were generated are evaluated for cost and performance. The fourth phase is one where the proposals are developed further and more specific costs are prepared and documented, and the final phase is to present the results to the clients.

EPA has prepared an interesting evaluation report of their experience in using value engineering (VE) in the wastewater treatment design process (B). They evaluated some 90 reports from value engineering workshops where teams were assembled to evaluate the designs of a wastewater treatment facility. In general, the process is viewed as very successful, and the report presents suggestions for getting the most from the process. Savings were on the order of 5% of the project costs, and the ratios of savings to the costs of the studies were from 10 to 25, representing high benefit–cost ratios for the program. Suggestions for implementing the program included the following:

Use standardized reporting formats.

Minimize the number of firms represented on a VE team to maximize interaction.

Avoid using engineers from the original design firm on the VE team.

Allow the recommendation of only ideas that the VE team members would use in their own designs.

Make sure that each recommendation is thoroughly developed.

Use trained VE team members wherever possible (there is a 40-hour workshop available).

Use at least two sanitary engineers on the team since they control over 70% of the spending decisions in the design.

Maximum benefits are obtained from a VE study at the 20–30% design point.

The report also describes some of the other interesting phenomena that would be expected, such as the pride of ownership and the tendency for design firms not to accept VE recommendations for various reasons having to do with human nature.

When the design process is completed, the outcome should be a set of workable construction documents and a firm decision to proceed if all of the permits, licenses, and other requirements have been met. When they have been met, it is time to advertise for bids and to enter the construction process.

The Construction Process

The construction process requires considerable documentation. Korbitz, in his text on public works management, lists the documents that are required as follows (9): bid advertisement, information for bidders, contract form, general conditions, detail specifications, plans and drawings, and bonds and other certificates. With all of the legal aspects of these documents, the importance of having good advice during this stage of construction is clear. Contractors and engineers must resort to protection through bonding and liability insurance.

The important concern from the standpoint of the urban water system manager is that the construction be performed on time, within the budget, without excessive disruption of the other urban systems, and in a quality manner. The importance of the project manager in this process is obvious, as well as experience in construction management itself.

If a system is accepted without the proper checks and guarantees, the result will be continual grief during the operation and management phase. The only way that the owner can be sure of the performance is to have adequate inspection, testing, and prove out processes. To illustrate this I would like to describe a joke I saw once about the importance of inspection of a sewer line. The completed sewer line was supposed to be straight, and to test it the resident engineer looked through it to see if he could see a light at the end. Since the line was not straight, but bowed, the contractor took a roller skate and lowered the light to the bottom of the bowed part, satisfying the inexperienced inspector. Imagine how this line would have operated!

Once all of the checks have been completed and all of the operating systems have been proved out, the system becomes the responsibility of the urban water authority, and the operating and maintenance phase begins. Chapter 9 describes it in detail.

REFERENCES

1. Timothy R. Lazaro, *Urban Hydrology: A Multidisciplinary Perspective*, Ann Arbor Science Publishers, Ann Arbor, MI, 1979.
2. Luna B. Leopold, Hydrology for Land Use Planning: A Guidebook on the Hydrologic Effects of Urban Land Use, USGS Circular 554, 1968.
3. M. B. McPherson, Need for Metropolitan Water Balance Inventories, *ASCE J. Hydraulics Division*, October 1973.
4. Wright-McLauglin Engineers Inc., Urban Storm Drainage Criteria Manual, as revised by the Urban Drainage and Flood Control District, Denver, 1984.
5. Bruce H. Bradford, Neil S. Grigg, and L. Scott Tucker, *Stormwater Management*, Vol. 1, *Urban Hydrology*, Water Management Science, Atlanta, 1979.
6. G. Paul Grimsrud, E. J. Finnemore, and H. J. Owen, Evaluation of Water Quality Models: A Management Guide for Planners, EPA-600/5-76-004, July 1976.
7. Environmental Protection Agency, Procedural Handbook for Value Engineering, EPA-430/9-75-020, December 1975.
8. Robert B. Williams, R. G. Gumerman, W. R. Whittenberg, and N. E. Heim, Recommendations from Value Engineering Studies on Wastewater Treatment Works, EPA-430/9-80-010, September 1980.
9. William E. Korbitz, Engineering Management, in *Urban Public Works Administration*, W. E. Korbitz, Ed., International City Management Association, Washington, 1976.

CHAPTER EIGHT

COMPUTER-BASED DECISION SUPPORT SYSTEMS

IMPACT OF COMPUTERS ON THE MANAGEMENT OF URBAN WATER SYSTEMS

So far we have been discussing the urban water systems themselves and techniques for planning, designing, and managing them. Now we turn to three chapters about the planning and operation of facilities in the computer age.

Every manager has at least heard about the value of computers in managing organizations and enterprises. The more complex the undertaking, the more the computer is needed. But for what exactly are the computers needed? Are they just expensive toys, soon to be put on the shelf? Of course, some, such as mainframe computers for writing checks and keeping up with finance, have proven their value, but what about all the recent publicity of personal computers and computer-aided design?

Many managers know that they can be sold a "bill of goods" in a computer system, and for them it is essential to know for what the computer is going to be used. To illustrate this point a recent humorous management article described a professional who went through all the process of buying a personal computer with all the promises of the things it was to do for him. When he finally finished learning how to use it, he realized that all he could use it for was to catalog his neckties! While this illustration probably does describe some overzealous computer fans who went out and bought computer hardware before they knew what they would use it for, it ignores the real potential impact of computers in transforming the workplace, including the urban water system workplace. I for one am a believer that work will be permanently changed at all levels by the computer, and we need to plan for the changes realistically.

There are many uses for the computer in water organizations, and the effective manager needs a clear idea of them. They seem to cluster in three categories, arranged in order of the number of applications: office automation, decision support, and automatic control of operations. Office automation is much the same in most organizations, and in water utilities there are many applications for word processing, spreadsheet analysis, data base management, and communications. This text is more concerned with operational management and will stress the latter two categories, that is, decision support and automatic control of operations. Some of the computer uses cut across the categories of applications, however, and will support each other. For example, data base management can be useful in the office to mail to customers and to the maintenance forces to schedule inspections and carry out related duties. This chapter describes the uses of decision support systems, and Chapter 10 covers automatic control of operations.

In general, the use of computers to improve productivity of organizations is growing rapidly. The main questions are how the machines will be used rather than how much storage they have and how rapidly they are able to "crunch numbers." A recent book on the transformation of work in the computer age addresses the question of how the computers will be used, and seems right on the mark (1). The author, Paul Strassman, a vice-president of the Xerox Corporation, stresses that improvements in efficiency and the speeding up of work are not the aims in using computers; the aim is to use them to simplify the organization, serve the customer better, and make the team and the individual worker more effective. This thinking by an executive of a major information company illustrates that thinking in some quarters is now shifting from how to make machines bigger and faster to how to use them to organize work better.

The principal challenge, according to Strassman, will be to handle the unstructured management problem, for which flowcharts and logic are not applicable. The routine tasks will be handled well by computers, so in effect, work is pushed up in the organization chart, and the workers who were performing routine work will have to do their bosses' jobs in order to continue their growth. The way to make management more productive is to measure the "value added" by each function. Strassman states that this can only be done when the value of a product to a customer can be measured, and the approach cannot be used in overhead situations, which can only be measured by taking away the value added of the individual other units from the total value added, and the residual is the value added by management.

Strassman is probably right that there is not a direct way to measure the value added in monopolistic situations such as utilities, but it appears to me that there will be some surrogate measures than we can use to measure the productivity. This is one of the main challenges to management in the computer age.

Another interesting observation by Strassman is that the individual work station will be the most important element in the office of the future, and this seems to be happening in water utilities as well as private businesses. This chapter closes by making some forecasts about an ideal water management organization of the future, one based around the individual work station as the main management element.

In searching for the model of how computers will impact the generalized organization of the future, we need to look for the evolving concepts for organizational structure and the use of information for management and decision making, one of the most important tasks of the manager. In this chapter we use the term *decision support system* (DSS) to refer to the use of computers to develop and display information to improve decisions. We use the abbreviation DSS in anticipation of being accused of making the use of computer jargon worse. The term is a partial successor to the terms *automated data processing* (ADP) and *management information systems* (MIS), and most will remember those terms from the beginning of the computer era. Decision support implies, however, more than just the processing of data; it includes considerable analysis, some using computer models that actually add value to the raw data. In other words, the DSS is the system that organizes the processing, analyzing, and delivering of information that is necessary for decision making.

The main points of the DSS are shown in Figure 8.1, which illustrates the flow of requests for decision support from the decision maker to the support staff. There are two main activities in the DSS, data management and studying alternatives. These are the activities that convert data or "information" into knowledge that is useful in the decision-making process. The role of the DSS is thus to bring the data together with the studies, using models if necessary, to produce the decision support.

How is the DSS different from the traditional staff activity to make recommendations to the boss? It really isn't that different; it is just a way to organize the concepts better and arrange to handle more information through the use of computers.

One problem (or opportunity) associated with using large numbers of computers is how to organize them so they will "cooperate." This humanlike problem is one that is attracting the attention of researchers already since computers work in organizations just as humans do. Later in this chapter and in Chapter 10 we discuss how computers can be organized to do work in a hierarchial fashion, which is the generally evolving pattern. The researchers are already asking questions about what will be the lessons of how computers can work together better (2). One re-

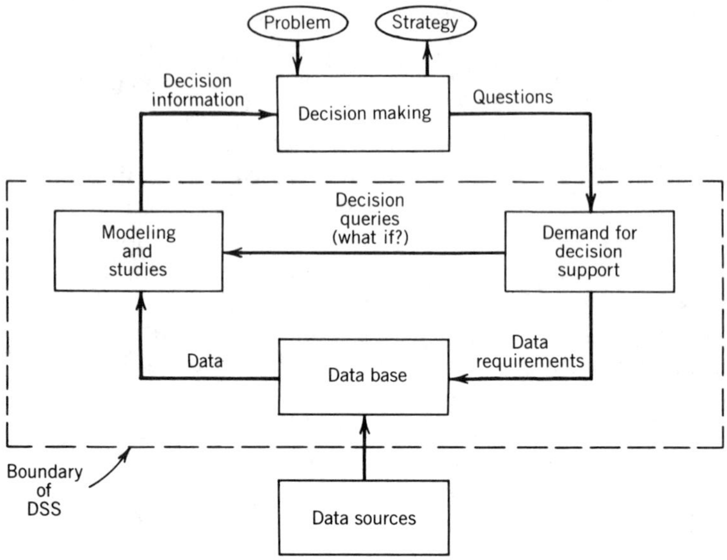

FIGURE 8.1 Activities of the decision support system.

searcher has been modeling groups of "agents," meaning computers in an organization, and he has found that when they do not know each other's goals they cooperate since that is the best way to achieve their own goals. When they are able to communicate with each other, they begin to take on some negative human traits such as manipulation, withholding information, and outright lying. They also establish ad hoc organizations. Certainly these actions are somewhat dependent on how the machines are programmed in the first place, but this illustrates some of the problems that are ahead.

TYPES AND STRUCTURE OF DECISIONS

In organizations both the management of data and decision making cluster around the organizational form. This is one reason that having a good organizational structure is important and a reason to be very careful before reorganizing an existing unit of workers due to the loss of data linkages and communications patterns. Before the computer, when data bases could be less formal, every department would have its own files, and both information management and decision making could be very customized. This, however, is a vestige of the past, and although we often desire to retain comfortable old patterns, the need to maximize the sharing of data and to make decisions that affect the public "in the open" both work toward improved organizational structures and data base management.

Decision making is one of the most important activities in management; some would say the most important. It is distributed throughout all levels and parts of organizations, many with dissimilar functions but with needs for the same information. For example, in an urban water utility almost all parts of the organization would need up-to-date statistics on the number and distribution of customers. Some say that in an organization all workers are engaged in one of two activities: "doing" or "deciding." It is always nice to be "deciding," as that implies a higher level of work, but we have to keep in mind that "decision making" refers to all levels of the organization, not just the top executive levels.

The strongest influence on changing organizational forms is the need for communication and for decision support, so as information flows become more sophisticated we will start to see them affect organizational forms more and more.

In water organizations the types of decisions will begin with the basic commodities that are dealt with: amount of water, rate of flow, pressure, depth, quality, condition of facilities, and money. In effect, for each of these system parameters we are interested in the level at any time and the rate of change. Decisions will eventually be made as regards the change of these parameters in time and at different places. These are the considerations that identify decisions and the need for data.

The main categories of functional decisions seem to have to do with operating and with finance. In the former case they would include to open or close a gate, to fill or empty a reservoir, to turn a pump on or off, or to maintain or repair a piece of equipment. Financial decisions would include rate setting, equipment purchases, contracting of services, and many other business decisions.

There are several ways to classify decisions according to the stage of management or the level of detail. Some like to refer to them as political, management, or operating, but this classification is about the same as that that refers to the stages of problem solving: problem identification and policy setting, planning, and implementing. Decisions have different "content" also, as Korbitz points out when he refers to the factual and the value content of decisions (3). This is a useful distinction since one would suppose that the factual content of decisions would be relatively nonnegotiable, whereas the value content will require trade-offs and political analysis. These distinctions would also be typical of policy versus operating decisions. According to Korbitz, the operating decision, as opposed to the management decision, would be subject to more definable rules, usually on a shorter time span, more repetitive, simpler, less risky and uncertain, and based more on knowledge of operational data.

The importance of good decision making lies in its link with action. It is only after a decision is made that action is possible. Management studies that look at effective organizations have concluded that a bias toward action is one of the characteristics of success, and a corollary to that might be that effective decision making will also be present.

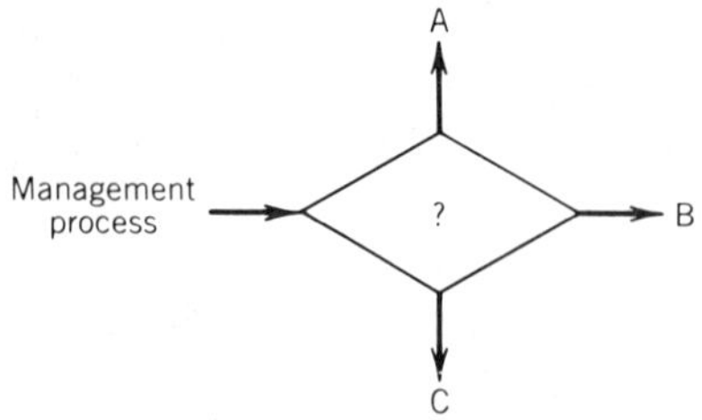

FIGURE 8.2 The structure of a decision.

A decision is the act of setting a course of action or making up one's mind. The act of making the decision is a singular point in the management process, as shown in Figure 8.2, and decision making and its aspects is part of the science of management. Many articles and books have been written on decision theory, decision making under uncertainty, and executive decision making. Most of these have to do with the quantification of decision making and with the use of statistics to estimate levels and place numbers on risk. It is very difficult, however, to teach anyone how to make the more difficult decisions that have so many political and subjective aspects, two characteristics that certainly apply to water management. For this reason decision making remains an art and is not amenable to pure quantification. Information is vitally needed, but it cannot be reduced to equations and numbers only.

In operational water management, decisions usually fit one of three types: how much, which one, or when. Examples would include planning the size of a treatment plant (how much); selecting a pump (which one); and staging an investment (when). Questions such as "what if" sometimes appear closely related to decisions, and may be subjected to the same analysis process, but they are not really the decisions themselves. They support the decision questions and are subordinate to them. The general question, what to do, is not really at the decision stage. It is at the earlier stage of the problem-solving process that has to do with formulating alternatives. In other words the supporting questions that relate to decision making are relevant at the different stages of the problem-solving process, and naturally lead to the clear decision question (see Figure 8.3). Decisions should be presented as clear choices and not muddled questions. This will improve planning and executive action.

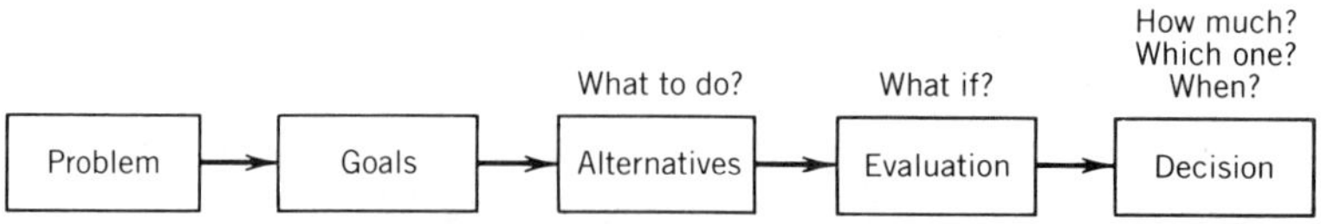

FIGURE 8.3 Sequence leading to decision.

Decision making in organizations is linked to the specific tasks or activities of the organization. An example of this can be seen in a recent study for the U.S. Army Corps of Engineers that developed recommendations for an integrated information system for the office of the Chief of Engineers (4). This study sought to link the business of the office to the information flows and data bases. More than 80 data bases were identified. They were following a process developed by the IBM Corporation for business system planning, which begins with a definition of the business that is engaged in. This is followed by a delineation of the business processes to be followed, independent of the position on the organization chart. These are related to clusters of data systems, called "data classes," that are mutually exclusive and independent of the organizational structure. The matrix that relates the business processes to the data classes is called the "information architecture" of the organization. This matrix identifies the key decision points of the organization and relates them to the information collected and utilized.

This approach can also be applied to a water utility, although a full study has not been done, but a general analysis can point out the basic decision-making structure of the organization. Figure 8.4 represents a hypothetical self-contained water utility with four levels in the organization; executive, staff, supervisory, and operator. Decisions are required throughout this structure, both as related to the level and to the tasks of management, planning, organizing, and controlling. Actually a matrix with at least three dimensions is required to illustrate these decisions since the management function must be considered as well as the level and the stage of management. Figure 8.5 illustrates a typical matrix of decisions and information needs as they relate to some of the activities of

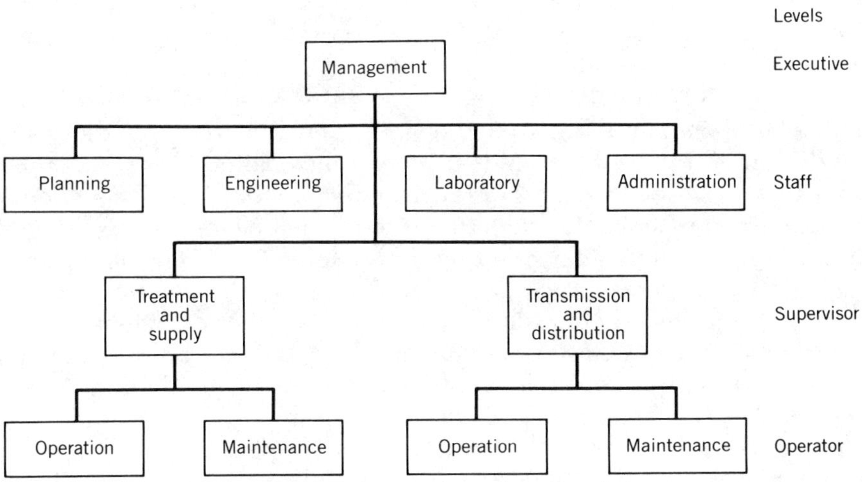

FIGURE 8.4 Hypothetical water utility.

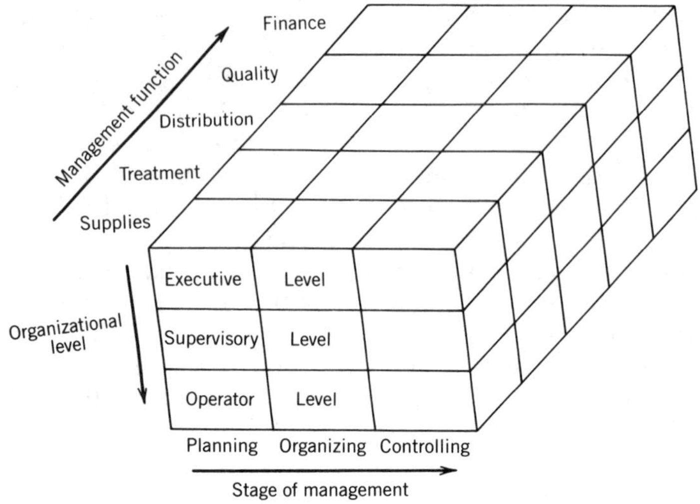

FIGURE 8.5 Three-dimensional space of decision making in water utility.

the water utility, but it is not a complete classification since the complete functioning of an organization cannot be present on such a simple display. The way to view Figure 8.5 is to see the decisions and actions that are required at the different stages of management for the different functions shown and consider that different decisions will be required for the different levels in the organization. Figure 8.5 can be said to define the decision-making "spaces" within a water organization. A complete presentation of the information shown would require far more space however.

The term *decision making* is also often used in management science and operations research. In those fields decision problems can be presented in terms of variables that describe the systems being managed. Hall and Dracup (5) present a good exposition of the basic concepts. A system to be controlled is subjected to controllable and uncontrollable inputs. As shown on Figure 8.6, the various characteristics of the system are described by "state variables," sometimes called "level variables" since they describe the states or levels in the system. The system boundary is defined in a manner such that the external factors are represented by the "environment" surrounding the system. The controllable inputs are called the "decision variables" since they represent the quantities that are controlled by decisions. The three types of decisions referred to earlier serve as examples: how much, x; which one, a or b; and when, t. These are typical decision variables in the language of operations research and systems analysis. "Constraints" limit the room to choose the decision variables, and choices of the decision variables are said to represent a "policy." A policy that does not violate the constraints is said to be a "feasible policy."

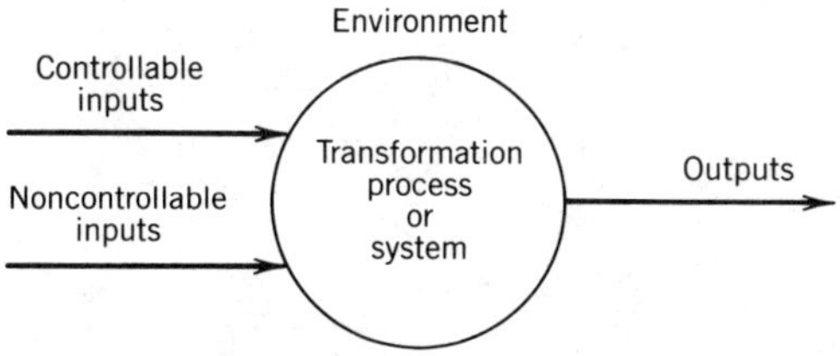

FIGURE 8.6 Decision making variables.

COMPUTERS AND THE DECISION SUPPORT SYSTEM

Although decisions refer to all levels in an organization, the most strategically important ones take place at the upper levels of the organization. This is normally white-collar work, and the most important challenge to improving productivity today in organizations is to improve the decision making and the use of time by managers. These managers may be working in the field and be engaged in very nitty-gritty work, but if they are managers, they are referred to as white collar in the jargon of the labor force. Their office can be a shack at a construction site and the desk can be a table there or their office can be at the top of a downtown highrise with the corresponding fancy surroundings. The challenge is summed up in the books and articles that are appearing about "white-collar productivity."

A few years ago I saw a diagram that explained white-collar work quite well. A worker sitting behind the desk, as in Figure 8.7, processes work in a standard fashion. The work came in through the in-box, was moved to the desktop where the worker applied logic, rules, procedures, and data from storage, and a decision or product was dispatched through the out-box. A calculator was available for computations, and a phone for communications with other workers. It is precisely these processes that computer hardware and software companies are rushing to automate. One classification listed the work in the office in four categories: document management, personal computing, support for decision making, and communications. All of these are shown in the "manual office" in Figure 8.7.

The role of the computer is to assist, not replace, the worker. Figure 8.8 is a depiction of the basic processes of the computer, showing how the inputs and outputs are arranged. Figure 8.9 shows how the computer memory relates to the inputs and outputs. The memories of computers are specified in kilobytes, normally eight bits of information, and one kilobyte is necessary to store one character of text. In some applications of computers, such as on–off process control decisions, the bits of information themselves are sufficient to store the necessary types of data. A personal computer might have, for example, 256 K of internal memory

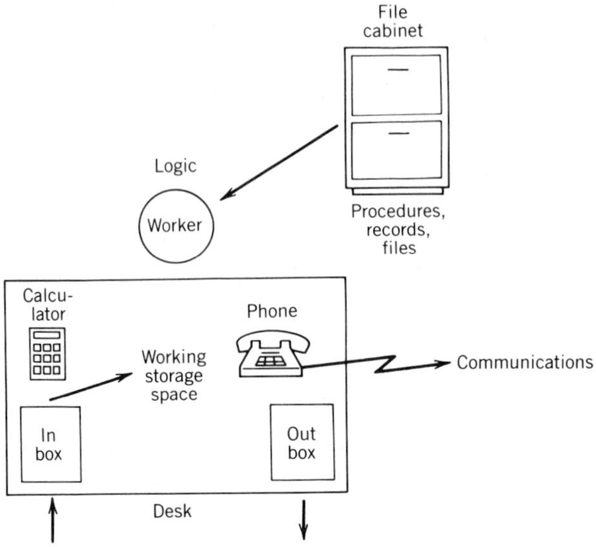

FIGURE 8.7 A desk-based working system.

and 10 megabytes of storage in a fixed disk. There are about 1500 characters of text on a page so the maximum capacity of the 256 K memory would be about 170 pages, and the fixed disk would have a capacity of about 6670 pages, with appropriate deductions for the operating system and the software, of course.

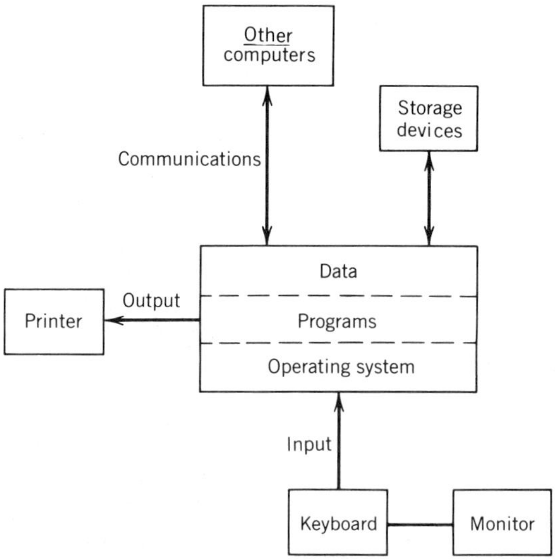

FIGURE 8.8 Elements of a computer.

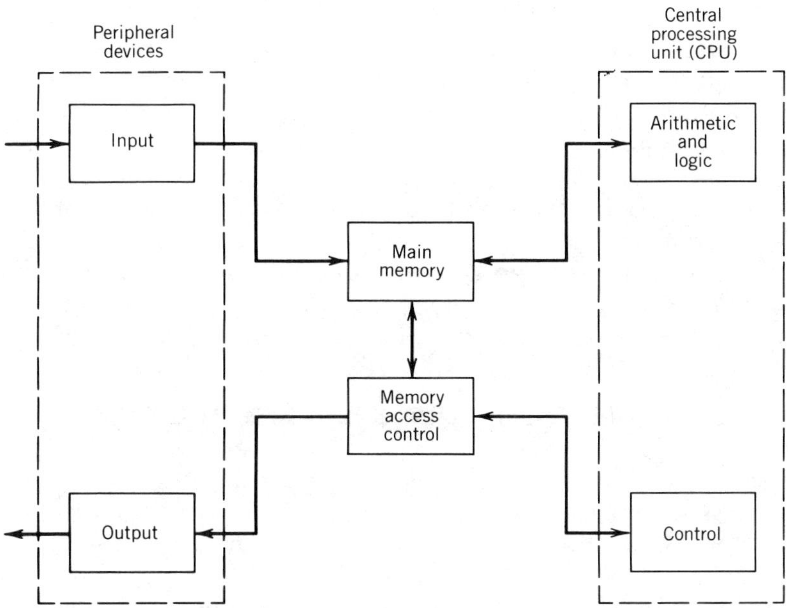

FIGURE 8.9 Roles of main memory and central processing unit in a computer.

The "outputs" shown in Figures 8.8 and 8.9 can be commands to print out results, or other outputs such as control commands for machines. These are similar to commands produced by the human brain, which has similar but far more complex functions than the computer. The functioning of the brain provides lessons for us toward improved uses of the computer for decision support and for commanding machines in the operational mode. Due to the similarities it seems helpful to understand the way the brain works to process information, make decisions, and issue commands; then the understanding of the computer possibilities is improved.

The similarities and differences between the human brain and the computer are startling. On the one hand the computer can do many of the simplest things that a person can do through reasoning, calculating, and communicating, but on the other hand the capabilities of the brain dwarf those of the computer.

The small computer described above would have 256(1024) = 262,000 bytes of storage capacity. Using the 8-bit byte, this would mean some 2 million storage registers with on–off information stored. These storage cells would not be able to provide any information other than 0 or 1, on or off. On the other hand the brain and central nervous system have some 10 billion to 1 trillion neurons, which are the basic units of information storage in the body, and many of these are different in character. Moreover, each of the neurons is connected to others with as many as 1000–

10,000 connections or neurotransmitters. Going beyond that, there is an electrochemical dimension to the transmission of information in the nervous system, meaning that in addition to the digital information capability there is an analog capability, with many dimensions corresponding to the different chemicals involved (6). With all of these complexities, it is no wonder the brain is the subject of so many inquiries and mysteries.

The interesting thing about the comparison with the brain for the support of decisions in water utilities and the use of computers is that computers and the use of information for decision making are evolving in the direction of the brain and its processes. This is the concept behind "artificial intelligence" and the "fifth generation" of computers.

ARTIFICIAL INTELLIGENCE AND EXPERT SYSTEMS

The rise of new concepts of using computers to think and help make decisions suggests that they will be able to assume more and more human tasks. As this was being written, I reviewed an article that suggests that a consulting organization has developed a hardware system that can reduce the number of required operators in a treatment plant by installing new computer hardware andsoftware at a very attractive price. This is an example of the emergence of automated control that relies on decision making and physical controls of valves and other devices. The controls have been available for a long time, and now the automated decision making is becoming more and more available.

Although it is clear that there is no danger of computers replacing humans completely, there are some very interesting parallels between automatic control systems, expert systems, artificial intelligence, and human decision making, especially the rather "automatic" human decision making that leads to the control of the body processes.

While the concept of artificial intelligence may seem remote to some, there are a number of very interesting potential applications for urban water systems. One of these applications, "expert systems," is something that has already been introduced in some water systems on a preliminary basis. Water organization has often been ahead of other fields in developing innovations, and some of the thinking about expert systems has been an example of this forward thinking. It is a natural arrangement since the expert operators who make the decisions about water are using experience gained over many years, and it is natural to try to utilize their knowledge in computer applications of expert systems. Another reason to use them is the fact that water decisions should be based on historical observations of rainfall and runoff data, and the data can be part of the expert system.

Some are not clear about the difference between artificial intelligence and expert systems. *Artificial intelligence* is an overall term that includes expert systems as one aspect, so it is the more general term. Expert systems can be defined as follows:

> An "expert system" is an intelligent computer program that uses knowledge and inference procedures to solve problems that are difficult enough to require significant human expertise for their solution. The knowledge necessary to perform at such a level, plus the inference procedures used, can be thought of as a model of the best practitioners of the fields.(7)

The report cited goes on to state that an expert system consists of facts and "heuristics," where the latter are rules of thumb and "know-how," such as that which is widely used in operating urban water systems. The report lists three parts of the expert system: a knowledge base of facts and heuristics, an inference procedure to use the knowledge base for problem solving, and a working memory.

There is an evolving research field called "knowledge engineering" which is used to prime the expert systems. Experts are interviewed to extract from them what they know about a particular problem. For example, a doctor can be interviewed about the knowledge used to diagnose an illness. Some say this field will soon overcome the processing of data as a main activity of the computer industry since so much of business consists of reasoning and problem solving rather than pure manipulation of numbers (8). This is probably overstating the case, but it seems clear that this will become an important part of the information economy. Already the commercial applications are beginning to appear, such as in medicine and geology (9).

We must be careful that expert systems are not oversold. One article in a recent edition of *Civil Engineering* suggested that expert systems are as promising as computers themselves in revolutionizing engineering practice (10), but the article reported few applications. Those reported, however, are familiar applications in water management: interpretation of data, diagnosis of problems, monitoring of systems, and design applications. The example given in the article is a water example. The expert being interviewed has a great deal of experience in selecting coefficients to calibrate the Stanford Watershed Model, and the expert system is called "HYDRO."

Artificial intelligence may seem too futuristic for a book on managing urban water systems, but the concept of using a computer to think and to help operate and make decisions will probably be attractive to most, and that is what the trend is all about. In fact, it is just possible that the terms *artificial intelligence* and *expert systems* will pass away and that we will continue making progress in using computers as we have for the last several decades. The concept of artificial intelligence is something to keep your eyes on, however, since a lot of effort is going into its development.

How specifically might artificial intelligence work in an urban water system? To answer that question, let's go to a simple engineering example of how it is evolving. Some professors at the University of Minnesota have begun using the concept to illustrate to students how to build an expert system (11). In this application we are using the expert system as our artificial brain, and our knowledge engineering application is to program the expert with the knowledge. The simple example given by the professors is of a student trying to decide how to use his evening for study so that he makes the best use of the time available. There are three general steps in the procedure: specifying the options, compiling a list of questions to elicit information needed to choose the right course of action, and building a set of rules that guide the decision. The rules are constructed in an "if–then" format.

The students in the class were able to use the procedure to build a more complex system that allowed the choice between techniques for blasting rock. The problem they selected was essentially one of making a design decision. We can formulate a number of similar problems in the urban water format. Let's choose one from operations and maintenance of a water distribution system. In the next chapter we will describe in some detail the problem of selecting repair and replacement strategies for distribution system mains. The problem is that the best replacement strategy is not based on age but on other factors such as frequency of repairs, structural loads, and corrosivity, just as examples. If we wanted to formulate an expert system to help us choose which pipelines to repair, rehabilitate, or replace, we could begin by formulating the options: for any portion of a pipeline in a given year, we could do nothing, test, repair, rehabilitate, or replace. There might be more options from combinations of these. The questions would be about the criteria for the decisions. For example, in the pipe being considered was there any problem in the previous year, is there a history of problems, is the corrosivity index above a certain value, and has there been any heavy traffic on the roadway? These are just sample questions, not meant to be complete at all. The rules for decision could be if a pipe had certain answers to the above questions then it could be placed into a pool for consideration for repair, rehabilitation, or replacement, depending on the budget available and the intangible factors. This is a very simple example, and we are tempted to dismiss it except if we consider that there are thousands of miles of pipes in a large system, and if they are divided into block-long sections for analysis, then the decision-making problem becomes too large for analysis without some system.

In Chapter 10 we look at the evolution of applications of expert systems in urban water systems as they are necessary to carry out real-time operation of systems under automatic control. They are not called expert systems by those developing the logic for the control, but advanced versions contain the same concept.

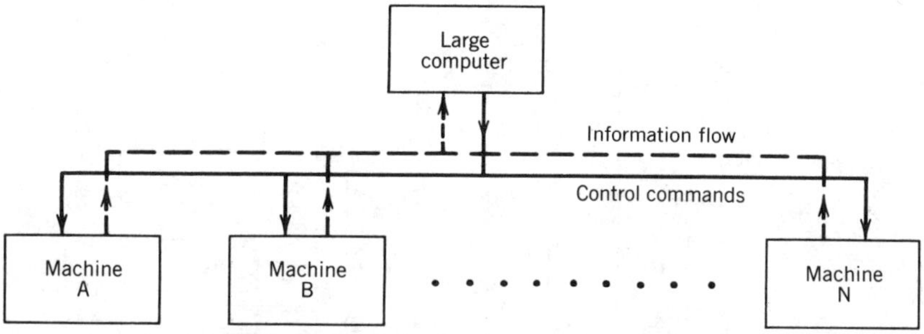

FIGURE 8.10 Centralized control system.

Although computers and control systems are still very crude compared to the human body, there are some very interesting trends, particularly in the distributed control of processes and in the automation of work in general. Figure 8.10 shows the general concept of a system of machines, all of the decision making being concentrated in one "large" computer. With the dramatic reduction in hardware cost, however, the trend has been toward distributed control systems such as those shown in Figure 8.11. This is similar to the way the body's control systems work with little distributed "computers" being distributed throughout the body in the form of different parts of the central nervous system. The brain is a fantastic central mainframe computer, and the nervous system provides a system for telemetry and control mechanisms that is very similar to the ones evolving on a crude basis in the automatic factory.

Since the brain is such a fantastic mechanism, I would like to show a little further how it compares to the working of control systems. The brain is actually part of a larger system called the central nervous system,

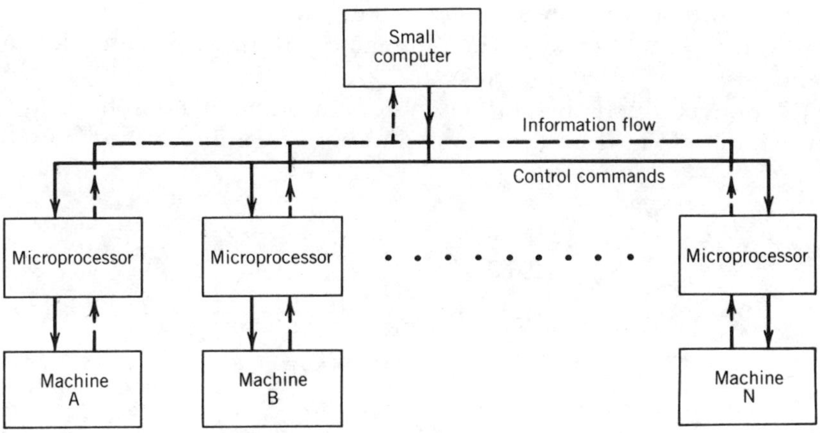

FIGURE 8.11 Distributed control system.

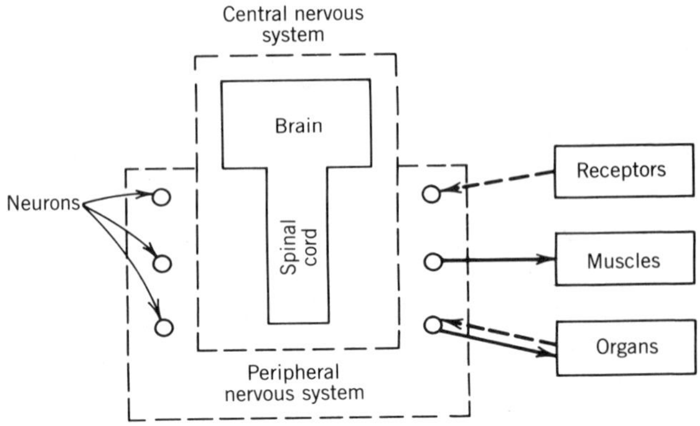

FIGURE 8.12 A simple model of the brain as a data collection and control device.

which includes the brain and the spinal cord. There is also a peripheral nervous system that includes the neurons that connect the central nervous system to the receptors, muscles, and organs at the periphery of the body (12). All of this is illustrated in Figure 8.12.

There are four aspects of the brain functioning that are particularly relevant to our discussion: the similarity to the concept of C^3I, the distributed processing, the connectivity, and the electrochemical operation.

Figure 8.13 shows a simple illustration of a neuron–connector–muscle arrangement. This illustrates the capability for decision making, command communications, and control of a single muscle control point. The neuron is connected to a vast number of other neurons in the brain and nervous system, and the decision making of this muscle can be coordinated with everything else that goes on. The comparable situation in an urban water system might be the operation of a single pump or valve arrangement, and as we know, we only can have the capability to coordinate their operation through human decision making at the present time. This illustrates the present gap between human capabilities and those of computers.

The connectivity in the nervous system through the neurotransmitters is truly impressive. If we accept the numbers given in Alexander (6) that

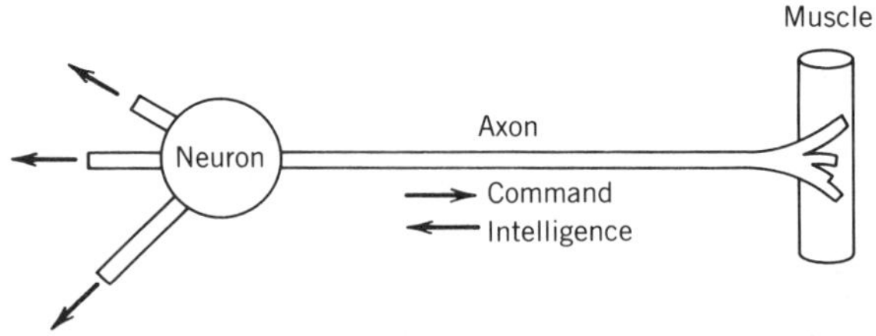

FIGURE 8.13 A simple model of neuron–muscle operation.

there are anywhere from 10 billion to 1 trillion neurons, each with around 10,000 connectors, then this would indicate that there are on the order of 10^{15} connection paths, a truly awesome number. The scale can be grasped somewhat by considering that each of the some 5 billion people in the world is connected to the others with 10,000 telephone lines, all going at once! And that example is some 100 times less than the maximum example of 1 trillion neurons given in Alexander's article. Think about the many possibilities for bypassing, decision making, and reponding to stimuli.

Since the structure of the nervous system also allows for electrochemical operation, we have the possibility of proportional or analog signals with qualitative changes due to the different chemicals involved. Computers only have the basic capability to generate on–off signals until programmed to do otherwise, so again the generic capability of the nervous system far outstrips the computer itself. Since man doesn't yet understand the brain very well (let alone the mind!) it seems that it will be a long time before the computer can be programmed to emulate it.

The really interesting thing about the control aspects of the brain is the arrangements for distributed control, as shown in Figure 8.12. This confirms that current directions toward distributed control systems in water systems are correct.

INFORMATION MANAGEMENT FOR DECISION SUPPORT

Computer companies have been eager to find ways to classify information uses in business to make the computer systems useful in managing the information. Five processes have emerged as useful in office automation and as compiled into today's integrated software: word processing, spreadsheet analysis, data base management, graphics, and communications. Other processes are in use, but these five seem to be the most frequently needed. My own current software has these features, and I find them to be very useful and complete.

However, the current software packages are only the leading edge of a cresting wave. *Scientific American* recently devoted an entire issue to the emergence of software, claiming that it is software that gives form and purpose to the computer, like a sculptor gives form to the clay (13). The analogy says that it is education and training that gives form and purpose to the human mind. The difference is that the human takes the training from all sources available in the environment, whereas the computer can only take it from the programs developed for specific purposes.

The software needed to manage water systems consists of the same facts and heuristics previously described. The need for facts in water management is undisputed, but the methods to get heuristics into the equation are still being developed. We approach the management of the facts through "data base management", but the management of heuristics must be through the development of expert systems. This is still mostly a research field rather than one that is ready for full application, although there are some applications ready in automatic control, as will be described later.

Data bases are mostly organized around the working units of the organization, beginning with the file drawers of each manager or supervisor and proceeding up to large central files. A "management information system" would refer to a coordinated arrangement of many of these, with the capability to access them when information is needed for management. The technology for developing and using the MIS is still in its infancy.

The data base builds on a hierarchy of information, beginning with the bit, or basic binary digital unit, and proceeding through the byte, the field, the record, the file, and the data base. Before the computer we were used to thinking about files in drawers, with the records in the files. When the computer "IBM card" came, the concept of fields of information became known as those which would go on one card. Now that cards are not used as much, we have a new set of concepts. The ultimate result, the data base, can simply be thought of as a related set of files.

There are substantial advantages to data base management. Consider two engineers sitting side by side in a water management district office. One of them, the reservoir and supply manager, has files needed to manage the reservoir, consisting of historical levels, releases, and inflows. The other, the treatment and distribution engineer, has information about customer demands, location of taps, treatment rates, and valve locations. If we enter all these files into a central data base, then both engineers can access all the information, and the capability for analysis can be greatly extended.

There is an interesting sociological aspect to the above example. The adage that information is power is true, and some are not willing to share the information they have. In some developing countries, for example, the governments are unwilling to make public even data such as weather information, fearing that it will be used by their enemies. In offices everywhere workers may be reluctant to give up information they have personally compiled to do their jobs, fearing that the loss of the information will give another an advantage over them. These are very real constraints on the development of effective data bases.

Urban Water Data Bases

The data bases that are needed for decision making and management in urban water systems are similar in water supply, wastewater, and stormwater system. As shown in Figure 8.14 the data can be classified as data on inputs, on the systems themselves, and as outputs. The system data

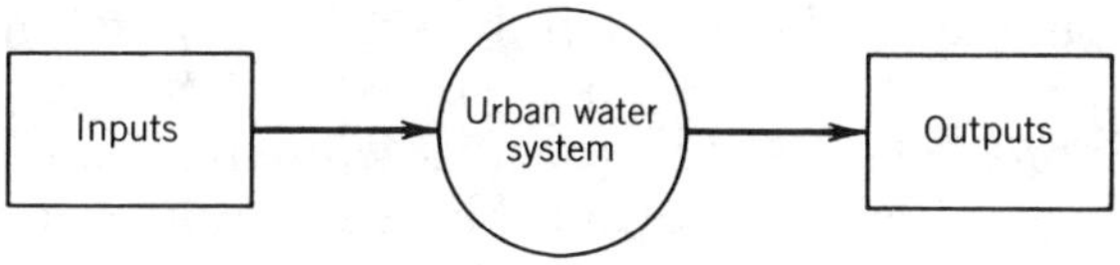

	Water Supply System	
Supply location Supply timing Supply quality	Facilities data Inventory Purpose Capacity Condition Maintenance history Operation data Flows Pressures Water Quality Levels	Demands by classes, location time and amount
	Wastewater System	
Input flows by classes, location, time, and amount	Facilities data Inventory Purpose Capacity Condition Maintenance history Operational data Flows Water quality	Receiving water, hydrology and quality
	Stormwater System	
Meteorology catchment data	Facilities data Inventory Purpose Capacity Condition Maintenance history	Receiving water hydrology
	Operational Data Event record Problems encountered	flows levels

FIGURE 8.14 Data needed for O&M of urban water system.

is further classified as facilities data and operational data. The focus on inputs, outputs, and the systems may be useful to engineers designing information systems, but these terms do not lend themselves to instant recognition by workers. As a result, we need to look for better ways to label the information systems.

Inputs are the noncontrollable variables that are imposed on the systems. The data management problem is one of keeping records and making forecasts. As shown in Figure 8.14, an example would be raw water supplies. The system data consists of variables that can be controlled, such as where pipes are located and how the flows in them can be managed. The output data is the result of the systems operating on the input data.

This simple presentation of the data structure does not tell the whole story, however, as there are several kinds of information that require different treatment. Some of them might be described as follows:

1. Historical data and records that go onto microfilm or microfiche. Examples would include as-built drawings and location maps.
2. Historical data that go into a data base and begin to form part of an eventual expert system. Examples include stream flows water demands, and problems with the system.
3. Inventory information that goes into a geographical data base for assistance in design and planning. Examples would include any utility data that can be drawn on a map by computer.
4. Real-time operational data that is used both for management decisions and to become part of the historical record and the expert system. Examples would include current system demands and meteorological data.
5. Real-time operational data that is only used for current decision making and is not needed in the historical record. Examples include distribution system water quality and equipment operating data.

One way to organize this data is shown in Figure 8.15, which has environmental data, operating data, and system data. The environmental data is independent of which system is of interest, but both the operating and the system data are dependent on the system, and subsystem, being considered. The components of the data base are interconnected and each part may supply or use information that can be used in other parts of the data base. For example, the rainfall component in the environmental part can be used to predict runoff, which then can be used to predict the supplies available. The demands for water supply provide data that leads to the prediction of demands for wastewater treatment. The whole data base is interconnected in this way.

Sometimes a forecast might need information from different parts of the data base; for example, supply as a function of time might be from an equation like this:

$$\text{Supply } (t) = F \text{ (climate, stream flow, watershed).}$$

More sophisticated models might consider lag times, such as that today's supply would depend on last week's rainfall:

$$\text{Supply } (t) = F \text{ (climate } (t - 7 \text{ days})).$$

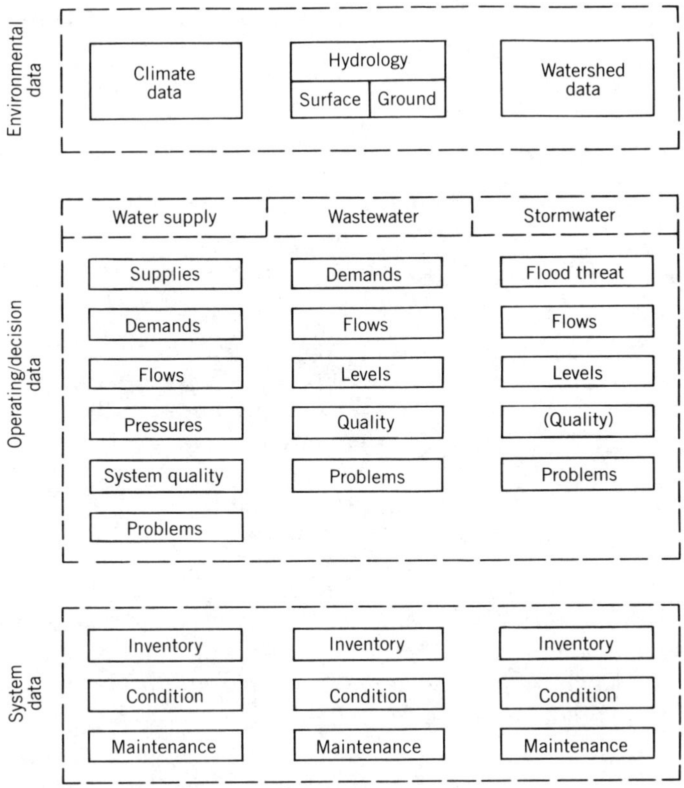

FIGURE 8.15 Data bases for urban water management.

As shown in Figure 8.1, which introduced the decision support system, the real use of data is to support decisions. The decisions can be "scripted," or in other words, there are decisions that occur over and over again. This is an approach that has been studied in the development of artificial intelligence systems (14), and one that merits consideration for urban water systems in the future, especially for operating systems. After they are scripted, they can be routinized and data bases can be brought to bear on the analysis. The starting place for routinizing decisions is with reservoir control, the subject that has received most attention from management scientists in research. However, there are many other important decisions in urban water systems, beginning with all the financial management questions.

For the reservoir control case, the script would be like this. An operational decision is needed to determine whether to release raw water, including when, how much, and at what place. The decision will have time, rate, location, and maybe quality dimensions. The data needed will consist of at least the following:

Climate data	To lead to forecasts of precipitation
Hydrology data	To learn about current inflows and outflows
Supply data	To balance current supplies and demands
Demand data	To know what demands will be
Watershed data	To relate precipitation to stream flow and to relate land uses to water demand
Wastewater flow data	To know what kind of in-stream dilution flows will be needed

As seen from this simple example, several of the data bases shown in Figure 8.15 are needed for this decision.

Another decision scenario is the planning problem of programming expenditures, say for a wastewater collection system. This would consist of the following data as a minimum:

Demand data	To know the flow characteristics in the wastewater system
Quality data	To know whether existing treatment facilities are adequate
Problem data	To know about the frequency and severity of operational problems
Inventory data	To know the characteristics of the existing system
Condition data	To be able to forecast the future needed repairs, rehabilitation, and replacement
Maintenance data	To study the economics of future maintenance versus replacement

The information systems needed to develop the kind of decision support needed for these examples are still evolving. System models are needed to aid in processing the information and determining what it means. This is the subject of the next section.

MODELS AND STUDIES

The second main element of the decision support system was the need to have a capability to study and analyze data. This infers the use of computer models but is not limited to them, as some studies and analysis requirements are simply too complex for the use of models by themselves. In effect, models and studies answer the "what if" questions that are required for decision support. The models and studies follow the organization chart just as the data base. The financial section will require financial models, and the engineering section will require engineering models of different kinds.

Studies that require models are, for the most part, planning efforts, and these have been described in the chapter on planning. In this section we will discuss models in some detail since they are important components of computer-based decision support systems.

There are many different kinds of urban water models. Table 8.1 lists some of them, classified as physical system or planning and management models. In effect, the physical system models are for simulating systems, and the planning and management models are for assisting in decision making. The physical system models should be considered to be in support of those for planning and management, although they can be used in detail level operation as well.

In a 1977 University of Minnesota study some 390 models were described, with 27 being selected for detailed study (15). I classified the 27 into five categories: general watershed models such as the Stanford Watershed model; flood simulation and stormwater models, such as the EPA Stormwater Management Model (SWMM); hydraulic and engi-

TABLE 8.1 Models for Water Supply Different Stages of Management

	Source	Storage	Treatment	Distribution	Use
Planning	Watershed models	Mass curve or balance models	—	Distribution systems models	Demand models
Design	Watershed models (greater detail)	Mass curve or balance models, (greater detail)	Component models (e.g., coagulation, sedimentation)	Distribution systems models (greater detail)	Demand models (greater detail)
Operation	Watershed models	Reservoir operation models	Process control models	Distribution systems model	—

neering analysis models, such as the Corps' HEC-2 that calculates flow profiles in open channels; groundwater models; and water quality models (16). These models fit, for the most part, into the category of physical system simulation models, and have to be coupled either with additional study or with other models before they can be used for planning and management.

Another way to classify models might be according to the following questions:

1. Which physical system does the model simulate? Which components are involved?
2. Does the model produce "facts" or "heuristics"?
3. Which stage of the management process is the model intended for—planning, design or operation?

This classification reveals that the available models are very heavily weighted toward physical system modeling and that their usefulness in direct decision support is limited.

Some of the applications of models in urban water supply systems are shown in Table 8.1, which shows the applications for different stages of management and different parts of the system. Examples given are from the system simulation viewpoint. The table shows that as the management process goes into more detail—that is, from general planning to operation—the detail required of the model increases. This illustrates one of the important aspects of modeling: the planning models must be consistent with the operational models, or there will not be consistency in the management process. This means that there is a "bottom-up" aspect to planning and management that should be recognized in water management, the same general principle that is behind the trend toward participatory management in organizations in other places.

There is a contradiction involved in modeling that should be recognized: on the one hand there is so much that models can do, and on the other there is not nearly as much application of models as there could be. There are reasons for the contradiction, of course, and models are like other parts of the early computer revolution; there has been some overpromising that has occurred. There seem to be five main reasons for the less than potential usage: lack of incentives, data problems, modeling difficulty, hardware–software confusion, and social barriers.

In the case of incentives my research suggests that where you find a lot of models and computers being used for other than office automation, you find someone in charge who relates to computers and models. Sometimes these persons are "odd balls" who are the same kinds who are inventing computers today, the "hacker" type. Much of the advances in models and computers is also due to good managers who recognize their value. Another factor in the use of computers and models is that they are sometimes required by contract. A flood insurance study may require a certain backwater program; a water quality study may require a certain waste load allocation model.

There have been many modeling studies that suggest that savings can occur, but few demonstrations that the savings have been realized. The future will undoubtably see more of them, and this will increase the incentives for using models.

In the case of data for calibrating models, only a few models have been well calibrated, and the data needed to make further improvements is very expensive and difficult to collect. The Stanford Watershed Model is an example. It has been around for nearly a quarter of a century in various forms, but there are still many parameters that would require guessing or the use of default values, which amount to someone guessing for you. The physical phenomena are admittedly complex, but that is true for all water systems. Take the urban water distribution system, for example. With all of the calibration needed, there is no way the model can ever be exact, and the cost for data to improve the validity of the model is very high. The biochemical properties of water also make water quality modeling very complex. There are many other examples from water management that illustrate the difficulty in bringing data to bear effectively on the solution of problems.

When modeling began in earnest about 1960, hardware limitations were severe. The mainframe computers available were less powerful than the small computers today, and advances were being made rapidly so that it was difficult to know what to expect next. This rapid development retarded the standardization of model efforts and left open the questions of proving the validity of applications through evaluation studies; in other words, there has not been much quality control in modeling. Software differences, different operating systems and computer languages, and the different approaches to problem solving by different programmers have also contributed to the situation today. However, much of this is being "shook out" through marketplace competition, and the profession is beginning to settle on some of the best models.

Human acceptance of models and computers is another large factor. The problem has several dimensions: the manager that rejects models without a fair hearing, the old line engineer that feels threatened by models, the operator who does not want his experience replaced by a machine, and the lawyer who would not accept the use of a model on the theory that it can be questioned in court.

In spite of all the difficulties in model acceptance, it is clear that models will be used in the future in important applications, and that the astute manager or engineer will pay careful attention to developments in the field, especially in the development of expert systems.

POTENTIAL URBAN WATER DECISION SUPPORT SYSTEMS

I would like to get more specific now by describing some potential decision support systems for urban water systems that are emerging. This will be followed in the next section with a design of a preliminary concept for locating the data bases and decision support capability throughout the organization.

This will not be an exhaustive list of potential decision support system aids, but represents what I consider to be the most logical and important emerging ones. The categories of decision support systems that seem to be emerging are the following:

1. A geographic-based system inventory data base
2. A condition index data base
3. A system water balance data base
4. A data base for real-time system studies and management
5. Data management systems for operating treatment plants and generating environmental information
6. Various analysis and design data bases
7. A financial data base

The first and most important might be the data base for locating and inventorying the components of the systems themselves. This is called the overall data management decision support system (DSS). An example of this kind of data base might be the effort that has been reported for Houston called "METROCOM." This data base required over 300 work years to develop and contains a continuous planimetric map of some 600 square miles of the city. Some 554,000 parcels of property are located and described, and there is location information on water, sanitary and storm sewers, roads, and bridges (17). There are other variations of this kind of data base appearing. One consulting firm has begun to market a lower cost system with limited graphics capabilities. Called the "Infrastructure Management System" it relies on Census Bureau maps with information on facilities added by the user organization. The City of St. Paul is planning to have such a system on line in 1985. They plan to take two years to create the data base at a cost of $250,000 excluding hardware (18).

The data base described above for St. Paul is planned to include the capability to add system condition information. It seems to me that this is really a different category of data base, however, one that should be linked with the location and inventory data. The condition data should be linked with the maintenance function, whereas the location and inventory is a more generalized data base.

Next, in the category of a hydrologic DSS would be a data base that would give us the capability to know the water balance at any time, both in the supply system and all throughout the urban area. This would be what McPherson called the "metropolitan water balance inventory," in real time. There are no complete examples of this kind of data base yet, but many utilities have part of it. For example, my own community maintains records of the water supply balances in a myriad of supply reservoirs and irrigation systems. An extension of this capability would involve water loss accounting and balances between water supply and wastewater discharges, showing consumptive use patterns, peak use factors,and other useful data.

A data base for the real-time management of systems would be very useful for applications such as flood and drought forecasting and for the control of water pressures during times of peak use. This is the general subject of Chapter 10 since it can lead to automatic control.

Data management systems that can be used to manage treatment plants and generate environmental reporting information are already coming into use. There are a number of examples of firms that are marketing information systems for these applications, and innovative plant managers are installing their own systems. The required functions range from maintenance scheduling to operations management to reporting for regulatory agencies. The use of these systems is described in more detail in Chapter 10.

Every system is unique and there are needs for specialized data bases for engineering purposes and for planning. As an example, consider a combined water and wastewater management district that had unique opportunities to trade water and to save money through innovative water management initiatives. They would need a specialized data base to study their opportunities. Another example would be the DSS developed for easier analysis and design. One of these for sewer system evaluation and design was recently described in *Public Works* (19). Implemented for the City of St. Petersburg, Florida, by the firm of Camp, Dresser and McKee, Inc., the system includes several linked models that run on microcomputers. The first model is a sewer system analysis model. It uses the input data about the existing system and potential new developments. The model predicts pipe peak flows and feeds information to the flow measurement analysis program. After an engineering analysis of the predicted flows, the information goes to the preliminary design program. There is also a cost program that provides a summary of the project costs. In effect, this series of microcomputer programs provides an automated analysis capability to really help in the engineering decisions made in St. Petersburg.

The water system manager is always going to be very sensitive about the financial condition of the system, and there needs to be a manager's financial model. This one, as distinct from the balance sheets and other accounting tools, should reflect the cost of doing business and allow the continuous tracking of productivity in the system. With this kind of decision support the manager will be able to provide good reports to the board and be able to have a complete perspective at all times of management and control actions that are taken. This DSS is not entirely separate from the financial management of the system itself, but will complement it, with emphasis on having financial information in real time for the chief operating officer. The need for this kind of financial decision support has been recognized and approached by researchers, but a lot of work remains to be done. The example I am familiar with is a system developed for the City of Tampa to integrate system cost data with the financial management. The system developers report some successes but more difficulties than envisioned and especially more work needed to integrate the cost data with the financial data base (20).

These data bases and decision support systems seem to represent the general categories that are evolving. Each organization is unique and the designs that are forthcoming represent the individual desires of the system developers as well as the nature of the physical systems that are being managed. The impact of commercial firms engaged in developing information systems will also continue to be a significant factor in the evolution of future decision support systems.

As a final word about the evolution of these data bases, there is a continuing and increasing cross fertilization between data base management for different applications. The evolution of data bases for facilities and building management seems to have many parallel requirements with that for infrastructure, and some emerging computer applications for facilities may apply to infrastructure. *ENR* had a cover story on the evolution of facilities management as a business opportunity area in early 1985 (21). It points out that facilities managers are seeing the following as computer opportunities: access to as-built drawings, space allocation, tenant records, utility information, and long-range planning. These are the same kinds of problems we face with infrastructure, and we will be seeing the same kinds of skills going into the solution of the two parallel problem areas. As buildings become larger and more complex, their management problems will begin to rival infrastructure in scale as well.

AN IDEALIZED MANAGEMENT ORGANIZATION

This chapter has suggested that the increasing use of computers will dramatically affect the organization of work and the structure of organizations themselves. This section makes some further suggestions about how this will affect water management organizations, with the design of decision support and information systems in mind. The observations in this section are not meant to go so far as to recommend or design individual data management systems but to demonstrate the principles of organizational and data base design that can be used by managers to design their own customized systems.

Strassman, referred to earlier, states that the individual work station will be a very important part of the way organizations restructure to accommodate the computer (1). I can see this in my own work, and am sure that the reader will identify with this observation, as we are seeing the fascination with which managers and engineers learn how to use their individual work stations. As this is being written, I am using mine to compose these ideas. I take this to be a given fact, then, that the individual work station will have a dramatic effect on organizations. It follows from this that the use of information and data to support management and operations will depend on the use of the individual work stations. Let's see how this will affect the structure of the organization and of management.

Chapter 6 briefly discussed the organization of water management agencies, and the next chapter discusses how operations and maintenance should be organized. These discussions emphasize that the fundamental purpose of the organization (that is, operations) must be given the highest priority. Without successful operations, all of the planning and administrative support in the world will be to no avail. Operations cannot succeed for long without effective maintenance. The fundamental purpose of maintenance, as a basic task, is to make sure that operations are successful, to support operations. This means that a basic organization chart such as shown in Figure 8.16 is called for, with operations being the most important element for the attention of the management.

The support for operations, as shown in Figure 8.17, comes in two basic forms, administrative and technical. These seem to be convenient organizing concepts for the management of data also. In Figure 8.17 administrative support is shown to include personnel, finance, and information systems. Technical support is shown to include planning, design, and construction. These activities taken together could be called the "engineering" section on the organizational chart.

Figure 8.17 represents a clean organizational chart, which is functionally straightforward. Most real-life charts are adapted to accommodate

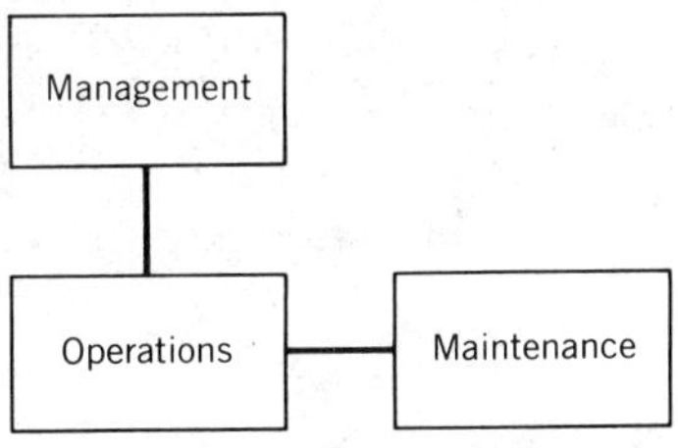

FIGURE 8.16 Central role of operations in the organization.

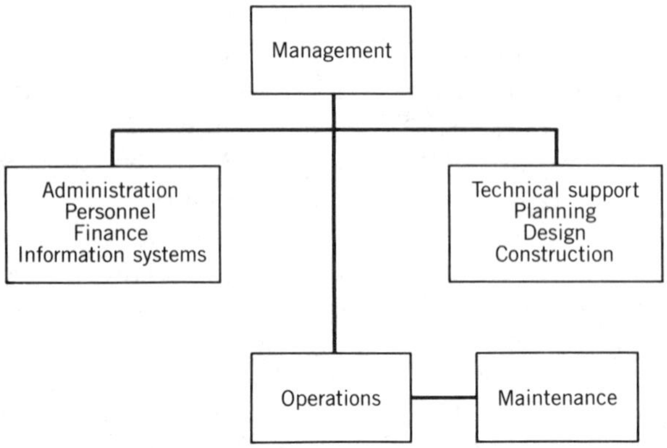

FIGURE 8.17 Organization with administrative and technical support.

the needs of individual organizations and look a lot more cluttered than Figure 8.17, but the underlying principles are usually the same.

The individual work stations that would be involved with the management of the functions shown in Figure 8.17 would be designed to process data and help in the normal decision making about operations and support. Since the organization itself involves distributed responsibilities, it follows that the data management responsibilities should be distributed throughout the organization. The closer the data management is to the managers and workers that need it, the better. This suggests that the individual work stations should be geared for full decision support; that is, to enable the collection, storage, and management of data, and the use of the data in management studies, involving modeling, financial studies, or engineering analysis.

To illustrate the layout for a more complete data management system, one keyed to one of the urban water services, let's look at the normal operations in a water supply utility: acquisition, treatment, and delivery. Figure 8.18 shows a sample concept of the organization for the data base design. The data bases are in the categories of facilities, operations, planning, and administration. Beginning with the lowest portions of the chart, each of the categories of facilities needs a data base for facilities management and for operations. Facilities management includes the information needed for minor and major maintenance, for condition assessment, for facilities tracking (inventory and location), and for other needs related to caring for the facilities themselves. These topics are discussed in more detail in the next chapter. The operational data bases would include the information needed to check up on operations and to report to the management levels and to the regulatory agencies. Each of the categories of the facilities would have their own data base. There are enough differences between the reservoirs, the well systems, the treatment plants, and the distribution system to call for individual systems.

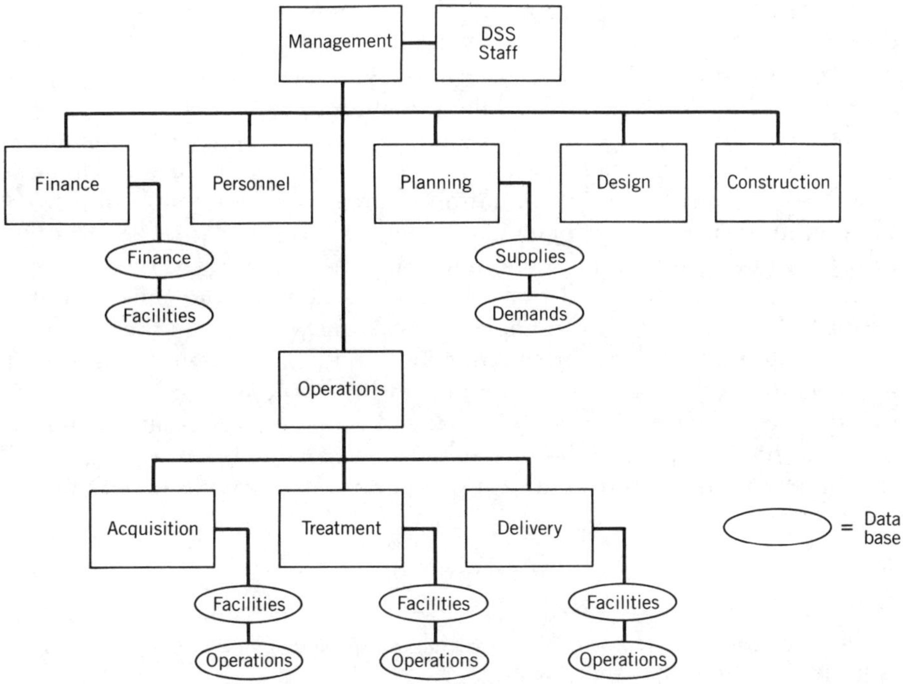

FIGURE 8.18 A conceptual water supply utility with distributed data bases.

In Figure 8.18 the information systems staff has been moved to directly support the system manager. The idea is that producing decision support information is important enough to deserve the direct attention of the operational manager. Not all managers will agree with the positioning shown, but the concept should be considered.

The data systems at the operational levels can be managed at individual work stations, assisted by programmable controllers and other automation devices (discussed in Chapter 10). The distributed network of work stations can be organized into a local area network, if it will improve management. There should be some upward reporting of data since management and the support functions can use the same data as the operational levels. How the data bases are assembled and coordinated is a task that requires a lot of care, and is very important to the effectiveness of the organization. It is the task of "data base design."

The support levels can profit by access to much of the same data as the operational levels, and so the data flow should be upward with some centralized management and coordination. This might be coordinated by the information systems staff function, although the extent of the management has to be coordinated carefully. I advocate caution here since the suppressive force of over-centralization on the productive use of distributed information systems is powerful.

In the example water supply utility, the planning data bases include data on the supply system, the demographics, and the demand parameters. The financial data is probably the most well developed in most water utilities at the present time, and the personnel data is not much used in operations, although it is obviously important in overall utility management. The planning data bases offer management the opportunity to gain a perspective on the entire operation at once, and need attention from the planning staff to be properly organized. Of particular interest in the future will be the compilation of planning data bases that allow the combination of hydrologic and financial studies at the same time. Combinations of hydrologic and financial forecasting will enable the manager to do a better job of decision making about rates, water acquisitions, and other matters that eventually show up in the "bottom line."

In the next two chapters we discuss the matter of operations more, first from the viewpoint of how operations and maintenance are organized and then from the standpoint of how operations can be automated.

REFERENCES

1. Paul A. Strassman, *Information Payoff: The Transformation of Work in the Electronic Age*, Free Press, New York, 1985.
2. The Intelligence of Organizations, *Science*, September 1984.
3. William E. Korbitz, Ed., *Urban Public Works Management*, International City Management Association, Washington, 1976.
4. Office of the Chief of Engineers, Information Systems Plan, Fort Belvoir, June, 1984.
5. Warren A. Hall, and J. A. Dracup, *Water Resources Systems Engineering*, McGraw-Hill, New York, 1970.
6. Tom Alexander, The New Technology of the Mind, *Fortune*, January 24, 1983.
7. William B. Gevarter, An Overview of Expert Systems, National Bureau of Standards, NBSIR 82-2505, Washington, May 1982.
8. Chronicle of Higher Education, Artificial Intelligence and Artificial "Smarts," *Notes on Computers*, Washington, November 21, 1984.
9. Richard O. Duda, and E. H. Shortliff, Expert Systems Research, *Science*, April 15, 1983.
10. Steven J. Fenves, M. L. Maher, and D. Sriram, Expert Systems: CE Potential, *Civil Engineering*, October 1983.
11. A. M. Starfield, K. L. Butala, M. M. England, and K. A. Smith, Mastering Engineering Concepts by Building an Expert System, *Engineering Education*, November 1983.
12. James W. Kalat, *Biological Psychology*, Wadsworth Publishing, Belmont, CA, 1984.
13. Alan Kay, Computer Software, *Scientific American*, September 1984.
14. Frank Rose, *Into the Heart of the Mind: The American Quest for Artificial Intelligence*, Harper and Row, New York, 1984.
15. C. S. Chu, and C. E. Bowers, Computer Programs in Water Resources, University of Minnesota, WRRC 97, November 1977.
16. Neil S. Grigg, *Water Resources Planning*, McGraw-Hill, New York, 1985.

17. Francis L. Hanihan, and C. A. Rivera, METROCOM: An Interactive Database for Managing a Modern Metropolis, *JAWWA*, July 1984.

18. Consultants Join to Market Municipal Mapping Service, *ENR*, December 20, 1984.

19. S. John, Calise, G. S. Walters, and D. T. Zimmer, Microcomputers and Software Provide Useful Tools for Sewer System Evaluation, *Public Works*, July, 1984.

20. David L. Tippin, E. Smith, J. I. Gillean, and Robert M. Clark, City of Tampa: Management Analysis and Report System (MARS), EPA report 600/S2-84-158, Cincinnati, November 1984.

21. Computer Maze Slows Managers Seeking Answers, *ENR*, April 4, 1985.

CHAPTER NINE

MANAGING OPERATIONS AND MAINTENANCE

In the overall engineering, construction, and management process described and illustrated in Chapters 6 and 7, the last step is "operations and maintenance," or "O&M." In a sense, O&M is like saying "and then they lived happily ever after" to end a story. The story is the project. People like building projects, and there is an enthusiasm that goes with it. After the ribbon is cut and the nitty-gritty of making it work day after day sets in, then interest sometimes wanes. This is true for businesses, which must grow or stagnate, for churches, which sometimes need to keep building to keep the interest of the congregation, and for water projects, which need a continuous injection of attention to avoid stagnation and neglect. The conclusion is that the operations and maintenance phase of projects requires special attention to retain the vitality of the original mission and goals. That is the subject of this chapter—how that vitality can be retained through an analytical approach to managing operations and maintenance. This is an important aspect to gaining control over infrastructure problems in general.

Operations and maintenance is a dynamic, not static, process. Some say that O&M should refer to "operations and management" rather than "operations and maintenance" due to the broad responsibilities implied. The model I prefer is to regard management during an operational phase of a project to consist of two parallel but equal tasks: operations and maintenance. In other words

$$\text{Operational management} = \text{operations} + \text{maintenance}.$$

Maintenance is a particularly important subject and should not be regarded as a minor task at all, although it is subordinate to operations in priority. The purpose of operations is to achieve the mission of the organization. The purpose of maintenance is to make sure that the operations can take place.

To refer to the importance of maintenance, the military has begun to use terms such as *maintainability engineering* and *sustainability*, which are meant to capture the intent of maintenance to keep equipment and systems ready to go at all times. The same emphasis on preparedness that the military has is also needed for industrial and for urban water systems, if they are to remain productive. In the case of urban water systems the consequence of a lack of preparedness is sometimes very costly failure, or even hazard to health and life.

In this chapter we will describe both the operating and maintenance functions, with applications from urban water systems. We will show the linkage with the planning, programming, and budgeting process through the needs assessment activity, one that continuously should identify needs for investment as a result of the system monitoring that is part of O&M. The credibility of budget requests depends on valid needs assessments, and the failure of public works to develop enough funds in the past has been largely due to a failure of this process. The most important aspect of needs assessment is developing the ability to make valid system condition assessments. These depend on having effective inspection and information systems.

GENERAL APPROACHES TO OPERATIONS AND MAINTENANCE

While urban water systems have not had too much attention from industrial engineers and efficiency experts, there have been improvements in recent years, spurred by pressure from the federal government and from better approaches to public works management. Still, there are things that can be learned from industrial approaches to operations. Operations and management is taught in business management and engineering schools under the general topics of production management and industrial engineering, and the objective of both these and operations in urban water systems is to find ways to improve the production and operational processes, whatever the objectives might be.

In preparing this chapter what might be transferred from the literature of production operations to urban water systems was researched, and I find that the concepts of production in industry and in urban water systems are very similar. If we take a basic text for operations management used in a business school, for example, we see immediately what the similarities are. I chose a text used at Purdue University with the title *Operations Management: Productivity and Profit* (1). The choice of this text was fortuitous since it illustrates the similarities between operations in a business and in a public utility.

The topics covered in Greene's text are generally related in the present text, with special attention to the unique needs of the urban water system. To illustrate, these are the principal topics found in Greene (1) as compared with the treatment in the present text:

Operations Management (1)	Present Text
Systems and decision making	Planning and management decision support systems
Communications for planning and control	Organizational aspects
Product and process information	Urban water system details
Facilities location and management	Urban water system information
Facilities recovery	Rehabilitation decisions
Human resources management	Topic not covered in detail
Manufacturing control	Planning and management essentials
Inventory management	Covered in this chapter
Production control	Covered in this chapter
Quality control	Covered in chapters on systems
Management controls	Covered in chapter on planning and management

As is evident from this list, there are several striking similarities between the coverage here and in the text on operations management. It appears that the objectives of the two works are very similar; that is, they strive to show how to improve management operations using the scientific approach. The present text does not go into any detail on human resources management, for reasons that are explained in the chapter on management, and it goes into much more detail than the operations text on the nature of the systems being operated. The other major difference is that we use the term *inventory* in a different way than does the Greene text. In manufacturing, inventory refers to the parts and materials that are necessary to manufacture the product. This is a very important aspect of cost control and eventually the profit of the organization. For example, a Japanese "just in time" inventory system refers to one where the parts and materials arrive just as they are needed, and this avoids the inefficiency of having a large inventory sitting around. In the case of the urban water system, the term *inventory* is used for the water system itself. That is, the pipes and physical facilities themselves. This distinction is necessary since we must know where the system components are and what their condition is before we can schedule the maintenance. This information is referred to later under the general topic of facilities data.

I will summarize by stating that there is a general model of operations and maintenance that is implied by all of the above. It is one that has three parts. The first part is to ask the question whether the system and the staff are organized properly. The second question is whether the work, all of it, is planned thoroughly, at all levels. The third question is whether the operation is adequately controlled. To ensure that it is, two checks need to be continuously made. The first is whether the product or the service is being adequately delivered, or in other words, is the quality of the product good enough. The second is whether the condition of the system is acceptable. These themes will arise later when we discuss how they apply to the specific parts of the urban water systems.

COMPLEXITY OF OPERATIONS AND NEED FOR DIAGNOSTIC ANALYSIS

In studying operations of enterprises in general, I am struck that they are all complex, and although it is possible to model them and make studies, the roles of experience, intuition, and human problem solving are critical to success. The goal of operations is to make a success of the venture that is undertaken. In urban water systems the venture is to deliver a service. In the terms of Greene's text (1), the urban water problem is like that of other service industries. Different ventures have different goals, and to achieve them takes different organizations and different processes. I have had opportunities to study operations in organizations in the military, government, university, church, small business, and water management organizations. Similarities exist in all of these in the way that operations are managed, and the complexity and uniqueness of all of them is evident.

To gain a perspective on how to improve operations in complex operating environments a technique called "diagnostic analysis" has been effective in different settings. It represents a general approach to studying performance improvement. In a sense it is a comprehensive performance audit.

The generalized procedure is like the evaluations that the military calls the "training test" or the inspection at the university called the "accreditation visit." In these the adequacy of the operations is judged by an outside evaluation team. In the case of water operations, a good suggested procedure is in the publication done for EPA by the Government Finance Research Center and Peat, Marwick, and Mitchell (2).

The EPA report recommends a diagnostic analysis of the performance of entire wastewater utilities to be carried out in four phases. The first phase is a preliminary investigation that identifies problems and sets the stage for the remainder of the study. The second stage consists of an on-site evaluation of the general management, the support services, and the operations and maintenance activities of the utility (see Figure 9.1). The evaluation of the adequacy of the general management would consider how the utility was organized, directed, and operated at the overall level. The support services that would be evaluated would consist of the finance, engineering, purchasing, personnel, and information services. The operations and maintenance, or facilities function, would be studied for adequacy in the following areas: policies and procedures, organization, staffing and training, planning, and management controls.

The third phase is the one where the individual findings are compiled into an overall utility evaluation, and evaluations are made of problems and potential solutions. The final phase is where the solutions identified are reviewed with the management of the utility and implementation plans are developed.

The general applicability of the diagnostic analysis approach can be verified by its use in other approaches. My university has been active in studying ways to improve the functioning of irrigation systems in developing countries, for example, and it has developed a "diagnostic analysis" approach to improving operations in the complex social-political arena of irrigation systems in developing countries (3). In this approach the developers stress the importance of the work of "interdisciplinary teams" and the involvement of the farmer in the analysis. Translated to the urban water system, this would mean to stress the involvement of

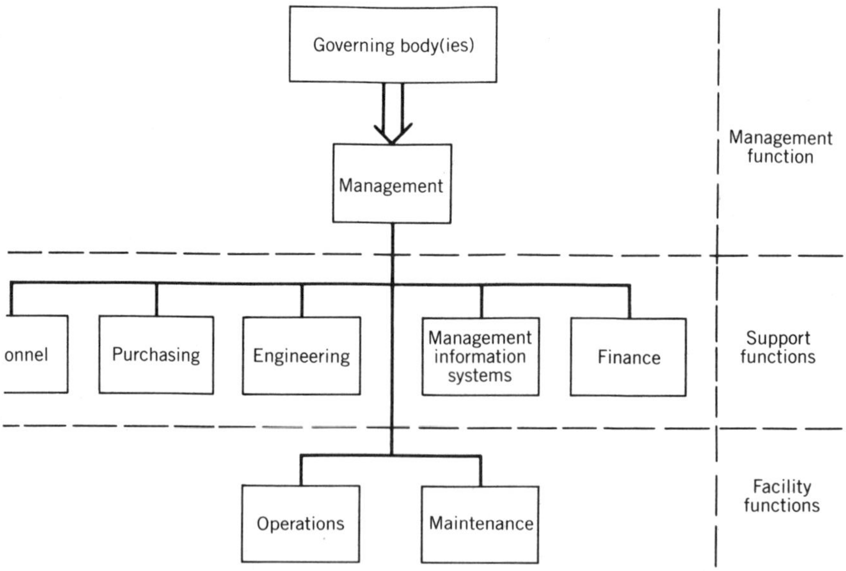

FIGURE 9.1 Wastewater management organization (illustrative) (1).

the system operators and to make sure that all points of view are incorporated.

The lesson from the above references and experiences is that to improve operations in complex systems is a tough management challenge, one that should not be taken lightly. The difficulties of making complex urban water systems work properly cannot be dismissed, but the total problem environment must be examined, using an approach that accomplishes the objectives of techniques such as diagnostic analysis. We will examine further an approach to this that is recommended for the wastewater treatment plant later in this chapter. The same techniques can be used for other parts of the urban water system.

OPERATION AND MAINTENANCE IN URBAN WATER SYSTEMS

Organizations that manage urban water systems are like others in the sense that they have missions, organizational structures, work processes, and products and services to be delivered. In the case of the water supply utility the result of the operation is a product: safe drinking water delivered to the point of use for about 25 cents per ton (when the rate charged is about $1 per thousand gallons). To achieve the desired results, the management organizations need line and staff functions.

In the operating agency O&M are the most important line functions. To illustrate that Figure 9.1 shows them as the direct operating functions of a wastewater agency, parallel on the line (2). The staff functions are in support of O&M, and they are shown where they belong, at the staff level. The illustration in the figure shows three levels of the organization: management, support, and facility functions, the latter being the line operations and maintenance. This is a descriptive chart, but I prefer one that shows a unified responsibility for operations and maintenance, as the chart shown in Figure 9.2 illustrates. The important aspect is that

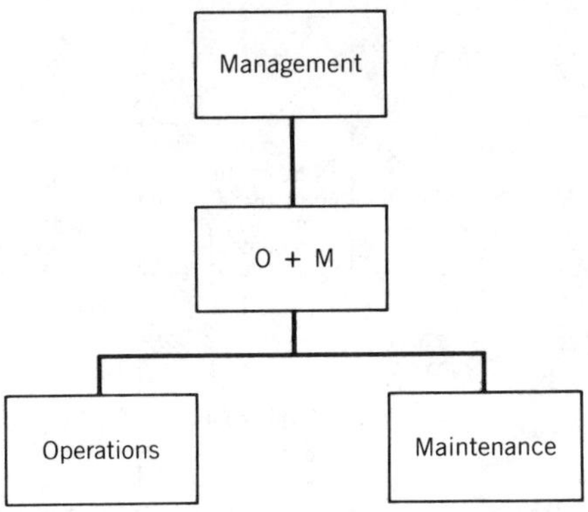

FIGURE 9.2 A unified view of operations and maintenance.

the responsibility at the operating level be unified for operations and maintenance. If they are split, it seems that the top management will have continued grief when maintenance does not properly support operations.

The functions of O&M shown in Figure 9.2 should be broken apart to show what is included in each of operations and maintenance. They link together in the role that maintenance plays in supporting the need of the system to be operational. Figure 9.2 shows a continuing check built into the operational function to determine whether the system is working as it should. Another check is built into the maintenance function to enable the scheduling of repair work. The definition of maintenance should be expanded to include condition assessment, inventory, and maintenance management, as shown in Figure 9.3.

In effect, Figure 9.4 shows four separate maintenance-related functions: condition assessment, inventory, preventive maintenance, and corrective maintenance. The latter is sometimes called major and corrective maintenance to stress the major nature of some of the corrective actions. The condition assessment activity is a linkage between the operation and the maintenance function and illustrates why the two functions must be unified. If the condition starts to worsen, the operation will be affected, and it will be time to schedule repairs and maintenance.

The result of the inventory and functions is a continuous determination of the location and condition of the system component of interest, these two pieces of information being taken together. There should always be an answer to the question, "Is the system OK?" If so, the operations and

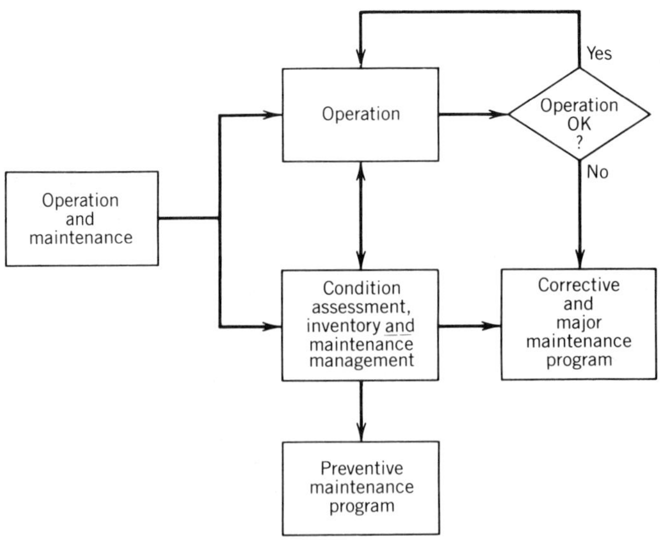

FIGURE 9.3 Condition assessment and maintenance management programs.

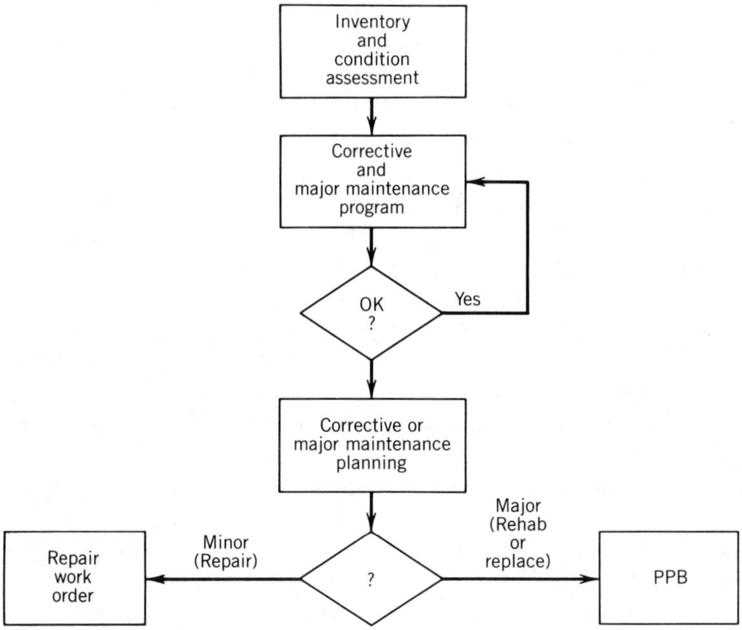

FIGURE 9.4 Corrective and major maintenance program.

surveillance continues. If not, we must enter the corrective and major maintenance activity. The need for an inventory seems obvious for valuable equipment, but we are just now starting to think about the need for inventories of full urban water systems. This means to have complete records of all the pipes, valves, manholes, and other appurtenances of the systems.

As shown in Figure 9.4, corrective maintenance requires a decision: Is the deficiency serious enough to warrant entering the planning, programming, and budgeting (PPB) activity, leading to a capital request, or is it minor enough to go ahead and take care of? If the problem is major, the PPB activity incorporates information about new standards and growth forecasts to lead to decisions about rehabilitation and/or replacement. The planning part of PPB also yields information to be used in the needs assessment process, which is the linkage between the major maintenance program and the capital budgeting program. Due to this linkage, it seems appropriate to have the same planning staff do the planning work for rehabilitation work as well as new facilities.

The sequence of activities that leads to a decision about minor or major corrective actions also explains the differences between the three "R's:" repair, rehabilitation, and replacement. These activities are part of a spectrum that covers the range from routine operating budget activities to major capital budget activities. This is an important distinction for management, since the operating budget should have enough funds for routine and important repairs, and there should be sufficient funds in the capital budget for major rehabilitation, without excessive deferral. The dividing point is somewhere between repair and rehabilitation, as shown in Figure 9.4, but each manager and organization must decide this individually.

Operation of any system follows a generalized model much as is shown in Figure 9.5, which illustrates the data collection, operator decision, and control command sequence that characterizes the operation of systems. That the operations function connects to maintenance through the condition assessment activity is also shown in this figure. As shown, the data collection or monitoring activity is part of operation since the system state (such as a water level) must be continuously known to make control decisions. The figure illustrates the connection of system control with operations, where the operator makes control decisions on the basis of information about the system. The information also feeds into the condition assessment file. The figure also illustrates a concept known to the military as C^3I, which means "command, communications, control, and intelligence." In other words, these are four key functions of managing any kind of system. The decision maker commands on the basis of communications and intelligence about the system, and the commands are transmitted to the system through controls.

Figure 9.6 shows an expansion of the system operation scheme of Figure 9.5 to incorporate expert systems and computer control. The only real change is that the expert system and system simulation model can be used to advise a decision model in choosing the correct control commands. This future addition to the operational model is described in more detail in Chapter 10, which is devoted to automation.

The automation implied by Figure 9.6 does not have to be closed loop, meaning without any human operator. It can be accomplished in phases. The first phase is the implementation of some supervisory control, meaning that system conditions or states are telemetered to the control

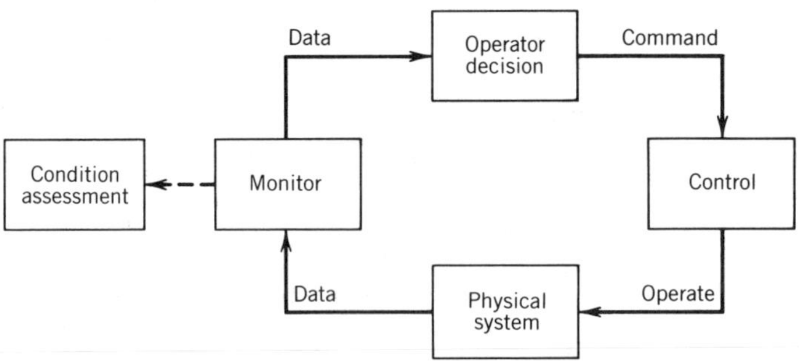

FIGURE 9.5 Operations sequence.

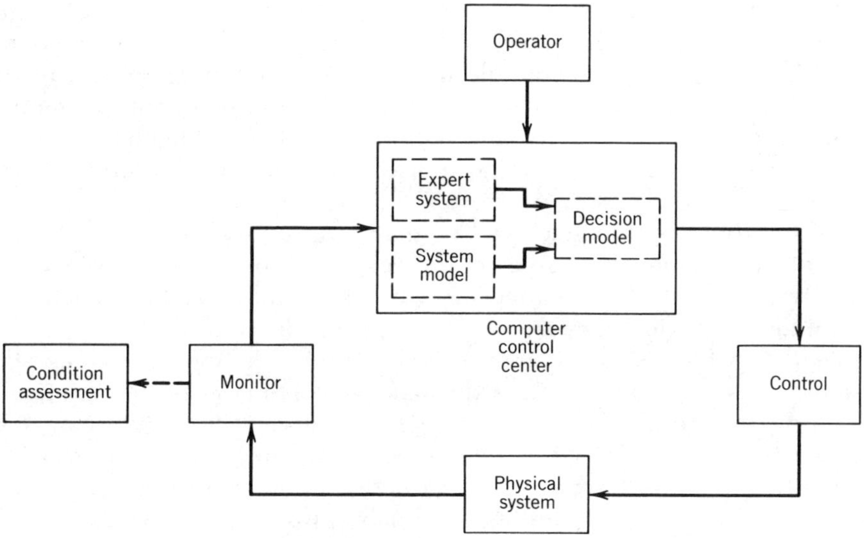

FIGURE 9.6 Operations sequence with computer control.

center. A second phase might be where the data is logged and stored automatically. A third might be where some computer models provide information about operating the system, and this information could be used by the operator to make decisions. The final or ultimate stage would be where a computer model makes all the decisions and the system goes on closed-loop automatic control.

Computer control systems such as those described above are being implemented in many urban water systems today. The automation has been in stages, a natural result with computer hardware evolving so rapidly. It appears that now the hardware has reached a level of cost low enough so that it represents no barrier even at the lowest levels of operation, and automation will proceed more rapidly. It will be following the distributed control route, rather than centralized control, this concept being explained more in detail in Chapter 10.

MAINTENANCE MANAGEMENT SYSTEMS

The concept of the "maintenance management system" (MMS) has arisen in recent years as an attempt to bring together the disparate concepts of maintenance activities into a holistic approach to caring for a system. In effect, it is the "systems approach" applied to maintenance. There are some consulting firms that specialize in maintenance management, and since there has been considerable attention given to the need to make wastewater treatment plants work better, conventional consulting firms have also gained recent experience in the concept.

An MMS is, in effect, a program for making sure that overall maintenance is managed adequately. It involves all the tasks of management: planning, organizing, and controlling, and it requires an effective decision support system. The MMS will include condition assessment, preventive maintenance, and corrective and major maintenance, and the decision support system will provide the information and data needed for these activities.

In studying maintenance, we find many variations of the MMS. There are not many comprehensive references, but those that do appear are usually based on maintenance in a specified industry. References are found again in the literature of industrial production, such as in Newbrough (4), which is based on experience in the steel industry. Generally, references such as this contain the same management principles found in the water industry experiences: information systems, work orders, preventive and corrective maintenance, and managing maintenance as a significant support activity of production. The same principles have been found in several references published in current journals. For example, one that specified procedures for vehicle maintenance referred to the importance of the maintenance management information system (5). Another referred to the need to have valid condition assessment data to develop a management system for roof maintenance (6). Another discussed the advantages of contracting out highway maintenance (7). All of these are current references, published to report interesting and new information to the public works profession.

The principles of developing maintenance systems are no secret in industrial and public works organizations, but it has been the lack of application of known principles that has been the problem. This is the general conclusion of a study by the Urban Institute where some 62 cities were surveyed about their practices and experiences in capital stock management and maintenance (8). One of the most significant findings of the study related to maintenance management: that large improvements could be made by paying more attention to selecting effective maintenance strategies. The selection of these strategies involves systematic identification and analysis of the strategies that are available. These include the following, as well as mixtures of them:

Do crisis maintenance only
Maintain the worst facilities first
Perform opportunistic maintenance, when related work is scheduled
Use prespecified maintenance cycles
Repair those components with the highest risk of failure
Use preventive maintenance
Reduce the demand for wear and tear on the facility
Compare the economic advantages of maintenance strategies

The generalized aspects of maintenance management systems are reflected in the apparently growing market for professional services in developing computerized MMS. One consultant, Roy Jorgensen Associates, has literature that suggests the need to develop comprehensive MMS that start with the basic operating procedures and work toward a complete system that includes work activities, documentation of system inventory, establishment of service levels, developing performance standards, and calculating work programs and budgets. The American Public Works Association has felt the need for and developed a set of guidelines for developing maintenance management systems (9). The Urban Institute study just cited found that when evaluation of maintenance strategies was done, it was usually by an outside consultant (8).

The MMS should comprise an integral part of the overall organization management control system. As a minimum, it will have an inventory control system, with numbering and identification for all equipment and components, and a record system for maintenance scheduling and completions. This record system will include equipment data, the preventive maintenance (PM) record, the repair record, and a spare parts stock card.

INVENTORIES AND CONDITION ASSESSMENTS

The inventory and condition assessment process is critical to both maintenance management and needs assessment. In effect, condition assessment is a process of measuring the physical condition of system components using objective criteria. According to O'Day and Neumann, the process should include such parameters as safety and structural integrity, capacity, quality of service, role, and age (10).

The National Bridge Inspection Program and the National Dam Safety Program are examples of condition assessment activities that were mandated nationally since safety was such an important factor. Later in this chapter we will discuss some condition assessment activities that are included in other activities such as corrosion monitoring, an activity that is a subset of condition assessment. There are two aspects of condition assessment that stand out for the management process. First, condition assessment is unique for each facility type because each one has to be monitored in the way that it is configured and operates. Second, the condition assessment activity requires good record keeping or an effective information system.

One of the volumes of the Urban Institute study was devoted to the assessment of capital stock condition (11). In this analysis the writers include inventory as one of the requirements of condition assessment and discuss applications to roads and streets, bridges, water, and sewer facilities. The inventory needs to be related to a geographical data coding system that will provide some basis for geographic location of facilities.

According to Godwin and Peterson (11), Kansas City has utilized the DIME system of the U.S. Census Bureau to define the location of streets. The DIME file is a dual independent map encoding technique that is based on the Census Bureau's mapping of 256 metropolitan areas at the block, census tract, and metropolitan scales. The DIME file provides a geographic base file, and, for this reason, is called a GBF–DIME file.

NEEDS ASSESSMENTS FOR MAINTENANCE MANAGEMENT AND CAPITAL BUDGETS

One of the most difficult parts of planning and budgeting is knowing how much of a system to build. The problem is to determine what the "need" is. The needs assessment process is the part of the PPB process that leads to a workable definition of "needs" as distinct from "wishes" or some other expression of demand for facilities. This is important because the needs statements must be credible, or they will be ignored. Opinions that are not based on objective data are not worth that much, even when they are somewhat informed. This is an important issue for public works managers.

Needs assessments are part of both the maintenance management and capital improvement processes. In effect, it is the needs assessment that links these two processes together. This is shown in Figure 9.7, which illustrates a planning, programming, and budgeting process for replacement and rehabilitation. It is natural to ask, how do we make the decision between rehabilitation and new facilities? The answer comes from the

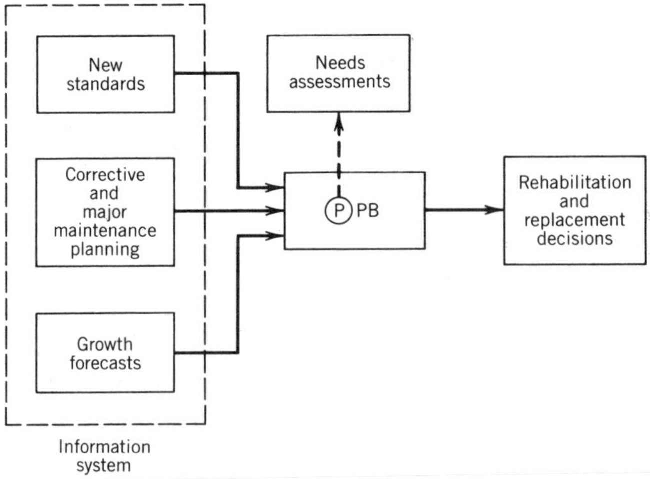

FIGURE 9.7 PPB for rehabilitation and replacement.

needs assessment and other evaluations, and must be an input into the PPB process. As shown in Figure 9.7, growth forecasts and new standards will come into the equation as well. The important point illustrated by Figure 9.7, and that comes out in the Urban Institute reports, is that maintenance decisions and strategies need planning. These are presently not much used by local governments for a number of reasons, and they need attention through the planning process. Valid needs assessments will be a product of improvements in this process.

The needs assessment process involves an inventory, a condition assessment, and an identification of desired levels of improvement and maintenance. These must be viewed in the light of present and expected future conditions. O'Day and Neumann present a framework for infrastructure needs assessments that integrates the inventory, condition assessment, and future expected conditions together. As shown in Figure 9.8, the process identifies deficiencies and evaluates alternative programs and projects to produce a needs report and selected programs and projects (10).

While Figure 9.8 presents a valid framework for needs assessments, many improvements are needed in the process. Many decision makers have seen so many poorly done needs assessments and plans that they consider a needs assessment as hardly more than a "wish list." This is a debilitating problem for public works managers, one that needs attention to improve the management of all kinds of public facilities.

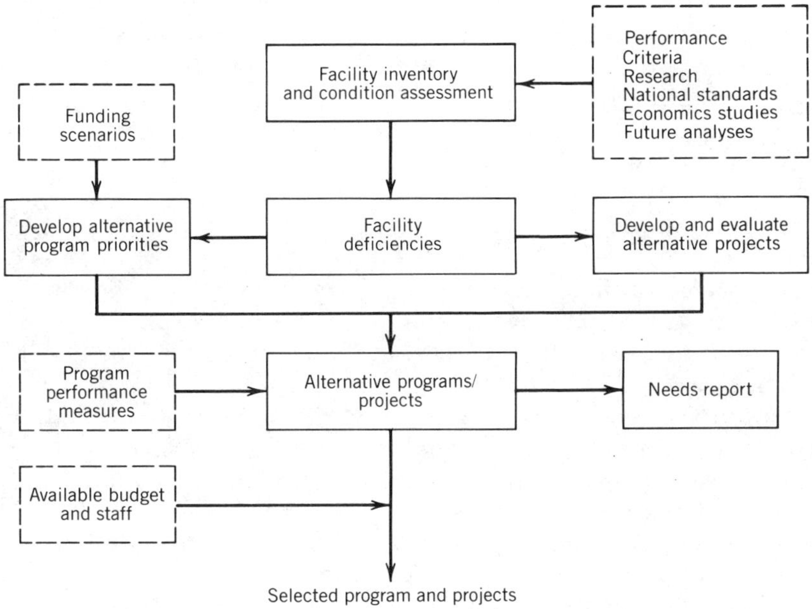

FIGURE 9.8 A framework for infrastructure needs assessment (10).

WATER DISTRIBUTION SYSTEMS

Caring for a distribution system requires close attention to water quantity and quality issues and the physical condition of the distribution system itself. Water quantity problems appear in the form of abnormal pressure, insufficient water supplies, leaks, and breaks. They can be caused by accidents, corrosion, system failures, and other causes. Water quality issues become more noticeable when new drinking water regulations are strenuously applied, and it has become apparent in recent years that water quality is a distribution system issue as well as a treatment plant issue. It also relates to the care and attention given to the distribution facilities inside the home or workplace. The physical condition of the distribution system is an important determinant of both water quantity and quality results; thus its management is a critical function of urban water managers. AWWA's annual attention to the issue through the distribution systems symposia is testimony to the increased concern over caring for distribution systems. These symposia were initiated in 1980 as a companion specialty conference to the annual water quality technology conference started in 1973. These conferences reflect the concerns of the water system manager: water quality and distribution systems. Since financial issues are also very important, perhaps we will see a conference on them in the near future as well.

Water Quantity Issues: Pressure, Flow, and Leakage

Managing the quantity of water in a system involves knowing the system's physical conditions at all times. This means, in effect, knowing how much water is going where at what time and what the pressures are. Pressures in closed conduits are analogous to depths of flow in open channels; they characterize the potential for flow. This direct relationship can be seen from a simple orifice expression,

$$Q = C\sqrt{2gH}$$

where Q = rate of flow, C = the orifice coefficient, g = gravitational acceleration, and H = the expression of head, or pressure. This general expression is the generic form of energy equation that relates flow to its potential, expressed in the form of pressure. This is important in managing water supply systems since the potential to supply water, whether for household, commercial, or fire flow use, is dependent on system pressure.

Water system managers must begin with a general plan for pressure distribution throughout a system. The pressure needed will determine design factors such as pipe size, pump capacity, and tank elevation. Once a system is properly designed, constructed, and initiated, the operation and management phase begins.

The first important consideration for pressure is that it remain within the expected bounds. If a system is working according to the design, the pressure would be expected to fluctuate smoothly between limits established by the design considerations. When system flows are high, the pressures will drop. When demands are low, the pressures should rise to maximum levels. In this framework it is the deviations from normal pressure that are important to observe. This underscores the importance of an adequate system monitoring plan. Also useful are water distribution system network models.

System Monitoring Plan

In earlier days a water utility would be expected to have a few pressure gauges around that could be used to check pressure from time to time, but today modern utilities are moving toward automated pressure monitoring. Such systems require pressure-sensing devices, recording equipment, and a way to collect and store the data. In smaller communities the placement of pressure recorders in critical locations provides a way to systematically check the system pressure for abnormalities.

Examples of benefits from placement of recording equipment include situations such as that reported by a small community where a rush problem arose from a pressure problem (12). The pressure problem was traced through study of the records to surges that were eliminated by valve adjustment. Another application reported by the same utility solved problems with pump station malfunctions by analyzing pressure fluctuations.

The extent to which pressure recording equipment should be telemetered depends on factors such as budget, size of system, and the management preferences of the system operators. We cannot ignore the fact that telemetry offers a lot of potential for automating systems in the future. We will shortly see routine centralized monitoring of pressure as a way to improve water utility operations.

Flow Monitoring and Use of Models

A network simulation model can tell you what the pressures ought to be at a given point under the conditions simulated. Although utilities are not using models for real-time management of systems, the models offer much promise for applications such as predicting problems, planning for expansion, and decision making for system operation such as pump operation, pressure reducing valves, and related decisions. In the future it seems inevitable that models will find application in the real-time control of systems. As stated later in Chapter 10, sometimes it is cheaper and more reliable to rely on models to tell what a system is doing than to try to instrument it at the level of detail needed for real-time operation.

It is important that distribution system models be reliable, and this means that the models must be operationally valid and properly calibrated. Working properly, a distribution system model will predict pressures at different points in a distribution system that will arise under the specific conditions simulated. This means that system calibration must be done under the same conditions that are modeled, otherwise the results of the modeling effort will not be valid. Walski has written several articles and a book about the importance of proper calibration, and referring to them will provide additional insight about the use of models and additional factors (13–15).

Water Hammer

One of the pressure problems that must be dealt with in distribution systems is pressure surge—water hammer. Not controlling this situation can result in serious damage to distribution facilities and water user property. I can personally remember three situations that illustrate this: one firsthand and two others secondhand. On a personal basis, my home was damaged by water leaks that resulted from a pressure surge from rapid closure of a hydrant valve near my house. One of the leaks was underground and not discovered for months. The saturation of the soil nearby led to settlement and damage to the house. A few years earlier I was studying valve problems in a water treatment plant and learned of a situation where the operators had witnessed a failure of a butterfly valve, resulting in sudden valve closure, and failure of a 30-inch high-pressure line. This was a life-threatening situation, and made the operators very conscious of any pressure fluctuation in their valve house. The third situation was a setup water hammer problem. It had to do with a cadet at West Point who had studied fluid mechanics (a true story) and learned about pressure fluctuations. He calculated that if a number of cadets simultaneously flushed toilets and closed valves that the pressure surge would rupture a main. Sure enough the experiment worked, and the cadet spent some time serving punishment tours after his deed was discovered.

Experienced water system operators have many stories about water hammer and surges. Force is required to change flow velocities (following the momentum principle), and these forces occur as hydraulic pressure transients, meaning conditions that occur rapidly and then disappear, hopefully without damage.

The causes and some remedies for water hammer are discussed in the November 1984 issue of the AWWA journal (16). Four common situations to watch are cited: pump startup, pump power failure, valve motion, and improper operation of surge protection devices. Solutions to water hammer problems involve both operational procedures and surge control devices, such as relief valves and surge tanks.

Leakage and Water Audits

The problems of leakage, breaks, and unaccounted-for water are important in water systems due to the loss in revenue involved, the higher operational costs, and the need for greater system capacity as a result of needing more water than is actually delivered. Fortunately, in recent years much progress has been made in solving the problems of locating breaks and leaks.

The place to start is with the water audit. According to Siedler, a water audit results in a reasonably accurate accounting for water in a system (17). This means that a water audit is like a financial audit; it checks the books. Siedler's version includes checking master meters for accuracy, testing industrial meters, checking for unauthorized use of water, and locating underground leaks through a systematic survey. A typical audit would result in a balance sheet that showed total accounted-for water, including known leaks, and total unaccounted-for water, including unmetered uses and unmeasured leaks.

Standard terminology for water auditing has evolved, and it is important to understand the terms involved. According to a report from the State of California (18), unaccounted-for water (UAW) is the difference between the total water produced and the water sold. This definition seems to be in common use in the industry. "Authorized UAW" refers to the uses that are unmetered but sanctioned, such as street cleaning, sewer flushing, park watering, fire fighting, construction uses, and other public uses. "Unauthorized UAW" would then refer to everything else, that is, losses, errors, and perhaps theft.

In the California report 10% of the water utilities reported UAW greater than 15%, and 30% reported UAW exceeding 10%. The average UAW was 9.3%, and the average unauthorized UAW was 7.2% of the UAW.

The 1981 AWWA water utility operating data report (18) reported California's average at 10.6% for 2.7 million service connections. The difference between the two reports is not explained.

In the Urban Institute study discussed earlier, some 49 communities reported reliable data on UAW. The percentages generally ranged between a low of less than 5% to highs about 40%. The highest figures were generally in the Northeast, with Albany reporting 40% and Boston 39% (19). The average percentage of UAW in the Urban Institute survey was about 15%, which seems to be a figure similar to the national averages in other reports, with consideration given to the difficulty in obtaining valid data on UAW.

The water audit must analyze leakage data, but there are a number of other losses that must also be considered. These are listed in the California report as:

1. Inaccurate meters
2. Unmetered use of fire hydrants (for all uses)

3. Unmetered connections
4. Unauthorized uses
5. Inadequate system controls
6. Incorrect meter reading or billing
7. Water system leakage
8. Reservoir seepage and evaporation

A critical part of a water audit is the leak detection program. An experience with the Westchester Joint Water Works in Mamaroneck, New York, showed that leak detection and repair can yield important benefits (20). Since this utility purchases water from New York City, the economics of saving water are readily apparent. In the project sonic leak detection was used and found to be effective, if it is operated by skilled personnel. The utility has acquired a leak detection van so the training of personnel should be facilitated. Unaccounted-for water was reduced to 16.3% by the program, and efforts to reduce it further continue. A key feature of the success of the program was the effort by the utility manager to keep good records. This is certainly an important feature of water auditing. The specific data recorded in the program are included in the article.

Leak detection and repair has gained considerable attention across the country in recent years. The State of California, where water supplies have always been a high priority, found that water audit and leak detection procedures could be beneficial even for communities with low values of unaccounted-for water (18).

Data collection is important for all categories of management in water utilities, and the leak detection program offers an opportunity to improve the data base while solving leak and breakage problems. O'Day writes about the importance of organizing leak and break data to aid in making main replacement decisions (21) and states that properly organized data can save substantial sums in main replacement decisions. This is an example of the need to have a valid condition assessment program for both minor maintenance (leak repair) and major maintenance (replacement of a main). O'Day states that a mile of main will cost between \$200,000–500,000 in 1980 dollars and that decisions on replacement by age are incorrect; they should be based on condition. Obviously a data base, along with careful analysis, will be of great help there. Since considerable savings can accrue from leak repair, the proper assembly of data is essential to good management. O'Day reports that a data base on main repair will cost \$50–100 per mile to develop and \$10–20 per mile per year to maintain. These figures are obviously insignificant compared to the stakes involved in main repair decisions.

Decision making on when to replace mains depends on careful economic analysis. Another article that presents methods to consider is by Walski and Pelliccia (22).

Valves

No discussion of maintaining a distribution system could be complete without mentioning valves. In older systems many of the valves may have not been exercised for years. In fact, some may not even be listed on the records. The importance of a good record-keeping system and a regular program of preventive maintenance should be emphasized to the maintenance forces. Some tips on developing preventive maintenance programs for valves are included in a recent article by Skorcz (23). The secrets to success are to have good records, to exercise the valves regularly, and to make sure that the need to maintain valves is a recognized part of the utility's management program.

Water Quality Issues

Operating and maintaining distribution systems also requires attention to water quality issues. The responsibility of the water system manager extends, to a certain extent, all the way to the tap water that the consumer takes from the faucet. While the legal aspect of this responsibility is debatable, most water system managers feel the responsibility to assure quality all the way through the private plumbing system whenever possible. In fact, the National Interim Primary Drinking Water Regulations (NIPDWR) specify that the maximum contaminant levels for inorganic chemicals such as cadmium and lead be met at the consumer's tap, and this action signals an increasing legal responsibility for quality all the way to the consumer (24). The subject of corrosion, to be covered later in this chapter, is but one of the quality concerns that distribution system managers face. Others have to do with quality control questions such as contamination, cross connections, and sabotage.

A recent AWWA survey showed that 32% of 314 utilities reporting found deterioration in water quality after the water reached the distribution system (25). Another report stated that distribution systems contamination accounted for 33% of the outbreaks of waterborne disease reported for the period 1971–1980 (26).

Two practices stand out as guards against contamination of drinking water after it reaches the distribution system. One is ensuring that a residual level of chlorine is present in the water while it is in the distribution system. This enables any minor bacterial contamination that remains to be treated. The other is the elimination of cross connections of all kinds.

The threat of cross connections that interject dangerous materials into water supplies have been recognized for a long time. There are two general ways that the materials can enter: pumping into the distribution system and being siphoned in when distribution pressures fall. All systems must have an active program to prevent these from happening (27). AWWA lists several situations that may be potentially dangerous sources of contamination through cross connection. These include unapproved supplementary supplies, industrial or fire protection water, premises that handle sewage or industrial process water, circumstances where there is a special possibility of backflow of sewage or contaminated water, and water supply lines that end at piers. Provisions recommended by AWWA to guard against these are plumbing changes, relief valves, backflow protection on all supply lines in addition to the primary ones controlled on a property, inspection of backflow protective devices, and discontinuing service in some cases (28).

In spite of all the preventative measures coliforms may still be present in distribution systems, requiring direct action to remove them. A case study of a remedial approach was reported by Springfield, Illinois, when bacterial contamination required a program to reduce bacteria in the treatment process and distribution system and purge the distribution system of chlorine-resistant bacteria. In this situation the distribution system was flushed for 8 days and several treatment actions were taken. Dead end situations and cul-de-sacs are a particular problem since residual chlorine may not last there until the water is discharged. This kind of problem is serious and long term. The problem in Springfield had been suspected as early as 1976, but it became serious in March of 1982. The actions reported lasted until July 28, 1982, when an order to boil water was lifted. It was March of 1983 before the coliform problem seemed to be fully under control (29).

Corrosion and Postprecipitation

Adequate maintenance of a distribution system means the avoidance of excessive corrosion or any deposition of unwanted materials on the inside of pipelines. The AWWA has a policy statement that refers to these: "water should not be corrosive or encrusting to, or leave deposits on, water-conveying structures through which it passes, or in which it may be retained, including pipes, tanks, water heaters, and plumbing fixtures" (30). Considering the enormous investment we have in the facilities named, the stakes involved in preventing corrosion and deposition are indeed high. Corrosion is estimated to cost $700 million per year in U.S. water supply systems, not including plumbing systems (31).

There are many benefits to controlling corrosion, some with direct economic measures and some with regard to health. Corrosion is of concern to public health since the metals leached from the pipes, such as lead, can be harmful to public health. Corrosion also causes financial damage such as staining clothes in washers.

Recognizing the importance of corrosion to health the NIPDWR requires that utilities determine the corrosivity of water and the materials in the distribution and home plumbing lines (32). This monitoring requirement calls for the design of a program to examine water quality and pipe material determinants of corrosion.

The water quality sampling program may begin with the collection of data necessary to calculate indexes. Common corrosion indexes include the Langelier Index, the Ryznar Index, the Aggressive Index, and the Driving Force Index, all of which are based on the soluability of calcium carbonate. Some of the characteristics used to calculate these indexes are pH, alkalinity, hardness, calcium, and total dissolved solids. To determine the corrosion that occurs in customer service lines, samples will be required inside the homes. The first flush from service lines after a night of no flow can help determine the metals present. These samples may reveal any of a large number of metals present. Other sources of data can include customer surveys and an analysis of staining.

Testing pipes from the distribution system or the plumbing systems can involve sampling and any of several tests. As discussed in Kirmeyer and Logsdon (32), this involves removing the pipe for the distribution system and using such standard procedures as visual examination, micrographs, weight loss, pitting potential, scale analysis, and corrosion probes.

The evaluation of the condition of home plumbing systems is more complex since the variety of situations encountered is greater and the range of materials is wider. A combination program of sampling, interviewing homeowners and local plumbers, and visual inspection of discarded water heaters and other equipment would be a first step.

For the most part corrosion in water systems is electrochemical in nature. The mechanism is basically the movement of electrons and ions from anode to cathode areas and vice versa. The result is the actual loss of metal from the system. Two general types are identified: uniform and localized corrosion, the latter including galvanic, crevice and pitting corrosion, and erosion. These are described in a little detail in Foster (33).

Corrosion reactions for iron pipe, galvanized pipe, lead pipe, and copper pipe are different in nature. There is also a variety of corrosion potential in treatment plants due to the many dissimilar materials used.

In an example given by EPA (34), the authors show that implementation of a corrosion control program had a benefit–cost ratio of 4.2 to 1, a very favorable ratio, if indeed the situation was correctly analyzed, and their analysis was deemed to be conservative. This indicates a direct financial reason for water utility managers to seriously consider corrosion control program, going beyond the requirements of the NIPDWR.

Another common problem associated with corrosion is tuberculation, or the formation of tubercles on the pipe walls as a result of the corrosion activity. These roughen the pipe and increase the energy required to pump the water.

The corrosion and encrusting phenomena in pipes is very similar to hardening of the arteries in humans where scales deposit on the walls of the "pipes" and reduce the "C factor" or eventually shut flow off altogether. The heart bypass operation, an expensive remedy, has been the answer to hardening of the arteries. In distribution systems pipe replacement is the response.

Postprecipitation is a related phenomenon that clogs pipes with deposits on their walls. Costello describes several types of postprecipitation reactions: calcium carbonate, iron, lead, zinc, aluminum, magnesium, manganese, polyelectrolytes, and microbial growth. The iron postprecipitation phenomenon is corrosion itself, with iron being the metal that is precipitated through ferric hydroxide or ferric oxide. Zinc and lead reactions are similar to iron, but they form a much more compact precipitate than iron and lead to a tough coating rather than tubercles as iron does.

Remedial action concerning postprecipitation requires careful analysis of the reasons for it, followed by a long-term adjustment of the treatment processes used for the particular source waters involved. Since the postprecipitation products discovered may have been in formation over a long period, the remedy ought to consider the history of source waters and treatment.

In the Urban Institute report, four indicators of water distribution condition are presented: unaccounted-for water, leakage, main breaks, and the adequacy of the maintenance program. Taken together, these categories also provide a general guide for the development of a total program of care for distribution systems. That is to say, if all of these condition indicators can be made to fall into the satisfactory range, then the total management of the system can be considered to be adequate.

WASTEWATER COLLECTION SYSTEMS

Wastewater collection systems are, for the most part, gravity systems meant to be installed and, hopefully, forgotten. That at least is the conventional wisdom in some places. The truth is, however, that operating and maintaining a wastewater collection system is an important and dynamic part of a total wastewater management program. The emphasis is on the maintenance since there are not many operating decisions associated with the collection system itself. However, the maintenance of the collection system affects the operation of the wastewater treatment facility greatly, since the flows into the facility are greatly affected.

Foster (33) states that there are five common maintenance problems associated with sewer systems: infiltration of groundwater, inflow of stormwater, clogging, breaks, and damage from unauthorized and improper waste material. The corrective measures called for fall into seven categories: replacement of damaged pipe with infiltration-resistant pipe, better installation of pipe, sewer cleaning, analysis of infiltration and inflow through flow measurement techniques, inspection and testing of sewers, grouting, and implementation of a safety program. These lists outline most of the challenges faced by the manager attempting to improve the performance of the sewer system.

The Clean Water Act of 1972 and its implementing regulations specified that new plants constructed through the construction grants program had to be preceded by sewer system evaluation and rehabilitation programs, complete with infiltration/inflow analyses (I/I analyses). This led to considerable use of new techniques such as video inspection to find places of leakage. The results, however, were not completely successful.

Figure 9.9 shows a definition of the terms used in I/I, with a correlation with a rainstorm event. It can be seen from the figure that infiltration is a rather constant additive, and inflow corresponds to the rain event (34).

EPA found that by 1978 it was obvious that original requirements for I/I analyses were too burdensome, and it issued revised regulations providing a technique for simplified I/I investigations. Moreover, a subsequent evaluation determined that expectations of the I/I program were not being fulfilled, in spite of the fact that the plants studied had significant problems with excessive I/I. EPA observed that most municipalities did not seem to be implementing long-term sewer system maintenance programs, and that operating budgets were the first to be cut in times of budget problems. They observed that operating budgets barely contained enough funds to handle emergencies, much less preventive maintenance. This means that more convincing arguments are necessary to convince public officials of the need for preventive maintenance, and that responsible programs for sewer system maintenance must then be undertaken.

As an example of a place where active maintenance is being consid-

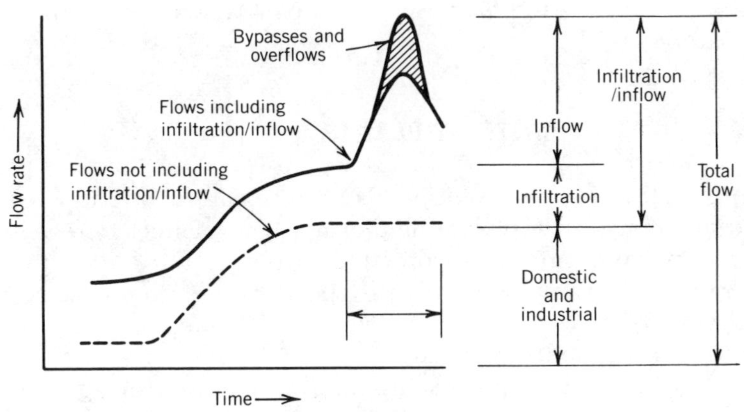

FIGURE 9.9 Infiltration/inflow (34).

ered, the Metropolitan St. Louis Sewer District recently reported on its program to identify and repair defective combined sewers (35). The district has about 1100 miles of sewer serving a population of 450,000. Some of the sewers are 125 years old. In 1981 it had 370 failures with repair costs ranging from \$5000 to \$500,000. Since the cost to detect and repair all of the defective sewers is too high, it is developing methods to zoom in on the high-risk areas using information technology. Specifically, the district is compiling maps with the risk factors that will enable a schedule of inspections with potential payoffs to occur. Detection technologies being considered include physical inspections, ground penetrating radar, gravimeters, and infrared photography. Some of the techniques are still considered experimental. The results of the detection program will be entered onto system maps using computer-aided design and drafting equipment. Rehabilitation methods to be considered include shotcreting, seal grouting, including chemical grouting, compaction grouting, sliplining, steel plate lining, and in situ forming.

Since sewer system rehabilitation is important, ASCE and WPCF are completing a new manual of practice on the topic. It was included in an article that reviewed the methods of sewer rehabilitation. Some 26 methods were identified, most developed during the past 20 years (36).

The Environmental Protection Agency has issued several reports concerning techniques for sewer systems evaluation and rehabilitation. One of them is called a "handbook" and may be of use to municipalities planning a program (37).

On an overall basis, O&M of wastewater collection systems is a challenging task that requires dedication and skill. Training and motivation of personnel is a principal challenge of management. EPA has sponsored the development of a comprehensive training manual by California State University at Sacramento, one that covers the important topics in considerable detail (38). The topics covered are the same that have been discussed: nature of the system, the operator's responsibility, organization for O&M, pipeline cleaning and maintenance, repair procedures, administration, and safety.

TREATMENT PLANTS

Both water and wastewater treatment plants require careful attention to operations and maintenance. Although neither has been perfect in recent years, water plants have a better record due to public interest in the quality of their drinking water. Certainly the specific operations of water and wastewater plants are different, but the principles of operations and management are similar, so they are covered here in the same section.

A treatment facility is like a processing plant in manufacturing. The product is treated water or wastewater. There are goals and objectives for operations, and they can be quantified for evaluation and testing. There can be automatic control of the processes. Inventory control is important to hold down the costs. The quality control function of the product is essential.

In the past few years there has been considerable attention given to developing procedures for O&M in treatment plants. One group developed a general model for a new treatment plant in Houston (39). They considered that the important factors in assuring a smooth startup and operation were maintenance management, training, startup operations, and administrative management. Their system contained nine elements:

Equipment technical manuals
Corrective maintenance systems
Preventive maintenance systems
Reporting systems
Inventory management systems
Spare parts stock records
Equipment histories
Routine operation/maintenance instructions and safety precautions
User guides

This is a rather complete statement of the general elements needed in a management system as outlined earlier in this chapter. The same elements appear in most operations and maintenance systems.

Water Treatment Plants

The operation and maintenance of a water treatment plant is guided by the need to produce drinking water within the quality guidelines of the regulatory authorities and the mandates of the local management. The quality of drinking water is usually a source of local pride (or it should be!) and the operators of the systems can be expected to be under close scrutiny. In Fort Collins, Colorado, the treatment plant recently experienced an increase in the incidence of the cyst Giardia, and treatment operations were modified to take care of the problem. This involved a slight change in dosage of chlorine. Some fish in aquariums died as a result, and the reaction was predictable—an outcry. It even earned a lead editorial in the local paper. This close attention is not normally given to wastewater treatment plants—a subject treated in more detail in the next section.

The operational problem for water treatment plants is basically one of operating the plant within the design guidelines to ensure adequate physical, biological, and chemical treatment. This means that the equipment must be installed and operated properly and be carefully maintained. The laboratory must function smoothly or there must be provisions for outside laboratory work.

Every locality has a unique source of water, and the treatment strategies must be adapted for the unique needs of the source. This will mean seasonal changes in treatment strategies and adjustments in the chemical dosages utilized. Source waters will not be uniform over a long period of time, since watersheds will undergo changes. This must be accounted for as well.

Wastewater Treatment Plants

Most of the recent furor in treatment plant operations has been about wastewater treatment plants. The General Accounting Office (GAO) reported that in a random sample of 242 plants in 10 states 87% were in violation of their permits, and, in the opinion of GAO, 31% were in serious violation (40). GAO's study showed that problems generally fell into one of the following categories:

Design deficiencies
Equipment deficiencies
Infiltration/inflow overloads
Industrial waste overloads
Operation and maintenance deficiencies

GAO elaborated on the last category, operation and maintenance, with an example from its studies. The plant studied had the usual problems: an inadequate budget and undertrained staff. Since it takes an adequate budget to attract and hold staff, these problems are not really two separate ones. The plant was a 5.5-mgd secondary treatment facility costing $4.2 million. The secondary digesters were inoperative due to poorly trained staff and ineffective maintenance. Also, the centrifuge and the chlorinators were inoperative. The plant had no system for scheduling maintenance, no operator training programs, and the operators were utilized on other duties for part of each week. After the city hired an independent contractor to run the plant, the problems cleared up. In a few months the plant was meeting its permit requirements.

These kind of problems are common. When I was working with a water quality regulatory program in North Carolina, we received many complaints from the public about water quality problems. One utility had problems with its centrifuges and took them all out of service for several months. This resulted in a high-visibility fine for the utility and negative publicity for its management. The incident could have been prevented with more attention to O&M.

The advent of computer control will mean improved attention to operation in treatment plants. While some of the processes do not lend themselves to automation yet, and while instrumentation limitations are a serious problem, automating means better information for management and almost always will result in better operations as a result of improved understanding. Computers as tools to improve operations and maintenance are already being installed in some plants. The firm CH2M-Hill has introduced a system it refers to as the Computerized Operations and Maintenance Programs (COMPS), which offers four features: a maintenance management system, a dataworker system, an inventory control system, and operator training programs. The dataworker system is to aid the operator to generate reports and make process checking calculations.

EPA has been very concerned about how to deal with the difficulties in making treatment plants operate better. It sponsored several studies and evaluations that began to show there was always a combination of factors that explain poor plant performance. In a sense, the EPA study was a compilation of 287 site visits that constituted limited scale diagnostic analyses as discussed earlier. The results confirm the generalized observation that difficulties are caused by complex interactions, and not by simple causes. EPA found that there were ten principal recurring difficulties: (41).

1. Operator application of concepts and testing to process control or operator error
2. Inadequate process control testing procedures
3. Infiltration/inflow problems
4. Inadequate understanding of wastewater treatment
5. Improper technical guidance
6. Sludge wasting capability not adequate
7. Inadequate control of processes
8. Inadequate flexibility of processes
9. Inadequate quality of O&M manuals
10. Poor design of aerators

As a result of the experience with inadequately operating plants, EPA has developed a program called the "composite correction program" that consists of a procedure for evaluating and correcting performance (42). In effect, the recommended program consists of the diagnostic approach, with provisions for prescribing and implementing improvement measures. The program focuses on the weak points in the system by identifying the performance limiting factors.

STORMWATER MANAGEMENT SYSTEMS

Stormwater systems differ from both water supply and wastewater in that no treatment is involved. However, the same water resources facilities are involved, including channels, pipes, ponds, instruments, and maybe pumping facilities. Maintenance programs are coming to the attention of management just now, since the costs of neglecting stormwater facilities is beginning to hurt in some areas.

For the most part stormwater facilities are not "operated." Most public works jurisdictions do not have staffs to operate them, and few if any discretionary or decision points are built into the systems. This means that they are "static" rather than "dynamic." However, modern computer control equipment and the lower-cost instrumentation and telemetry that is now becoming available mean that more control will become feasible in the future. Control of stormwater systems can bring significant benefits in terms of better performance, less flooding, better maintenance, and lower cost. Some cities are pioneering it, and of course, where combined sewers are involved, there is more awareness of the need for control.

When it does become possible to control stormwater systems, the places where decisions can be made will be releases from reservoirs, pumping stations, and overflow points to receiving waters. These will call for the right instruments, telemetry, gates, pump stations, and other devices. Stormwater agencies across the country are already studying these possibilities.

Maintenance of stormwater facilities is practiced by some public works organizations, but reports about it are scanty since the maintenance function may be submerged with other organizational missions such as street maintenance.

The Urban Drainage and Flood Control District in Denver has developed a maintenance program for its major drainageways (43). It classifies the work required as routine work, restoration work, and rehabilitative work. Each is approached differently in the contracting.

Example of routine work include vegetation mowing, trash and debris cleanup, weed control, and revegetation efforts. Examples of restoration work include detention pond mucking, trash rack cleaning, rebuilding steep rundowns, tree thinning and clearing operations, extending trickle channels, repairing local erosion problems, and doing local channel grading and shaping. Examples of rehabilitative work include rebuilding or replacing drop structures, installing trickle channels, reshaping channels, installing riprap to correct or prevent erosion, establishing maintenance access into drainageways, and providing protection for existing box culverts, retaining walls, or road crossings.

The three types of maintenance work require different approaches to contracting. The district has found that routine work can be packaged and bid to obtain competitive prices. Restoration work is difficult to package for bidding so the district reviews the work with potential contractors and has them submit proposals. Rehabilitative work requires design and construction in the same way that new facilities do.

DuPage County, Illinois, has prepared a comprehensive manual for operation and maintenance of stormwater storage facilities (44). The manual has received compliments for its comprehensiveness and its innovation. Since the manual is a pioneer, these compliments are really an expression of pleasant surprise that such a manual has become available.

The DuPage manual addresses several interdisciplinary topics: inspection and monitoring, record keeping, economic studies, sedimentation control, aquatic vegetation control, mosquito control, pollution and erosion control, structural maintenance, safety measures, and multiple-use management.

QUALITY CONTROL

Quality control (QC) is a term that is becoming very important to U.S. industry since we are learning that high product quality is not only the best way to be competitive it is essential. U.S. industry is learning that the best way to have high quality is to do things right the first time. There is a lot of discussion about quality control today, and there will continue to be emphasis on it in urban water organizations as well.

A discussion of quality control belongs in this chapter on operations and maintenance since it is really an operational concern. Basically, QC means to assure that the quality of the product, whatever it is, is within acceptable limits of quality, as defined for that particular product. Sometimes the words *quality assurance* are used in place of QC. This is common in the water industry, when the word *quality* already has so many important ramifications relative to water quality.

Again, the concept of production operations offers a place to find ideas about quality control. Greene (1) devotes most of two chapters to the subject, with one of them covering the statistical aspects of QC alone. His discussion of the internal organization for quality control offers food for thought for water managers since it offers a test as to whether our organizations are adequate to make sure the quality checks are there. The parts of the organization Greene identifies are inspection, administration and record keeping, quality control engineering, and the gauge room. Most of these functions have familiar rings relative to the water or wastewater treatment functions since they have become necessary to meet regulatory requirements. The principles can be applied even more vigorously than the regulations require, however, and there are also other opportunities for the application of the principles. As an example, if the goal of the water utility is to deliver service that involves water of certain quality, with certain pressures, during certain time periods, how can it make sure that it does? It would need an inspection program, a quality control sampling effort, laboratory work, and record keeping. Another example might have to do with the stormwater department. If the goal is to prevent a certain level of flooding, how do we know we are being successful? The normal indicator might be the absence of complaints, but the organization could plan and execute an inspection program.

There are many aspects to QC programs that can be considered by the managers of water organizations. It can be considered as taking the offensive, rather than always being on the defensive, in water management.

REFERENCES

1. James H. Greene, *Operations Management: Productivity and Profit,* Reston Publishing Company, Reston, VA, 1984.
2. Government Finance Research Center and Peat Marwick, Mitchell & Co, Wastewater Utility Management Manual (draft), prepared for the EPA, July 1981.
3. C. A. Podmore, Ed., *Diagnostic Analysis of Irrigation Systems,* Vol. 1, *Methodology, Water Management Synthesis Project,* Colorado State University, December 1983.
4. E. T. Newbrough, *Effective Maintenance Management,* McGraw-Hill, New York, 1967.
5. Peter Ward, The Fleet Maintenance Management Cycle, *Public Works,* November 1984.
6. Edward S. Lindow, Maintenance Management for Roofing Systems, *APWA Reporter,* October 1984.
7. John C. Whitman, Contracting Highway and Street Maintenance, *APWA Reporter,* March 1984.
8. Harry P. Hatry and Bruce G. Steinhal, *Guide to Selecting Maintenance Strategies for Capital Facilities,* Urban Institute Press, Washington, D.C., 1984.
9. APWA, Guidelines for Developing a Maintenance Management System, Chicago, 1980.
10. D. Kelly O'Day and L. A. Neumann, *Assessing Infrastructure Needs: The State of the Art,* Hanson, Royce, Ed., *Perspectives on Urban Infrastructure,* National Academy of Sciences Press, Washington, D.C., 1984.
11. Stephen R. Godwin and George E. Peterson, *Guide to Assessing Capital Stock Condition,* Urban Institute, Washington, D.C., 1984.
12. John Gorish, Pressure Recorders in the Right Places Make a Better Distribution System, *Water Engineering Management,* October 1984.
13. Thomas M. Walski, Why Calibrate Water Distribution System Models, *Water Engineering Management,* October 1983.
14. Thomas M. Walski, Using Water Distribution System Models, *JAWWA,* February 1983.
15. Thomas M. Walski, *Analysis of Water Distribution Systems,* Van Nostrand Reinhold, New York, 1984.
16. Joseph R. Kroon, M. A. Stoner, and W. A. Hunt, Water Hammer: Causes and Effects, *JAWWA,* November 1984.
17. Moses Siedler, Obtaining an Analytical Grasp of Water Distribution Systems, *JAWWA,* December 1982.
18. Boyle Engineering Corp., Municipal Leak Detection Program: Loss Reduction, Research and Analysis, State of California, Department of Water Resources, Sacramento, August, 1982.
19. George E. Peterson, M. J. Miller, S. R. Godwin, and C. Shapiro, *Guide to Benchmarks of Urban Capital Condition,* Urban Institute, Washington, D.C., 1984.
20. Ellen E. Moyer, J. W. Male, I. C. Moore, and J. G. Hock, The Economics of Leak Detection and Repair—A Case Study, *JAWWA,* January 1983.

21. D. Kelly O'Day, Organizing and Analyzing Leak and Break Data for Making Main Replacement Decisions, *JAWWA*, November 1982.
22. Thomas M. Walski, and Anthony Pelluccia, Economic Analysis of Water Main Breaks, *JAWWA*, March 1982.
23. Dan Skorcz, Maintaining Distribution System Valves, *JAWWA*., November 1983.
24. Committee on Control of Water Quality in Transmission and Distribution Systems, AWWA, Determining Internal Corrosion Potential in Water Supply Systems, *JAWWA*, August 1984.
25. Committee on Disinfection, AWWA, Disinfection, Water Quality Control, and Safety Practices of the Water Utility Industry in 1978 in the United States, *JAWWA*, January 1983.
26. Joseph A. Cotruvo and Craig Vogt, Development of Revised Primary Drinking Water Regulations, *JAWWA*, November 1984.
27. Donald M. Somers, The Potable Water System, in Korbitz, William E., Ed., *Urban Public Works Administration*, International City Management Association, Washington, D.C., 1976.
28. American Water Works Association, Water Utility Management, *Manual of Practice* M5, Denver, 1980.
29. LaVerne D. Hudson, J. W. Hankins, and M. Battaglia, Coliforms in a Water Distribution System: A Remedial Approach, *JAWWA*, November 1983.
30. James J. Costello, Postprecipitation in Distribution Systems, *JAWWA*, November 1984.
31. American Water Works Association, Curbing Corrosion, Theme Introduction, *JAWWA*, August 1984.
32. Gregory J. Kirmeyer and G. S. Logsdon, Principles of Internal Corrosion and Corrosion Monitoring, *JAWWA*, February 1983.
33. William S. Foster, Ed., *Handbook of Municipal Administration and Engineering*, McGraw-Hill, New York, 1976.
34. EPA, 1990 Preliminary Draft Strategy for Municipal Wastewater Treatment: Operations, Washington, D.C., January 1981.
35. John F. Koeper, K. L. Koble, and J. Coll, Combined Sewers: Maintaining Yesterday's Infrastructure with Today's Technology, *J. Water Pollution Control Federation*, September 1982.
36. Allen Morrison, Sewers: Repairing Beats Replacing, *Civil Engineering*, September 1983.
37. Environmental Protection Agency, *Handbook for Sewer System Evaluation and Rehabilitation*, Washington, D.C., December 1975.
38. California State University, Sacramento, Operation and Maintenance of Wastewater Collection Systems, for the Environmental Protection Agency, 1983.
39. Leo M. Weinberg, and R. D. C. Lloyd, Operations and Maintenance Assurance and Management Systems (OMAMS), *J. Water Pollution Control Federation*, October 1983.
40. General Accounting Office, Costly Wastewater Treatment Plants Fail to Perform as Expected, CED 81-9, November 14, 1980.
41. EPA, Summary of National Operational and Maintenance Cause and Effect Survey, Technology transfer program, Cincinnati, July 1979.
42. EPA, Handbook for Improving POTW Performance Using the Composite Correction Program Approach, EPA-625/6-84-008, Cincinnati, October 1984.
43. Mark R. Hunter and L. S. Tucker, Contract Maintenance for Drainage and Flood Control Facilities, *Flood Hazard News*, Urban Drainage and Flood Control District, Denver, December 1982.
44. DuPage County (Illinois) Development Department, Stormwater Storage Facilities Operation and Maintenance Handbook, August 1983.

CHAPTER TEN

AUTOMATING OPERATIONS OF URBAN WATER SYSTEMS

Chapter 9 emphasized the importance of improving operations, and one way to accomplish the improvements is with automation. The purpose of this chapter is to discuss how automation can be applied to improve operations, with emphasis first on what is happening in the industry and second on what the future may hold.

To "automate" generally means to convert to self-operation. A machine or system that is automated is able to run itself. There must be, however, provisions for how this is done, and it is the design and implementation of these provisions that constitutes the challenge to automating urban water systems.

Some do not realize that the use of the term *automation* is not very old. John Diebold helped to make the term an everyday one, and its use is still evolving. Diebold recently republished his 1952 book explaining the concepts involved, and upon reading it, we definitely see that we are still learning about the way that computing and data management machines affect work in general and in the water industry in particular (1). Actually the term probably evolved in the 1930s at General Motors rather than with Diebold's 1952 book, but the point is the same: the term is a relatively new one for water managers to deal with, and they will have to stay up to date with automation just as their colleagues in manufacturing must.

The use of the term *automating* sometimes introduces confusion. You might walk into a modern treatment plant with panels of instruments and blinking lights and ask if the system is automated. The answer may be "yes," while the truth is that the plant is under remote surveillance with a capability for remote control, but with no automatic self-operation invovled at all. The plant may be able to run unattended, with the operators located at the remote control center, but there is still no automation in a strict use of the word. There may even be computers involved, but no automation. The computers can be used as data collection devices, for example. In that sense you could say that data collection is part of operation, and so the plant is partially automated.

The term automation has good and bad connotations. To a manager it offers the possibility that higher productivity may result from replacing operators with computers and instruments. To a worker, however, automation may raise the threat of the loss of a job. Regardless of these considerations, it now seems apparent that automation is here to stay, and ways must be found to accommodate it rather than to resist it, but to do so in a responsible and wise manner. This view is behind the objective of this chapter: to outline the possibilities and procedures for automation and to point out ways for it to be implemented successfully.

Responsible automation offers improvements for everyone, including workers and responsible and effective programs of implementation offer help for all classes of workers. The careful and incremental approach to automating offers an educational experience to all workers and managers, for the system has to be understood better under automatic control than under manual control. In other words an automation exercise offers the chance for a step up in improved operations, both due to the automation and due to the new insights gained in the process.

Automation of processes and manufacturing seem to be increasing rapidly. Many examples of application are reported; including the "factory of the future," air traffic control, urban street traffic, nuclear and conventional power plants, aircraft autopilot, elevators, "smart buildings," and spacecraft.

There seem to be a number of excellent opportunities for automation in urban water systems. Examples could be classified as follows: storage reservoirs, water treatment plants, water supply distribution systems, wastewater collection systems, wastewater treatment plants, and stormwater systems. The latter could include flood warning systems, detention storage facilities, and combined sewer systems. We will discuss automation opportunities for these different components of the urban water system in more detail later in the chapter.

CONCEPTS OF AUTOMATION

Since automation means different things to different people, it is useful to define it in greater detail. It is convenient to refer to automation in its stages of development. There are different classifications, but four seem clear as they apply to urban water systems.

1. Data collection, processing, and reporting
2. Conventional supervisory control without automatic control
3. Automation of parts of systems using set point controls
4. Complete computer control using prediction models

These categories describe the situation today pretty well, and observations over the past 10–15 years verify that there is an incremental approach to automation along the sequence indicated above. For example, the improved availability of small computers since 1971 has made them very popular for data logging and processing. There have been reductions in labor required by implementing supervisory control, such as in water treatment plants or distant pumping stations. We are also seeing set point automation of parts of systems, particularly treatment plants, but we do not yet see examples of full closed-loop computer control of linked complex systems, the final category of automation. We will certainly see examples in the future, however, for the water management field has many innovators and they are ready to apply the latest technology when they can.

Control theory has many complicated concepts and theories. The theory begins with the basic control loop. This means the lowest level of control, such as an individual machine. The control of a pump in a pumping station would be an example, as shown in Figure 10.1. The pump in the figure is controlled by some arrangement of digital or analog devices that respond to stimuli from the measuring instruments, in this case flow and/or pressure. The control program involved will allow for instructing the controls about how to keep the pump within allowable limits. These may involve either being on or off or, in the case of a variable capacity pump, different states of pumping. The disturbances in the diagram could be changes in demand or supply, or some system problem such as a leak. Figure 10.1 is an example of what is called "feedback control." There

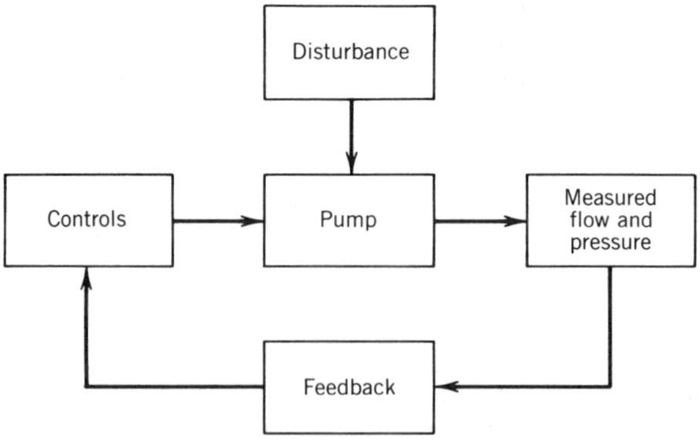

FIGURE 10.1 A pump control loop.

is also a version called "feedforward control," where the measurement is reported to the next stage of the process. The control loop illustrated in Figure 10.1 represents the simplest level of control. Other water examples of single control loops might be chemical feed rates, valve operation, or tank storage level. In the case of wastewater treatment plants, some control loops reported in Milwaukee were as follows (2):

Primary sludge pumping
Dissolved oxygen process control for aeration
Mixed liquor suspended solids control
Automatic final clarifier wasting
Dissolved air flotation thickener loading
Digester feed and withdrawal
Acid feed for phosphorus control
Sampler flow pacing

These control loops are those in the plant that are amenable to measurement and control, and these requirements are basic necessities to make automation successful.

The next level would be combinations of control loops. These might be called the unit control level (3). Terminology here will vary, but Corbin describes the loops as being connected with a data "highway" and an operator station, as shown in Figure 10.2. This figure shows two microcomputers, each operating one control loop. Extending this concept, multiple controllers or local area control networks can be linked to a data highway as shown in Figure 10.3. These concepts seem to be followed by several manufacturers who offer control equipment for the water industry, and the equipment and approaches seem to differ mostly in component configuration and programming language and concepts. The fea-

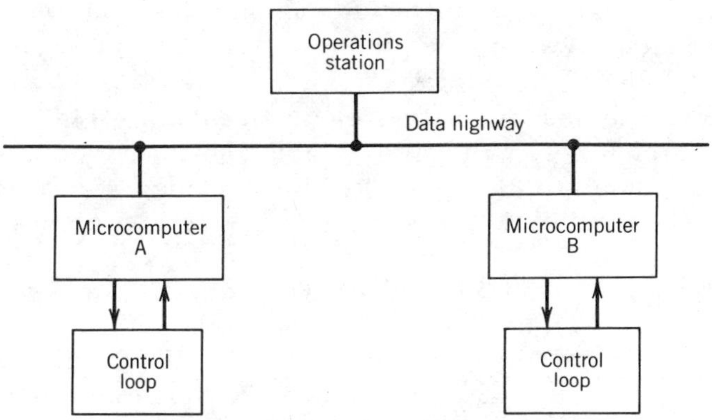

FIGURE 10.2 Unit level processing.

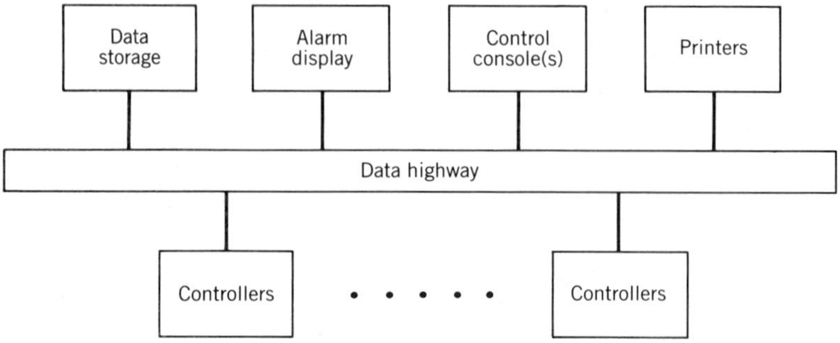

FIGURE 10.3 Concept of the "data highway."

tures shown in Figure 10.3 are generally self-explanatory. The data storage devices can be tapes, disks, or drums. The alarm display is a necessary feature of any control center. The control consoles can be designed for on-line control, design, or data analysis, and the printers can be of different applications. The capacity of the data highway will be given in terms of bits of information per unit time, such as one megabit per second.

The next level of automation might be the "plant level." This might have a configuration such as that shown on Figure 10.3, and it could involve numerous control loops or other connections to the data highway. Westinghouse's system, called the Westinghouse Distributed Processing Family, provided for up to 254 connections or "drops" in 1982, and with hardware capabilities increasing, these figures will surely be expanded (4).

Automating a plant brings to mind a manufacturing plant, processing plant, or, in the case of water facilities, a treatment plant. In the case of industrial applications advances made have been phenomenal, especially in industries where automation is necessary to meet the competition. Water systems are ideal for the kind of distributed processing shown in Figure 10.3, however, and "plant automation" concepts could be used, for example, on a citywide water distribution system, just as it can on an electric power distribution network.

There are some phenomenal developments going on in factory automation today, and these will continue. Imagine a factory of robots, controlled by computers, building things 24 hours a day with practically no downtime. The possibilities for improving productivity seem almost limitless.

An interesting illustration of plant automation is one reported from General Motors, which has implemented a "numerically controlled paint system" (5). In this example there are a number of machine controllers for different stages of the auto painting work. Each of the painting devices has considerable flexibility and can be regarded as a semirobot controlled by the computers. All of the assembly line painting is controlled by the supervisory computer, which allows for painting different types of cars as they enter the assembly line. This, then, is not a single model assembly line such as that introduced by Henry Ford (who is reported to have said, "you can have any color you like as long as it's black"), but it is a flexible and adaptive approach to the assembly line. The supervisory computer is backed up with a redundant computer, and both receive signals from a model recognition sensor. There is another computer in a "teaching booth" where a worker can go in and program painting instructions as needed.

A deeper understanding of the computer's role in control is seen in an analysis of how the software works in control applications. Spector (6) provides some explanations that are helpful. He takes the general components of the operating system that we have already discussed and gives some different names to them, names that might be considered more general (see Figure 10.4). First, he describes the links between the computer and the process (or system) to be controlled as sensors and actuators. These are the data collection devices and the control devices. The computer uses a model of the actual system or process, which has three components: a model of the state of the system or process, a state update function, and a predictor function. It is the use of this predictor function that offers the most possibilities for making advances in computer control in the future, and we will discuss this more later. The computer also needs instructions to tell it how to sequence its checks and control signals, and it requires a set of goals or a "strategic plan" as Spector calls it, to establish the framework or strategy for control. In effect, the state model keeps up with the state variables, the state update model transforms data received into new information about the state of the system, and the predictor function uses the rules for operation to predict how the controls (or actuators) must be operated.

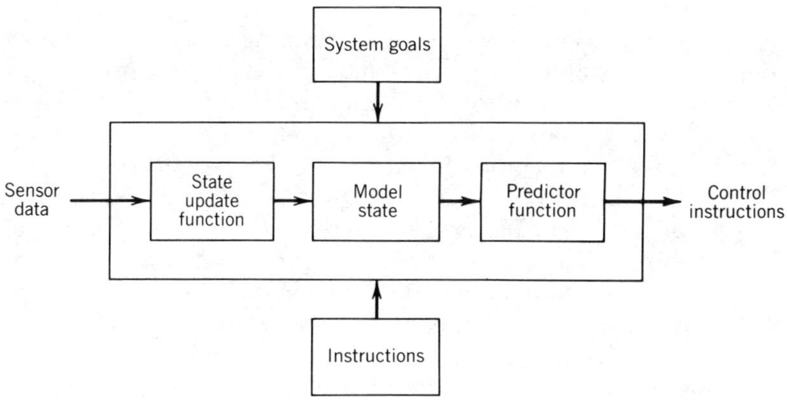

FIGURE 10.4 Components of the control center.

The models used by process control systems can be simple or complex. A simple one might be a temperature control system, while a complex one might be a multitask control system with many variables and system states. Complex multitask control systems might be appropriate for applications where operations are tightly coordinated, but in water systems components are often geographically separate and a different approach is required.

The approach that is evolving, as illustrated before, is based on distributed processors that normally do not share memory and that send messages back and forth along the communications channel, or data highway. This distributed processing offers several advantages, such as the minimization of data communications, with most computing being done by the computer in its own local network. Protection against failure is an obvious advantage to distributed processing, and in water systems it offers the possibility for lower-cost stage construction of the automation network. There is no other choice.

When discussing automatic control, it is important to understand the developments that are taking place in "control theory," which is the basic scientific area responsible for the research advances driving automatic control. Control theory is a scientific area falling between electrical and mechanical engineering, and it is concerned with how to measure and control the performance of devices of all kinds. It becomes highly mathematical quickly, and I will not present much of it here except to point out the linkages between it and the commonly used devices in water systems.

Basically, control theory deals with the response of a system to commands or disturbances. The system can be as complex as an entire treatment plant or as simple as a single pump. There is a controller that controls a manipulated variable, resulting in control of the process. The controller makes decisions on the basis of the outputs of the system, so that there can be corrections to compensate for deviations away from the norms that are set for the process.

There are three basic approaches to controlling processes that are used in commercial controllers: proportional control, rate control, and integral control. In proportional control the output and input are required to have a continuous linear relationship, or to be proportional to each other. In rate, or derivative, control the output is set to be proportional to the rate of change of the input. In integral control the output is proportional to the integral of the input or, said in reverse, the rate of change of the output is proportional to the input. These three approaches are found in commercial controllers that are called proportional (P), proportional–integral (PI), and proportional–integral–derivative (PID). The controls literature makes frequent reference to these three approaches (7).

Although there are presently not many extensive computer control applications in urban water systems, the future will see them, and they will operate reliably and in real time, two essential characteristics for important services such as urban water.

EMERGING AUTOMATION CONCEPTS

Automation is really exploding everywhere, and so much is happening that it is really difficult to stay current with all of it. Some of the trends that seem evident in the water industry are identified with the use of programmable controllers and personal computers, the need for improved instruments, and the need for predictor function displays. We shall discuss these trends in general and then give some specific illustrations in the next sections where the different water applications are discussed.

Improvements in programmable controllers (called PCs in the controls industry) have been especially important to advances in treatment plant operations. This is a different "PC" than the personal computer, but the controls industry is very used to calling these controllers by that acronym. Basically, a programmable controller is a replacement for the timer and relay arrangement for controls that has become so familiar. It is not a computer per se, since it has not been designed to "compute," but it has many of the features of the computer and is certainly part of automation equipment (8). The real advantage of the programmable controller is its programmability: it can be reused even when the logic changes. It has the capability to perform a number of the functions needed in a control center: calculations, comparisons, and analysis of data, and it can be connected with a computer to enhance its usefulness in on-line decision making.

The trend seems to be to increase the use of personal computers for industrial data logging and control decision making, and firms are now coming out with systems that combine personal computers and programmable controllers (PCs with PCs, if you will). These developments are described in a series published by *Control Engineering* in 1984, with a key reference by Miller (9). An obvious reason for the marriage between computers and controllers is to make the new computing equipment work with existing control equipment.

It seems that the use of the programmable controllers, in tandem with a growing availability of personal computers, is setting the stage for some real growth in automation in the coming years. Along with this action, we will see two general trends: the increasing popularity of computer "control" for functions in all four of the categories of automation defined earlier and the increasing use of models to predict the future state of systems so that control decisions can be improved.

The limiting factor for the success of automation is the availability of instruments to measure what is going on so that corrections and control decisions can be made. There has been a great deal of discussion in the past few years about the lack of availability of good instruments, and it is true that the instruments that are available leave something to be desired. There are instruments presently available, however, and practical operation will have to rely on them for the time being. Their importance in automation cannot be overemphasized, since if the instruments do not function, the control decisions cannot be made. In a recent article about instruments, Knudsen and others reported that their studies showed that the instruments worked, but it was the misapplication, misinstallation, and the lack of maintenance that were the problems (10). The authors had studied liquid and airflow measurement devices and the in-line measurement of dissolved oxygen. These basic quantities can be used to compute other quantities of interest, and control decisions can be made from the calculated values when decisions are necessary. The authors give an example that shows that some $200,000 per year (at $0.04 per kWh) in power cost savings would be realized in a 100-mgd plant from the automation of the decisions regarding airflow to an aeration basin.

The use of set point controls and the general use of programmable controllers to replace controls based on operator logic is the order of the day. The future use of prediction based on mathematical modeling is a tempting prospect, however. The use of prediction and predictor displays is not widespread in controls applications anywhere yet. The chemical process industry does not use it in spite of the fact that the concept has been around for over 25 years (11). It just may be that the water industry will pioneer the practical application of prediction.

The predictor display, as explained by Banks and Cerven (11) and as I understand it, would provide the operator with simulations of what the system will do if certain things happen. The operator can ask what would happen if the temperature hit a certain level, or if it rained, or if the chemical feed rate changed to a certain level, or any of a thousand things that will affect operations. The simulation model would then predict and display the consequences. This could help the operator to make complex decisions and to operate even simple systems under stressful situations.

APPLICATION OF AUTOMATION TO URBAN WATER SYSTEMS

There has been considerable interest in applying automation to urban water systems for years. Murray B. McPherson, director of ASCE's Urban Water Resources Research Program from about 1965 to 1982, made it a point to keep up with automation and other innovations, and his report on metropolitan water intelligence systems (12) documented the status of the thinking in the early 1970s. As far as specific applications, McPherson saw the automation of water distribution systems as having advanced the furthest (this was in 1971) and foresaw that actual applications of automatic control to distribution systems would be feasible soon (after

1971). Since this has not happened, the reasons need examination, and this subject will be covered later in this chapter. McPherson saw the automation of "integrated metropolitan water systems" to be an elusive goal since the incentives for integration by service and by jurisdiction were generally missing. Applications such as control of combined sewer overflow seemed to be the most difficult technically due to the absence of control logic or strategies. McPherson's insights on these matters led to several research projects I participated in at Colorado State University, as well as scores of others at various universities and agencies.

In fact, distribution systems are not yet automated, although the hardware capabilities for data collection, transmission, and analysis are much improved since 1971. The reasons for lack of implementation are managerial, financial (or economic), and sociopolitical, rather than technical. There are numerous examples of treatment plant automation, however, as well as automation of water system components, where considered appropriate by managers, but many problems still inhibit total system automation.

Opportunities for automation generally follow the configuration of systems: water supply reservoirs, water treatment plant operations, water supply distribution systems, wastewater collection systems, wastewater treatment plants (including sludge treatment) and stormwater management systems (especially detention storage and combined sewer systems). There are examples of all of these in operation across the United States at the time of writing, and the successes are mounting.

Progress in automating operations is proceeding incrementally, and the best way to study it is to look at examples. The purpose of the next sections is to present examples of the different applications and to infer from them the benefits, procedures, and pitfalls of the different ways to automate.

Water Supply Reservoirs

The reservoir offers a very direct application of automatic control since water management decision making can be seen easily there. The control variable is to store or release water, and decisions can be based on forecasts and other hydrologic information.

The reservoir illustrates the three model categories described as part of control systems: a state model, a state update model, and a predictor function. The level of the reservoir or amount of water in storage at any time is the most important state, and disturbances come in the form of inputs or outputs. The state update model allows for the recalculation of water level after disturbances, and the predictor function predicts what the future states will be as compared to desirable states, as a result of control actions and/or predicted future disturbances.

Reservoir operations are complicated by the normal multipurpose nature of the water uses, meaning that reservoir control actions result in impacts on water supply, pollution control, flood control, recreation, hydropower, irrigation, fish and wildlife, and other uses such as navigation. The "predictor" model and the "stragetic plan" can be very complex. In fact, the state update model can be complex if it involves mathematical modeling and watershed analysis.

Water quality modeling can be incorporated into reservoir decision making as well, since the quality of raw water can affect decisions about treatment levels or the selection of water from different levels of a reservoir or mixtures of water from different sources. In fact, automation allows the consideration of these different alternatives, whereas manual operation might not allow the consideration of operational schemes that enable the best use of existing facilities and lowest cost operation to achieve the best performance.

Quite a bit has been written about the automatic control of reservoir systems, and they have been analyzed more than most other parts of water systems. In a recent study of the use of models for optimization of reservoir operations, Helweg and others showed that no major reservoir systems were using optimization models for real-time operation (13). Their article pointed out that the reasons for the delays in implementing optimization models were the same as for other aspects of automation: lack of management support, lack of trained personnel, and problems with software and hardware. Although there have been delays in using optimization models in reservoir systems operations, there has been progress in automation per se, since the control of the gates and other structures of reservoirs is a straightforward application of supervisory control systems. There are also opportunities for the use of computers in the operation of turbines and gates in reservoirs, such as the use of the computer-controlled fixed turbines that has been reported in the Pacific Northwest recently (14).

We will see more computer control of reservoir systems in the future, since they are the water facility most amenable to large-scale optimization, and it seems promising to squeeze a great deal of additional yield from them through the use of automation combined with good predictor and optimization models.

WATER TREATMENT PLANT OPERATIONS

There has been considerable interest in automating water treatment plant operations. The unit processes of a water treatment plant are similar to the industrial processes of refineries and process industries and a great deal is known about how to automate them. In reviewing the progress in automation we must, however, distinguish between plans and actual accomplishments. Much of the published literature deals with plans or initial installations and not much with long-term experiences with automation.

AWWA has published a number of papers and reports on treatment plant automation. Since the field is progressing rapidly, advances in implementation are outpacing the published reports. AWWA's report, "Water Plant Instrumentation and Automation" (15), contains a good summary of reports, but so much has happened since then that the document cannot be expected to be up-to-date. Automation is like the rest of the computer field, very difficult to stay up-to-date due to the rapid pace.

Nevertheless, AWWA's report is excellent to track activities and topics of interest. For the most part these focus on the following topics: control strategies and software for process elements, instrumentation problems and solutions, case studies, and human factors involved in automation. AWWA also has published a manual of practice on automation and instrumentation, one that is oriented toward the plant operator (16). It covers instruments, controls, electricity, telemetering, computers, and the basics of automatic control.

Philadelphia began to report a comprehensive automation project in 1976. It envisioned a hierarchical scheme that consisted of the following five levels of control: (1) local manual control, (2) remoted multiplexer instrumentation, (3) operator information systems, (4) operator information system supervisory level, and (5) the management information system (17).

The idea behind the hierarchical approach was similar to the plant automation concepts described earlier. Level 1 was to include the basic control processes of the plant, such as chemical feed, pumping, valve control, and related processes. Level 2 was to implement the distributed control approach with controllers and multiplexers located near the processes with data communications to Level 3 where the process computer would be located. Level 3 would be a decision making level with the basic computing capability located there. Level 4 was envisioned to have a treatment control headquarters that could eventually exercise supervisory and surveillance control over the level 3 controllers and the water quality laboratory. Level 5 was to have a large general-purpose data processing computer to implement the management information system and provide overall supervision for the system.

Since the Philadelphia plans were published in 1976, the reports leading to the published paper were at least 1 or 2 years old, and this means that the concepts reported were 1974 vintage or earlier. In the automation applications I studied the hardware limitations of that period have been very evident, and if the Philadelphia plan were redone, the concepts would be different, probably more distributive in nature, with fewer levels and more computing capacity located nearer the processes.

The current thinking in Philadelphia and other large cities must be along those lines since the water industry follows industry in general in implementing innovations.

The automation of water treatment plants offers opportunities to use programmable controllers and personal computers to operate the valves and pumps and to generate the required reports, all automatically. In my own community of Fort Collins, the utility is automating step by step with its own forces. It has purchased a programmable controller for the treatment plant, and the city's forces are programming it to take over the supervision and controls in the water plant. There is a gradual reduction in personnel needed, and it is very happy with progress with the controller. The utility is phasing in the use of personal computers as data loggers and report generators, and soon the plant will be quite automated. The computers and the graphics displays that come with the programmable controller are allowing the phase out of the old control consoles that are so common in plants.

Automation is such a pervasive concept in water treatment plants that one must be careful what is meant by the term. For example, a recent article about the Norwich, Connecticut, plant stated that simple automation was used (18). Upon reading the account one discovers that no computer is involved, and what is meant by simple automation is that the plant runs unattended. No doubt uses will be found for computers and programmable controllers in the future.

Water Distribution Systems

There is less application of computer control and automation to water distribution systems than to treatment plants. One of the reasons is the spatially more difficult problem of adapting automation to the distribution system. Another is the general lack of control going on today in distribution systems operations. They are more static, than dynamic, and although there are possibilities for energy and water savings through real-time control of distribution systems, they have not been constructed with such control in mind. Also, data is generally not available to know what is going on at different parts of the systems at different times, although there are trends toward the acquisition of the data especially with leak control and the control of unaccounted-for water in mind.

There are some reports of applications of computer control in distribution systems, although not nearly as many as predicted from the interest 10 years ago. Actually, the concept of automating a distribution system was described over 20 years ago by Dan Brock, a scientist with the Dallas Water Utilities (19). However, recent reports have not suggested much implementation, with there being more papers about modeling techniques and new ideas than actual applications. In a recent paper Brock (20) states that the loop has not been closed yet, at least not for any large water systems, but he believes that the prospect is attractive enough to be tried.

Applications will come in future years, however, as water conservation, energy conservation, control of distribution system quality, and water auditing become more important in everyday operations.

Wastewater Collection Systems

The wastewater collection system is probably the most static of all the urban water subsystems. There are opportunities for automatic operation of pumping stations, but for the most part, the systems are gravity operated and function according to design until something goes wrong. At that time there may be opportunities for automation in the sense that operational levels can be sensed or failures located. I am not aware, however, of much interest right now in automating the pure wastewater collection system. Combined sewers are, however, another matter. They will be discussed under the section dealing with storm sewers.

Advanced applications of automation in wastewater collection systems will be like those reported by the Upper Occoquan Sewage Authority, where 22 selected points in the collection system are monitored for flows, and 5 remote pumping stations are monitored. The pump stations are under local automatic control. This system will be described in more detail in the next section under wastewater treatment (21).

Wastewater Treatment Systems

Wastewater treatment plants are essentially like water treatment plants except for the severity of the operating conditions and the nature of the unit processes. By the same token, progress in automation of plants has been roughly parallel. The complexity of some of the biochemical reactions in wastewater treatment plants has made them difficult to instrument and control, and water treatment plants are, in some ways, a little simpler. The instrument problem is especially severe in the wastewater field.

A good starting point in the discussion of computer control of wastewater treatment is a 1970 report by the American Public Works Association (22). This comprehensive report involved many of the leaders in the plants operations field of that day, and it summarized progress of that time very well. While there was a lot of interest in automating plants, only 1 plant of some 22 that responded to the survey reported any automatic control being performed. A closer investigation showed that there was a pneumatic controller for the influent at the plant responding, but no computer was being employed. There was, of course, considerable activity in using the computers for off-line work such as reporting and analysis.

The APWA report thus sets the stage for the progress that has been reported since about 1970. With so much being spent on wastewater treatment plant construction through the 1970s, there should be a great deal to study at this time. However, much of the success in automating operations of wastewater treatment suffers from the same difficulties of wastewater treatment in general—low budgets and inadequate attention—so the situation today is not one of great advances but of small steps in automation.

The APWA report offered some excellent suggestions that constitute a starting point for analyzing progress. The following were conclusions and recommendations from that report:

CONCLUSIONS

1. While conventional treatment systems do have serious disadvantages and limitations, they will continue to be the most broadly applied for the next 20 years.

2. While no new large-scale innovation threatens to replace conventional treatment, modification to the processes and supplementary processes are being implemented. Examples are chemical coagulation and oxygenation.

3. Personnel pose a real problem in the operation of plants. Difficulties arise in obtaining qualified personnel, in training operators and maintenance men, and in reliable follow-through of operational instructions.

4. Interest in computer applications and automatic control is widespread among plant management.

5. A recognition of the advantages of off-line computer uses was unanimous with general appreciation of automated management reporting, maintenance scheduling, inventory control, payroll, and accounting.

6. Skepticism was apparent among plant management of on-line computer control of plant processes.

7. Computer hardware and communication equipment are available and reliable for use in industrial process control and are suitable for application to wastewater treatment as well.

8. Off-line computer implementation for plant data handling and management information reporting is being successfully practiced; it can serve a wide range of beneficial applications and is fully feasible today.

9. On-line instruments to sense some of the variables considered essential to continuous monitoring and control of wastewater treatment (such as BOD, bacteria, and volatile acids) are not on the market.

10. Computer control of the more critical treatment processes, activated sludge, and anaerobic digestion is possible in a limited sense based on

experienced operator control, and will require further study before optimization via computer control is achieved.

RECOMMENDATIONS

1. Plant management should consider using off-line computer data processing for all areas of plant data handling: reports, maintenance files and scheduling, inventory control, and business areas. Computer availability for off-line applications may exist via a municipal facility, a nearby service bureau, time sharing, or the plant's computer.

2. Studies are recommended to develop the mathematical–logical relationships of the plant physical processes, notably the activated sludge and anaerobic digestion processes. The studies should result in the development of control strategies for supervisory or closed-loop control.

3. Further work must be done in the development of reliable, maintenance-free on-line sensors. An in-depth survey of user's experience with analytical instrumentation and sensors in wastewater treatment plants should be conducted. The survey would provide guidelines for improvement and innovative approaches to solving the problems of continuous monitoring of wastewater treatment processes.

4. Plant management should give serious consideration to computer control of plant processes, despite the lack of both knowledge and sensors in some areas of wastewater treatment. Today's impetus in plant controls is toward increased application of local closed loops with advanced instrumentation. Some examples are return sludge on the basis of blanket level, aeration rate to maintain a dissolved oxygen level, and variable speed pumping for uniform plant flow. These local processes can be integrated to achieve greater operational and economic benefits via computer control of centralized plant monitoring, automated analyses, timely reporting, and improved plant controls. *Only by application of computer control can the wastewater treatment process be optimized as a total integrated system.*

5. Design of new plants or modifications of medium size (15–50 mgd) existing plants should incorporate automatic controls of the highest degree available. This should yield economies of scale, operation and management and prepare the plant for future conversion to full automatic control.

The above quotes from the APWA report have proved remarkably durable over the past 15 years, and as such, seem to constitute valid guidelines for the future, subject to the continuing accumulation of experience and new innovations.

The automation of the Upper Occoquan Sewage Authority in Virginia is representative of the state-of-the-art and will be described next (21). It is a 15-mgd plant with conventional primary/secondary treatment and five advanced treatment steps: chemical clarification with two-stage recarbonation, multimedia filtration, carbon adsorption, ion exchange for nitrogen removal, and breakpoint chlorination.

The designers sought the simplest possible control scheme and this led to the choice of four levels of control: manual, analog hardware, direct digital control (DDC), and DDC with analog backup (CAM).

The manual level provides the operator with the manual capability to control a device such as a valve. The analog hardware level provides the option of continuous analog control of a device that requires continuous proportional control, and has a manual override feature. In the DDC loop level the computer control performs all the functions of an analog controller, but there is a computer/manual (CM) station that provides for the direct control of the device, with both automatic and manual capabilities. In the final level the DDC with analog backup, or computer/auto/manual (CAM), the operator has three choices: DDC, analog, or manual.

The overall system has two major parallel control provisions: the central computer system (CCS) and the analog system. The CCS provides for data management, process monitoring, alarm functions, process control, energy management, and data processing. The analog system supplements the CCS with some process measurement, analog control, hardwired alarming, and data collection.

The CCS uses a special process control language called Control Diagram Language (CODIL). The basic computing equipment consists of two CPUs each with 48K of core memory and a fixed-head 3-megabyte disk. One of the computers is always in a standby mode, provided as a safeguard against failure.

There are three main control centers in the plant: conventional and chemical treatment, advances waste treatment, and solids handling. Local control stations are distributed throughout the plant.

Most of the processes in the plant are under computer control. For example, the aeration basin dissolved oxygen is maintained by controlling weir elevation through adjustments signalled by the CCS. The CAM control arrangement is used to maintain pH levels for lime treatment and in both stages of recarbonation. Six mixed-media filters are controlled by computer, with CAM control for filter flow rate and automated backwash scheduling. These are only a few of the examples.

Stormwater and Combined Sewer Systems

When describing the challenges ahead for automation of urban water systems, McPherson identified the combined sewer overflow problem as the toughest because of the lack of "control logic." By this he meant that the vagaries of storm hydrology, combined with the complex structure of combined sewers, removed any regularity from the decision making that would enable control of combined sewers (12). Considerable research ensued after that, and some control logic has been developed, but the goal of automatic control, or even much control at all, of combined sewers remains elusive.

One of the public works features of the 1960s and 1970s in the United States was the discovery that stormwater and combined sewer systems contributed to nuisance flooding and water quality problems that needed fixing. This discovery was followed by research and discussion about what policies and systems to implement. After more than 15 years of attention to these problems few comprehensive solutions are in place, mostly because the costs, compared to the benefits, are considered too large. The advent of computer-based automatic control offers help, however, and the promise that better performance can be wrung from existing or low-cost new systems. This could help solve stormwater and combined sewer problems without the billions of dollars of investment that pure structural solutions demand, and in the process, it could create business opportunities for service and manufacturing firms.

Automatic control was first proposed and considered in the late 1960s and early 1970s. Its implementation has been slow, however, and there really exist today only a few examples of cities that have implemented it on a large-scale successfully. There do exist, however, pockets of innovation that promise to show the way for the future.

Interest in the stormwater and combined sewer problems rose rapidly in the 1960s and 1970s fueled by the environmental movement, the rapid urbanization of the United States and the funds that became available through the Clean Water Act. Control concepts such as detention storage using rainfall forecasting and computer control had been suggested to deal with the problem. Expensive structural solutions such as sewer separation and stormwater treatment were also suggested, as well as low-cost "best management practices." In the absence of firm federal regulatory policies, still missing today, little was done about these problems, however. Still, the prospect of better performance of existing systems through computer-based automatic control attracted innovators in U.S. cities and firms.

In the late 1960s there was the promise that federal funds could be made available to help with demonstration projects for automating combined sewer systems. At that time there was even the prospect that the federal government would finance some stormwater systems, but both of these prospects have now disappeared. While they were present, however, a number of U.S. cities embarked on planning and development projects. To mention a few, they included Minneapolis–St.Paul, Cleveland, San Francisco, Detroit, and Seattle, as well as "super projects" such as Chicago's deep tunnel plan.

Whether these projects were in the planning or implementation stage, they included a common element of plans to use some computer-based automatic control. During this period computer hardware was expensive and had limited capabilities, and the software available was still in the developmental stage, and much of it was extremely complex for the machines of that day.

As an example, the EPA Stormwater Management Model (SWMM) was being developed and some of the modules were not fully de-bugged. Even when they were, it took a large machine to run the model. However, the computers available for actual control applications had very limited memories such as 8K to 45K, the size for small microcomputers today.

Ten years ago I wrote about the possibilities and barriers for implementing automatic control of combined sewer systems (23). At that time our research team foresaw substantial difficulties in overcoming the barriers to automation, even in situations where the funds were available. It is interesting to review the points made at that time.

First, we listed some "sociopolitical constraints." These were as follows: the manager's limited capability to affect the social system in his organization; the inability to quantify benefits and costs of automation; the lack of political skills to implement automation; the impacts on the organization; communication difficulties in organizations; institutional barriers; timing problems; financial constraints; and personality problems. These problems are not unique, of course, to water organization, but apply to automation in general. They represent a formidable array of barriers to implementation.

The situation today bears out the prediction that automation would face barriers. In fact, there hardly exist today large-scale successful applications of computer-based control of either stormwater or combined sewer systems.

A review of the current situation finds that there are a few combined sewer systems with real-time operation, but their scope of operation relies heavily on human supervision, and even the successful systems are cautious about expansion into systemwide automatic control and giving away the control task to a computer.

Our experience in studying automatic combined sewer and stormwater control convinces us that there is a real opportunity for future applications. However, some of the management and political factors seem to be at least as important as the technical ones.

Wherever automation is successful, there is a small group of enthusiastic innovators who have stayed with a system over a period of time, usually through the planning, design, and implementation stages. These innovators are backed by higher management, and they are willing to look across their own fields to others in different parts of the organization. Communication within the organization is good vertically and horizontally. All of these seem to be characteristics of healthy organizations with good leadership, whether in the public or private sector.

These characteristics of the organization that are observed in the successful ones relate closely to the sociopolitical factors reported 10 years ago, and it will be no surprise to behavioral scientists that they have been observed. The conclusion seems to be that good management and healthy organizations can produce innovations such as automation, even in the absence of strong regulatory constraints. The opposite conclusion also seems to hold: a messy management might be able to run a conventional system but not a complex areawide integrated setup.

FUTURE DIRECTIONS

The future for automation in operations seems clear: there will be increasing trends toward it in urban water management, and it will bring opportunities for professional growth, better service, lower costs, and new business creation. As the trend continues, however, it will have some problems to overcome, both managerial and technological.

In the managerial area the sociopolitical problems already described are the major stumbling block; that is, how to implement automation successfully. The advent of automation has brought with it some trauma to workers, much as the industrial revolution did, and the trauma has to be dealt with. Some suggestions for this are made in the next section dealing with procedures.

In the area of technology there seem to be three trends at work. First, there seems sure to be a continued decline in the cost of computing capability. At the time of this writing the industry is announcing breakthroughs in one megabyte chips, ones that will replace the standard 256-kilobyte chips that just replaced the 64K chip that has only been around for a short while. At this point it seems futile to forecast any end to the miniaturization that is occurring. Also, there are continued improvements in hardware in general: better configurations and more reliable systems, at lower cost. This trend has some definite limitations, but it seems assured that we can expect a continued decline in the cost of high-powered machines.

Software availability seems assured to improve. Computer operating systems are improving, and the general quality and availability of software is receiving considerable attention. However, the improvement in software has been mostly aimed at the general-purpose business user and not many really useful generalized packages have been developed for specialty water applications. Due to the limited market available, this situation seems destined to continue.

However, the complexity of the water problems to be solved will continue to increase. New quality problems, the complexity of infrastructure systems and the need to understand systems better will drive the awareness of the complexity. For this reason the challenges to automation will not abate at all, and it is a subject that needs a lot of attention.

Instrument problems will be a major limiting factor for urban water automation. The APWA report cited earlier identified the problem well, and the authors of that report have continued to work on the problem. The APWA report could have been interpreted to be the harbinger of new business opportunities for wastewater instrument companies, but the general business climate of the 1970s brought with it a shakeout in the wastewater equipment business, and instrument availability is still not satisfactory. The instruments that are available are expensive and not reliable enough for some of the complex applications, especially under the severe conditions encountered in wastewater plants. The situation in water supply systems is somewhat better, but there do not exist low-cost and reliable instruments for many of the newer applications encountered, such as trace organics and metals.

APWA has taken the initiative to start up an instrument Testing Service to help improve the quality of the instruments available, but in the meanwhile the market suffers from not being large enough to attract much innovation, and the regulatory climate and general public orientation of the agencies, with the attendant procurement rules, has inhibited a solution to the instrument problem.

Finally, the fifth-generation computers will certainly impact process control. The advent of artificial intelligence offers the application of "knowledge engineering" to automatic control, and the heuristics involved in learning to operate a treatment plant, for example, seem ideal. There are already signs that developments are on their way. For example, a company recently announced a system that can be installed in the control room of a complex plant, say a manufacturing plant, to work with the operators to learn the system. In effect, it would be like the "learning room" in the paint operation cited earlier, but the machine will improve its knowledge continuously.

The expert system can be put into operation after a model of the operation is constructed. Then the machine uses two processors: one that handles the "expertise" side of the operation and the other that handles the calculations and the monitoring of alarms and sensors. In effect, the same general operations take place here that were described earlier in the state model, state update function, and predictor functions; the difference will be the extent to which the new systems can continue to learn. The system provides for operators to check the recommendations of the expert systems and query it to find out why it thinks certain strategies are called for. There are no applications of this kind of expert system in automatic control yet, but they seem sure to arrive (24).

Suggested Procedures

The lessons learned during the past few years should have taught us how to improve operations through automation, and we have learned a lot. I have tried to put down here some general lessons that seem clear from the above discussion. In general, these constitute additions to management philosophy in the operations field.

Use Automation to Improve Operations. This seems obvious, but in the 1970s with some of the legislative push for innovation, there was a tendency in some quarters to try automation (and some other new approaches) for its own sake. Some applications just did not make sense, and for that reason they either failed or they raised costs excessively. As an example of legislatively pushed innovation, I mean programs such as programmatic "innovative and alternative" approaches. I am not criticizing them, but pointing out that leveraged new technologies such as automatic control have to be built on solid technological and management foundations if they are to succeed, and hasty approaches are to be avoided, as are approaches that are implemented as a way to get a government grant. Another way to state this principle is to go back to the sage management consultant advice—"know the business you are in and do it well."

Innovate Through Good Organization and Management. One way to succeed in innovation is to have it come from the ranks. If automation, for example, is planned and implemented by a utilities own forces, rather than by an outside consultant, there will be some "ownership" that is very valuable in making it succeed. Innovation should come through a careful approach to organization and management, including supervision and coaching; then it will emerge rather than be forced. Your own forces should know and accept what is happening. Security and innovation can be built into the system at the same time. The requirement for redundancy and fail-safe operation will also be taken care of.

Enter into Contracts Carefully. The general principle that it is necessary to have your own forces feel ownership of the automation effort also leads to the conclusion that it is important to enter into contracts carefully, including both hardware and service contracts. It is easy to be swept off our feet by demonstrations of new hardware and techniques, and for the boss to say, "I want that," even against the advice of the rank and file (who may feel too intimidated to give advice), but this will be a mistake. Again, it is a symptom of good management when the organization will unite behind an automation effort because the organization is so well managed in the first place.

By following the principle of entering into contracts only carefully, the organization also avoids the pitfall of being "sold a bill of goods." In the 1960s and 1970s there were examples of public utilities of all kinds making procurement mistakes due to miscalculations, and the fiscal climate today does not allow for that margin of error.

Build on Strength. This principle means that everything should be running well before considering automation. Do not try to solve problems through automation, unless the problems are related to general confusion and the inability to handle all the information involved. Proceed carefully up the automation ladder: manual operations, automated data logging, supervisory control, computer-assisted control, and finally, closed-loop automatic control. Succeed at all the stages, but do not try to leap too far too fast.

REFERENCES

1. John Diebold *Automation,* AMACOM, New York, 1983.
2. Joseph R. Grinkler and R. F. Meagher, Computer Controlled Operation of an Activated Sludge Plant, *J. WPCF,* July 1984.
3. Robert L. Corbin and T. Turai, Concepts of a Distributed Digital Process Control System, Proceedings, AWWA Seminar on Water Plant Instruments and Automation, New Orleans, June 20, 1976, AWWA, Denver, Colorado.
4. Westinghouse Corporation, Westinghouse Distributed Processing Family, Westinghouse Electric Corporation, January 1982.
5. Eli Ginzberg, The Mechanization of Work, *Scientific American,* September 1982.
6. Alfred Z. Spector, Computer Software for Process Control, *Scientific American,* September 1984.
7. J. L. Cleland, G. L. Kingsbury, and F. O. Mixon, Automated Control: A Review and Application in Industrial Environmental Protection, EPA-600/S2-83-126, February 1984.
8. Anthony M. Arnone, Understanding and Applying the Programmable Controller, *Water/Engineering Management,* January 1984.
9. Timothy J. Miller, Data Acquisition and Control with Personal Computers—A Blossoming, *Control Engineering,* May 1984.
10. Dag I. Knudsen, W. W. Schuk, and R. C. Manross, Instruments You Can Rely On, Parts I and II, *Public Works,* November and December 1983.
11. W. W. Banks and F. Cerven, Predictor Displays: The Application of Human Engineering in Process Control Systems, *Plant/Operations Progress,* October 1984.
12. Murray B. McPherson, *Feasibility of the Metropolitan Water Intelligence System,* ASCE, New York, 1971.
13. Otto J. Helweg, R. W. Hinks, and D. T. Ford, Reservoir Systems Optimization, *J. Water Planning Management Division,* ASCE, June 1982.
14. Computer Cuts Hydro Cost, *ENR,* October 25, 1984.
15. Water Plant Instrumentation and Automation, New Orleans Conference, *AWWA,* Denver, 1976.
16. AWWA, Automation and Instrumentation, *Manual of Practice* M2, Denver, 1983.
17. Carmen F. Guarino F, J. V. Radzuil, P. R. Cairo, R. E. Day, and J. J. Smith, Philadelphia's Water System Automation Plan, ASCE, *J. Environmental Engineering Division,* October 1976.
18. Grant Weaver and Richard E. Desroches, Good Water from Simple Automation, *Water/Engineering Management,* July 1984.
19. Dan A. Brock, Closed-Loop Automatic Control of Water System Operations, *AWWA,* April 1963.
20. Dan A. Brock, Predictive Modeling for Optimization of Water System Operation, Instrument Society of America, International Conference, Houston, 1983, ISA, Raleigh, N.C.

21. Millard H. Robbins, Jr., Tight Effluent Specs Met with Digital Control, *Water/Engineering Management*, October 1983.
22. Feasibility of Computer Control of Wastewater Treatment, Chicago, *APWA*, December 1970.
23. Neil S. Grigg, J. W. Labadie, and H. G. Wenzel, Metropolitan Water Intelligence Systems, Phase III Completion Report, Colorado State University, July 1974.
24. Expert Systems, *Inter. J. Knowledge Engineering*, October 1984, Learned Information, Oxford, England.

CHAPTER ELEVEN

FUTURE ISSUES IN URBAN WATER MANAGEMENT

It is an unavoidable temptation to look ahead and speculate on what the future issues will be for urban water managers, and in organizing and writing this book, I wondered how they can be resolved. I decided to assemble in this last chapter a discussion of the most important of them from my viewpoint in an attempt to identify business opportunities, research needs, and management improvement projects for system managers. The discussion can also be interesting for policy research, as it needs always a list of what the policy "issues" are. This is the way government works: we identify the issues first, and then solutions can be found through the political process. In the case of urban water systems the issues are not so much political on a grand scale as they are local in nature, with a need for the national policy to allow a climate where solutions can happen.

The issues fall into five main categories: management, decision support, finance, technical, and regulatory. The reasons for classifying them this way spring from my own views, and others would have different classifications. My reasoning is that if we are to really solve the problems of the urban water infrastructure then certain improvements will be necessary. The first category of improvement will have to be in the area of management, including planning and control. To be successful in management and control, good decision support and information systems will be required. Only if systems can be adequately financed is successful management possible. Technical issues support management since they must be solved if mangement is to make progress, but water systems can limp along without much technological innovation. For this reason the technical issues support the management issues rather than vice versa. However, if skilled management is in place, the opportunities to solve problems and improve effectiveness through innovation is really impressive. The final category of issues is regulatory, an important category since regulations issued by the federal and state governments often call the tune that urban water systems must dance to.

PLANNING, MANAGEMENT, AND CONTROL ISSUES

The first and most important issue for water management organizations is to have skilled management at the executive levels and professionalism throughout the organization. One of the greatest difficulties facing water management everywhere is to attract and retain good managers. This means that the water organization must have the same capabilities as other private and public organizations to motivate and take care of their employees. This also means that there can be no corruption in the management of the organization. If there is corruption and a lack of good manners, we can forget about seeking solutions through finance and technological innovation.

Assuming that skilled management is in place, there is a need to adopt and effectively utilize modern management approaches for the critical and unique problems faced in urban water systems. The first of these is to have a life cycle approach to managing capital facilities. This requires the capability to plan, program, and budget for both new and replacement facilities. The preparation of valid needs estimates and capital budgets is a must. The organization may need to enhance their problem-solving capabilities and develop techniques such as diagnostic analysis to find out where improvements are necessary in the organization.

The proper organization structure of the water management organization is important since effectiveness and efficiency are dependent on good flow of information and task assignments. The advent of different ways to organize information will mean more attention to organizational structures in the future. Just as we have organized around people and jobs in the past, future organizations will pay more attention to organizing around data bases and the need to support decisions.

Planning to solve problems will demand more consideration of the benefits of cooperation and integrated management. One problem that can be aided by this will be drought management. Better forms of risk assessment can be used to aid in planning new water supplies, including the consideration of the benefits of working together in regional organizations. In some cases the regional cooperation may have to be mandated by state legislatures. Demand management will also come to be a recognized part of urban water supply management, and may enter into wastewater management as well. Water conservation will become institutionalized as a good management practice, rather than an emotional and environmental issue.

It will be necessary to do a better job of cost control to hold rates down and begin to internalize the costs of water management so that the enterprise approach to management can become effective. There may be more rate regulation in the future, even for public entities offering water and wastewater service. To avoid problems with facilities after they are constructed, better quality in construction will be required.

The control aspect of management will rely more and more on automation and computer control in the future. Facilities management such as military bases or large buildings will begin to be automated, and the urban water infrastructure will follow the same pattern.

EVOLVING DECISION SUPPORT SYSTEMS

The advent of more and more computer control in urban water systems will require attention to the development of really effective information systems. Seven categories mentioned in Chapter 8 seemed likely. The 'nventory data bases are already coming on-line. We are going to see some more failures before some low-cost and practical approaches to this problem are found. The information system market right now is in the same position as the general market for software, with some shakeouts due. The condition assessment data base will be a natural successor to the geographically based inventory data base, since once we find out where a facility is, we need to know what its condition is and whether we can rely on it. Reliability is one of the most important characteristics needed in urban water systems, and the constant search for reliability explains many of the management actions in urban water systems.

The water balance data base is a natural one for planners and engineers to desire since so many of the facility planning questions require data on water flows, levels, quality, and variation with time. We are going to see a great deal of additional research by local water agencies on accounting for water losses and inflows, as well as about hydrologic balances and overall water auditing.

Data bases to support real-time control will be necessary in the future, especially to enable the build-up of expert systems that can aid in decision making for treatment plants in complex water distribution, stormwater, and wastewater collection systems. Treatment plants will become more and more automated, and the information systems necessary to improve operations and maintenance will improve. Better understanding of overall systems will lead to the development of data bases and information systems to aid in planning and design of specific facilities, and these will result in the development of improved system simulation models. Financial data bases and models can be used in small systems.

FINANCIAL ISSUES

To understand the importance of financial issues in urban water management, you must only consider the extent of the press coverage to them. For example, the AWWA journal devotes regular issues to financing water supply systems. As the infrastructure issue has developed, the attention given to financing solutions has dwarfed all other subjects. It is natural to expect that this attention to financial questions will continue.

The area where most improvement can be made is probably the general financial management capability of utilities and urban water systems in general. As discussed earlier in Chapter 6, this includes all of the reporting and control aspects of financial management, as well as the critical issues of raising enough money to support necessary capital as well as operating and maintenance requirements. The techniques of planning, programming, and budgeting must be improved if we are to overcome the past shortcomings in financial management.

Improving financial management is especially important to solve the problems of small water and wastewater systems. As discussed earlier in several places, solving small system problems is one of the critical issues facing urban water systems, and the heart of this problem is financial, meaning raising and managing enough money to care for the systems and meet their objectives.

Whereas the financing of water supply has received considerable attention due to the absence of government subsidies, wastewater finance is just beginning to emerge as a new professional challenge. The problem in the future as subsidies continue to shrink will be to internalize the cost of wastewater management, including the expensive treatment processes that are mandated by law. We have learned that these processes can cost more than $1000 per capita to construct, and that they include high O&M costs, and these public burdens come to more than some poorer communities can bear, especially in a time when utility costs are also rising in other areas. We need to find some way to have effective wastewater management at lower cost.

As a final financial issue we need to look much closer at the possibilities that can come through privatization. This newly rediscovered technique for building and operating water systems has a lot of possibilities that have not been heretofore considered, and some new experiences will be forthcoming. The possibilities for solving the financing problems of wastewater treatment should be especially interesting to evaluate.

EMERGING TECHNICAL ISSUES

A study of the water infrastructure issue areas would convince the non-technical person that all of the problems are financial, and there are not really any technical problems to be solved. This is absolutely not true, and my study of the issues convinces me that the reason for this perspective is simply the high interest level in financial and political issues as compared to technical ones.

In the area of water supply there are many new technical issues relating to the health effects of synthetic organic chemicals, and the detection and treatment of them raises significant challenges. These chemicals are carried throughout the environment as diffused pollutants as a result of some past disposal practices, as well as being carried in the point source effluents of cities and plants that have not controlled them yet. Solving this problem will be a real challenge for years in the future.

The general problem of nonpoint source pollution, especially stormwater quality, continues to be unresolved. Although much is known about the problem, the technology to handle it at a reasonable cost continues to elude us. The linkages between stormwater quality and toxic chemicals control are just beginning to be identified, and it is my prediction that we will see some renewed emphasis on stormwater quality control in the future.

Stormwater management presents the additional challenge that the results of urbanization are not controlled. We are seeing increases in runoff volume and rate and aggravated flooding damages as a result. The normally prescribed solution, detention of runoff, may not be working, but our research is inadequate to tell, and affordable techniques for control continue to elude us.

Many of the problems that we face from stormwater require better simulation models for their solution. It will never be possible to collect enough data to analyze all of the problems or to control them in real time. For this reason the availability of valid models is especially necessary for planning and design. This is true for rainfall problems, for stormwater rate problems, for the design of detention storage, and for solving the stormwater quality problem.

Problems of physical facilities loom large as they decay and begin to fall. Corrosion in water distribution systems and infiltration/inflow problems of wastewater collection systems have the common characteristic that they need better detection and in situ repair methods if improvements are to be made. The associated problem of accounting for water losses also needs to be solved. We also lack technological methods to assess the condition of our water facilities so that plans can be made to repair, replace, or rehabilitate them.

REGULATORY ISSUES

Ever since environmental legislation began to increase in the 1960s, we have heard complaints about the changes and difficulty in meeting all the requirements. Although this situation will continue to exist, the regulatory climate will begin to settle down.

The Safe Drinking Water Act will continue to be the driving force in drinking water programs, but we will probably also see greater attention given to rate regulation in the future. Technical issues such as the proper standards for the synthetic organic chemicals and limits for the products of disinfection will begin to settle, but there will be new technical issues to study. The importance of the attention to finance will be to give us the opportunity to balance health effects and environmental regulations with realistic timetables and costs, especially considering the real state of knowledge about levels of risk.

The Clean Water Act has undergone significant revisions to accommodate the pressures and discoveries of the 1970s, and there will continue to be incremental changes. The trend toward assigning to the states more and more responsibility for meeting clean water goals, including financial responsibility, seems sure to continue, however, and this should mean more stability in the requirements of the program. There are many technical issues still outstanding, such as the allowable levels of certain pollutants, and the related environmental effects. Some consider these to be "details," but the scientific challenges are awesome.

In the wastewater program area pollution from industrial discharges will continue to be a challenge, and the regulation of industrial wastes will become a separate issue area that serves as an organizing concept for several regulatory efforts having to do with toxic materials and hazardous wastes. The field of waste management engineering will become more and more specialized due to the increasing reagulatory and technical complexities.

Stormwater regulation will continue to be a question mark, and the requirements that do appear are likely to be in response to local problems with development and pollution of desirable facilities such as lakes and beaches. In the State of Florida, for example, development has been and continues to be so rapid that the solution of stormwater problems becomes a practical necessity and no longer an optional area of public policy. The results of the new requirement for permits for storm sewers in urban areas will have to be evaluated.

CONCLUSION

The issues described above will constitute a significant challenge for the years ahead. The overwhelming characteristic that has shaped the development of urban water systems in the past has been the need to have safe and reliable service that meets regulatory requirements. This set of forces will continue to shape the urban water industry. A new factor that will set in during the remainder of the century will be to find cost-effective approaches to maintain service levels in the face of the new regulatory climate. This new emphasis on management and finance, along with the continued engineering and technical challenges, will produce interesting tests for urban water managers in the future.

APPENDIX ONE

COST INDEXES

Managers and planners often deal with costs, and they must be expressed in equivalent terms to be comparable. This appendix explains a little about the use of cost indexes and presents some of the numerical values that are intended to be helpful. In my own professional work I have found it helpful to maintain a file on cost indexes, and this appendix may provide similar help to those just beginning to keep records on costs.

Cost indexes are intended to be used by applying a ratio of the indexes to the costs. This procedure allows the conversion of, say, a 1980 cost to a 1985 cost according to the equation.

$$\text{COST (1985)} = \text{COST (1980)} \times \text{(1985 INDEX/1980 INDEX)}.$$

There are a number of widely used indexes. The difference between them is the group of cost items that is used to make up the index. For example, the widely quoted Consumer Price Index (CPI) is intended to measure the general cost of living, and it includes everyday items. The Producer Price Index is intended to measure the costs of production rather than consumption. The most useful indexes to engineers trying to estimate construction costs are those that deal with construction. *ENR* (formerly *Engineering News-Record)* has published the ENR Construction Cost Index since 1905. It includes 200 hours of common labor and prices of steel, cement, and lumber (ENR, Construction Cost Index History, *ENR*, March 18, 1982). The ENR Building Cost Index is similar, but it includes different quantities of labor, steel, cement, and lumber costs.

There are many other special-purpose indices. The Bureau of Labor Statistics publishes producer prices and price indexes with some 3000 product groupings considered, and annual indexes of output per work hour for selected industries. The Environmental Protection Agency publishes complete urban sewer system (CUSS) construction cost indexes

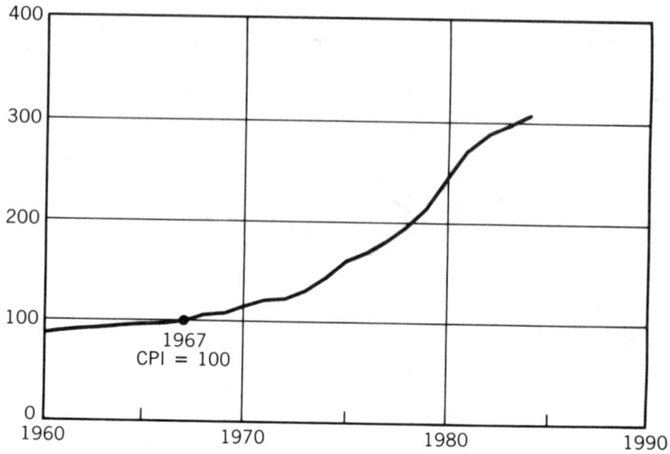

FIGURE A.1 Consumer price index.

for different cities. *American City and County* publishes a municipal cost index, and other federal agencies such as the Corps of Engineers and the Bureau of Reclamation have their own.

Figure A.1 shows the CPI since 1960 for general reference. The July values since 1967 are as follows:

Year	CPI
1967	100.0
1968	104.2
1969	109.8
1970	116.3
1971	121.3
1972	125.3
1973	133.1
1974	147.1
1975	161.1
1976	170.5
1977	181.5
1978	195.4
1979	217.4
1980	246.8
1981	272.4
1982	289.1
1983	298.4

Yearly average values of the ENR Construction Cost Index since 1906 are as follows:

1906	95	1926	208	1946	346	1966	1019
1907	101	1927	206	1947	413	1967	1070
1908	97	1928	207	1948	461	1968	1155
1909	91	1929	207	1949	477	1969	1269
1910	96	1930	203	1950	510	1970	1385
1911	93	1931	181	1951	543	1971	1581
1912	91	1932	157	1952	569	1972	1753
1913	100	1933	170	1953	600	1973	1895
1914	89	1934	198	1954	628	1974	2020
1915	93	1935	196	1955	660	1975	2212
1916	130	1936	206	1956	692	1976	2401
1917	181	1937	235	1957	724	1977	2577
1918	189	1938	236	1958	759	1978	2776
1919	198	1939	236	1959	797	1979	3003
1920	251	1940	242	1960	824	1980	3237
1921	202	1941	258	1961	847	1981	3533
1922	174	1942	276	1962	872	1982	3825
1923	214	1943	290	1963	901	1983	4055
1924	215	1944	299	1964	936	1984	4148
1915	207	1945	308	1965	971		

The reader is encouraged to consult the quarterly cost roundup editions of *ENR* and other business periodicals to update these figures and to improve his or her own understanding of cost-estimating procedures, especially under conditions of inflation.

APPENDIX TWO

PUBLICATIONS OF ASCE URBAN WATER RESOURCES RESEARCH PROGRAM UNDER MURRAY B. McPHERSON

The "Identifying Number" cited for nearly all entries is that of the National Technical Information Service, U.S. Department of Commerce, Springfield, Va., 22151, from which reprints are available for a cost-recovery charge).

PHASE I (1967–1969)

Urban Water Resources Research, First Year Report to the Office of Water Resources Research, U.S. Department of Interior, ASCE, New York, September 1968. (Identifying No.: PB 184 318).

Basic Information Needs in Urban Hydrology, a Report to the Geological Survey, U.S. Department of Interior, ASCE, New York, April 1969. (Identifying No.: PB 185 442).

T.M. No. 1, L. S. Tucker, Northwood Gaging Installation, Baltimore—Instrumentation and Data, August 1, 1968. (Identifying No.: PB 182 786).

T.M. No. 2, L. S. Tucker, Oakdale Gaging Installation, Chicago—Instrumentation and Data, August 15, 1968. (Identifying No.: PB 182 787).

T.M. No. 3, J. C. Schaaker, Jr., Response Characteristics of Urban Water Resources Data Systems, August 1968. (Identifying No.: PB 182 788).

T.M. No. 4, H. G. Wenzel, Jr., A Critical Review of Methods of Measuring Discharge Within a Sewer Pipe, September 1968. (Identifying No.: PB 182 789).

T.M. No. 5, M. B. McPherson, The Nature of Changes in Urban Watersheds and Their Importance in the Decades Ahead, December 1968. (Identifying No.: PB 182 790).

T.M. No. 6, M. B. McPherson, Some Notes on the Rational Method of Storm Drainage Design, January 22, 1969. (Identifying No.: PB 184 701).

T.M. No. 7, L. H. Clem, A Study of the Expenditures for Urban Water Services, The Travelers Research Corporation, February 1969. (Identifying No.: PB 184 702).

T.M. No. 8, L. S. Tucker, Availability of Rainfall-Runoff Data for Sewered Catchments, March 3, 1969. (Identifying No.: PB 184 703).

T.M. No. 9, L. S. Tucker, Raingage Networks in the Largest Cities, March 17, 1969. (Identifying No.: PB 184 704).

T.M. No. 10, L. S. Tucker, Sewered Drainage Catchments in Major Cities, March 31, 1969. (Identifying No.: PB 184 705).

PHASE II (1969–1971)

M. B. McPherson, *Prospects for Metropolitan Water Management*, ASCE, New York, December 1970. (Available at $2.00 per copy from ASCE, 345 East 47th Street, New York, NY 10017).

Report by Water Resources Engineers, Inc., Systems Analysis for Urban Water Management, Walnut Creek, CA, prepared for OWRR through the ASCE Urban Water Resources Research Program, September 1970. (Identifying No.: PB 197 677).

T.M. No. 11, L. S. Tucker, Non-Metropolitan Dense Raingage Networks, January 1970. (Identifying No.: PB 191 709).

T.M. No. 12, T. B. Prawdzik, Environmental and Technical Factors for Open Drainage Channels in Milwaukee, February 1970. (Identifying No.: PB 191 710).

T.M. No. 13, L. S. Tucker, Availability of Rainfall-Runoff Data for Partly Sewered Urban Drainage Catchments, March 1970. (Identifying No.: PB 191 755).

T.M. No. 14, M. B. McPherson, Management Problems in Metropolitan Water Resource Operations, September 1971. (Identifying No.: PB 206 087).

T.M. No. 15, M. B. McPherson, Feasibility of the Metropolitan Water Intelligence System Concept (Integrated Automatic Operational Control), December 1971. (Identifying No.: PB 207 301).

PHASE III (1971–1974)

T.M. No. 16, L. S. Tucker, Jaime Millan, and Wilford W. Burt, Metropolitan Industrial Water Use, May 1972. (Identifying No.: PB 212 578).

T.M. No. 17, M. B. McPherson, Hydrologic Effects of Urbanization in the United States, June 1972. (Identifying No.: PB 212 579).

T.M. No. 18, M. B. McPherson, Urban Runoff, August 1972. (Identifying No.: PB 212 580).

T.M. No. 19, James E. Goddard, An Evaluation of Urban Flood Plains, December 1973. (Identifying No.: PB 227, 337).

T.M. No. 20, Richard F. Lanyon and James P. Jackson, A Streamflow Model for Metropolitan Planning and Design, The Metropolitan Sanitary District of Greater Chicago, January 1974. (Identifying No.: PB 232 181).

T.M. No. 21, M. B. McPherson, Innovation: A Case Study, February 1974. (Identifying No.: PB 232 166).

T.M. No. 22, Noel Hobbs and Jay D. Britton, Computer Modeling Applications in Urban Water Planning, Denver Board of Water Commissioners, March 1974. (Identifying No.: PB 233 673).

T.M. No. 23, L. A. Roesner, et al., A Model for Evaluating Runoff-Quality in Metropolitan Master Planning, Water Resources Engineers, The Hydrologic Engineering Center/Corps of Engineers, and Department of Public Works/City and County of San Francisco, April 1974. (Identifying No.: PB 234 312).

T.M. No. 24, by Water Resources Engineers and The Hydrologic Engineering Center/Corps of Engineers, Management of Urban Storm Runoff, May 1974. (Identifying No.: PB 234 316).

1974–1980

T.M. No. 25, M. B. McPherson, G. F. Mangan, ASCE Urban Water Resources Research Program, 1967–74, New York, October 1974. (Also in July 1975 ASCE Hydraulics Div. Journal).

T.M. No. 26, R. P. Shubinski and S. N. Nelson, Effects of Urbanization on Water Quality, ASCE UWRR Program, ASCE, New York, March 1975. (NTIS No. PB 242 297).

T.M. No. 27, J. B. Wolff, F. P. Linaweaver, Jr., and J. C. Geyer, Commercial Water Use, ASCE UWRR Program, ASCE, New York, December 1975. (NTIS No. PB 250 998).

T.M. No. 28, M. B. McPherson, Household Water Use, ASCE UWRR Program, ASCE, New York, January 1976. (NTIS No. PB 250 879).

T.M. No. 29, N. S. Grigg, J. W. Labadie, G. R. Trimble, and D. A. Wismer, Computerized City-Wide Control of Urban Stormwater, ASCE UWRR Program, ASCE, New York, February 1976. (Available from Colorado State University, Dept. of Civil Engrg., Fort Collins, CO 80521).

T.M. No. 30, C. T. Earl, et al., Urban Flood Warning and Watershed Management Advances in Metropolitan Melbourne, ASCE UWRR Program, ASCE, New York, June 1976. (NTIS No. PB 259 549).

T.M. No. 31, M. B. McPherson, Ed., Utility of Urban Runoff Modeling, ASCE UWRR Program, ASCE, New York, July 1976. (NTIS No. PB 261 460).

T.M. No. 32, William H. Espey, Jr., D. G. Altman, and C. B. Graves, Nomographs for Ten-Minute Unit Hydrographs for Small Urban Watersheds, ASCE UWRRP, December 1977.

T.M. No. 33, Jiri Marsalek, Research on the Design Storm Concept, ASCE UWRRP, September 1978.

T.M. No. 34, Jess Abbott, Testing of Several Runoff Models on an Urban Watershed, ASCE UWRRP, September 1978.

T.M. No. 35, V. T .V. Nguyen, M. B. McPherson, and Jean Rouselle, Feasibility of Storm Tracking for Automatic Control of Combined Sewer Systems, ASCE UWRRP, November 1978.

T.M. No. 36, Philip D. Straffin, Jr., Introduction to Social Choice Theory for Environmental Decision Making, ASCE UWRRP, February 1979.

T.M. No. 37, M. B. McPherson, Challenges in Urban Runoff Control (Planning, Simulation and Control), ASCE UWRRP, August 1979.

T.M. No. 38, M. B. McPherson, Ed., International Symposium on Urban Hydrology, ASCE UWRRP, September 1979.

T.M. No. 39, M. B. McPherson, Ed., Urban Runoff and Section 208 Planning, ASCE UWRRP, October 1979.

IHP–TM SERIES

M. B. McPherson, Urban Hydrological Modeling and Catchment Research in the U.S.A., ASCE UWRR Program Technical Memorandum No. IHP-1, ASCE, New York, November 1975. (NTIS No. PB 260 685).

A. P. Aitken, Urban Hydrological Modeling and Catchment Research in Australia, ASCE UWRR Program Technical Memorandum No. IHP-2, ASCE, New York, May 1976. (NTIS No. PB 260 686).

J. Marsalek, Urban Hydrological Modeling and Catchment Research in Canada, ASCE UWRR Program Technical Memorandum No. IHP-3, ASCE, New York, June 1976. (NTIS No. PB 262 068).

M. J. Lowing, Urban Hydrological Modeling and Catchment Research in the United Kingdom, ASCE UWRR Program Technical Memorandum No. IHP-4, ASCE, New York, July 1976. (NTIS No. PB 262 069).

V. V. Kuprianov, Methods for Calculating Maximum Flood Discharges for Natural Watercourses and Urban Areas in the U.S.S.R., ASCE UWRR Program Technical Memorandum No. IHP-5, ASCE, New York, August 1976. (NTIS No. PB 262 070).

H. Massing, Urban Hydrology Studies and Mathematical Modeling in the Federal Republic of Germany, ASCE UWRR Program Technical Memorandum No. IHP-6, ASCE, New York, September 1976.

Gunnar Lindh, Urban Hydrological Modeling and Catchment Research in Sweden, ASCE UWRR Program Technical Memorandum No. IHP-7, ASCE, New York, October 1976.

M. Desbordes and D. Normand, Urban Hydrological Modeling and Catchment Research in France, ASCE UWRR Program Technical Memorandum No. IHP-8, ASCE, New York, November 1976.

Nils R. Selthun, Urban Hydrological Modeling and Catchment Research in Norway, ASCE UWRR Program Technical Memorandum No. IHP-9, ASCE, New York, December 1976.

F. C. Zuidema, Urban Hydrological Modeling and Catchment Research in the Netherlands, ASCE UWRR Program Technical Memorandum No. IHP-10, ASCE, New York, January 1977.

P. Blaszczyk, Urban Runoff Research in Poland, ASCE UWRR Program Technical Memorandum No. IHP-11, ASCE, New York, February 1977.

S. Ramaseshan and P. B. S. Sarma, Urban Hydrological Modeling and Catchment Research in India, ASCE UWRRP TM IHP-12, New York, May 1977.

M. B. McPherson and F. C. Zuidema, Urban Hydrological Modeling and Catchment Research: International Summary, ASCE UWRRP TM IHP-13, New York, November 1977.

OTHER REPORTS

M. B. McPherson, *Regional Earth Science Information in Local Water Management*, ASCE, New York, July 1975.

M. B. McPherson, Urban Hydrology Research Needs, a report for Division of Advanced Environmental Research and Technology, Research Applied to National Needs Program, National Science Foundation, Washington, D.C. 20550, April 1976.

M. B. McPherson, Urban Water Resources, *EOS*, Transactions of the AGU, Vol. 57, No. 11, pp. 798–806, November 1976.

M. B. McPherson, Urban Runoff Control Master Planning, ASCE UWRRP, June 1977.

M. B. McPherson, Challenges in Urban Runoff Control, Fourteenth Annual Henry M. Shaw Lecture Series in Civil Engineering, North Carolina State University, March 1979.

M. B. McPherson, Integrated Control of Combined Sewer Regulators Using Weather Radar, report for EPA, 1980.

INDEX